Heat Exposure and HUMAN HEALTH IN THE CONTEXT OF CLIMATE CHANGE

AF795086

Heat Exposure and HUMAN HEALTH IN THE CONTEXT OF CLIMATE CHANGE

Edited by

YUMING GUO

Climate and Air Quality Research Unit,
Monash University, Melbourne, Australia

SHANSHAN LI

School of Public Health and Preventive Medicine,
Monash University, Melbourne, Australia

ELSEVIER

Elsevier
Radarweg 29, PO Box 211, 1000 AE Amsterdam, Netherlands
The Boulevard, Langford Lane, Kidlington, Oxford OX5 1GB, United Kingdom
50 Hampshire Street, 5th Floor, Cambridge, MA 02139, United States

Copyright © 2023 Elsevier Inc. All rights reserved.

No part of this publication may be reproduced or transmitted in any form or by any means, electronic or mechanical, including photocopying, recording, or any information storage and retrieval system, without permission in writing from the publisher. Details on how to seek permission, further information about the Publisher's permissions policies and our arrangements with organizations such as the Copyright Clearance Center and the Copyright Licensing Agency, can be found at our website: www.elsevier.com/permissions.

This book and the individual contributions contained in it are protected under copyright by the Publisher (other than as may be noted herein).

Notices
Knowledge and best practice in this field are constantly changing. As new research and experience broaden our understanding, changes in research methods, professional practices, or medical treatment may become necessary.

Practitioners and researchers must always rely on their own experience and knowledge in evaluating and using any information, methods, compounds, or experiments described herein. In using such information or methods they should be mindful of their own safety and the safety of others, including parties for whom they have a professional responsibility.

To the fullest extent of the law, neither the Publisher nor the authors, contributors, or editors, assume any liability for any injury and/or damage to persons or property as a matter of products liability, negligence or otherwise, or from any use or operation of any methods, products, instructions, or ideas contained in the material herein.

ISBN: 978-0-12-819080-7

For information on all Elsevier publications visit our website at https://www.elsevier.com/books-and-journals

Publisher: Candice G Janco
Acquisitions Editor: Kathryn — Candice G Janco
Editorial Project Manager: Andrae Akeh
Production Project Manager: R. Vijay Bharath
Cover Designer: Victoria Pearson Esser

Typeset by TNQ Technologies

ELSEVIER Book Aid International Working together to grow libraries in developing countries
www.elsevier.com • www.bookaid.org

Contents

Contributors

Katherine Arbuthnott
The Oxford School of Public Health (OxSPH), United Kingdom

Paul L.C. Chua
Department of Global Health Policy, Graduate School of Medicine, University of Tokyo, Bunkyo-ku, Tokyo, Japan

Rosario Di Corleto
Principal Consultant, Ashgrove, Brisbane, QLD, Australia

Naoto Fujii
Faculty of Health and Sport Sciences, University of Tsukuba, Tsukuba, Japan

Antonio Gasparrini
Department of Public Health Environments and Society, London School of Hygiene & Tropical Medicine, London, United Kingdom

Donna Green
Climate Change Research Centre, University of New South Wales, Sydney, NSW, Australia

Suying Guo
Climate, Air Quality Research Unit, School of Public Health and Preventive Medicine, Monash University, Melbourne, Australia

Yuming Guo
Climate, Air Quality Research Unit, School of Public Health and Preventive Medicine, Monash University, Melbourne, Australia; Department of Epidemiology and Preventive Medicine, School of Public Health and Preventive Medicine, Monash University, Melbourne, Australia

Masahiro Hashizume
Department of Global Health Policy, Graduate School of Medicine, University of Tokyo, Bunkyo-ku, Tokyo, Japan; Department of Global Health, School of Tropical Medicine and Global Health, Nagasaki University, Nagasaki, Japan

Yasushi Honda
Faculty of Health and Sport Sciences, University of Tsukuba, Tsukuba, Japan; Center for Climate Change Adaptation, National Institute for Environmental Studies, Tsukuba, Japan

Jouni J.K. Jaakkola
Center for Environmental and Respiratory Health Research (CERH), University of Oulu, Finland; Biocenter Oulu, University of Oulu, Finland; Medical Research Center Oulu, University of Oulu and Oulu University Hospital, Oulu, Finland; Finnish Meteorological Institute, Helsinki, Finland

Ho Kim
Department of Public Health Science, Graduate School of Public Health, Seoul National University, Republic of Korea

Hyewon Lee
Department of Health Administration and Management, Soonchunhyang University, Republic of Korea

Shanshan Li
Climate, Air Quality Research Unit, School of Public Health and Preventive Medicine, Monash University, Melbourne, Australia; Department of Epidemiology and Preventive Medicine, School of Public Health and Preventive Medicine, Monash University, Melbourne, Australia

Rahini Mahendran
Climate, Air Quality Research Unit, School of Public Health and Preventive Medicine, Monash University, Melbourne, Australia

Ana Maria Vicedo-Cabrera
Institute of Social and Preventive Medicine & Oeschger Center for Climate Change Research, University of Bern, Switzerland

Mariana Matera Veras
Department of Pathology, University of Sao Paulo, School of Medicine, Sao Paulo, Brazil

Paulo Hilario Nascimento Saldiva
Department of Pathology, University of Sao Paulo, School of Medicine, Sao Paulo, Brazil

Nazife Pehlivan
Department of Public Health Science, Graduate School of Public Health, Seoul National University, Republic of Korea

Sarah Perkins-Kirkpatrick
School of Science, UNSW Canberra, Canberra, Australia

Dino Pisaniello
School of Public Health, University of Adelaide, Adelaide, Australia

Niilo R.I. Ryti
Center for Environmental and Respiratory Health Research (CERH), University of Oulu, Finland; Biocenter Oulu, University of Oulu, Finland; Medical Research Center Oulu, University of Oulu and Oulu University Hospital, Oulu, Finland

Xerxes T. Seposo
Department of Hygiene, Graduate School of Medicine, Hokkaido University, Hokkaido, Japan; School of Tropical Medicine and Global Health, Nagasaki University, Nagasaki, Japan; Ateneo Center for Research and Innovation, Ateneo de Manila School of Medicine and Public Health, Ateneo de Manila University, Pasig, Philippines

Micheline de Sousa Zanotti Stagliorio Coelho
Department of Pathology, University of Sao Paulo, School of Medicine, Sao Paulo, Brazil

Bo Wen
Climate, Air Quality Research Unit, School of Public Health and Preventive Medicine, Monash University, Melbourne, Australia

Yao Wu
Climate, Air Quality Research Unit, School of Public Health and Preventive Medicine, Monash University, Melbourne, Australia

Zhengyu Yang
Climate, Air Quality Research Unit, School of Public Health and Preventive Medicine, Monash University, Melbourne, Australia

Pei Yu
Climate, Air Quality Research Unit, School of Public Health and Preventive Medicine, Monash University, Melbourne, Australia

Yiwen Zhang
Climate, Air Quality Research Unit, School of Public Health and Preventive Medicine, Monash University, Melbourne, Australia

Qi Zhao
Department of Epidemiology and Preventive Medicine, School of Public Health and Preventive Medicine, Monash University, Melbourne, Australia

CHAPTER 1

Introduction

Yuming Guo and Shanshan Li
Department of Epidemiology and Preventive Medicine, School of Public Health and Preventive Medicine, Monash University, Melbourne, Australia

The world has been experiencing a marked increases in the frequency, magnitude, and duration of heatwaves over the past century, especially in the region of West Asia, Eastern Europe, and Northern Africa (Fig. 1.1). Heatwaves are sensitive indicators of changes in climate and are closely related to climate change. The global mean surface temperature anomaly (GMST) has increased at an average rate of about 0.01°C per year since 1920 and over twice that rate (over 0.02°C) since 1990 (Fig. 1.1). From 1920 to 1990, the changes in heatwaves were relatively small; however, since 1990, the increasing trend of frequency, maximum duration, and extra heat of heatwaves have accelerated over tenfold to 0.48 days,

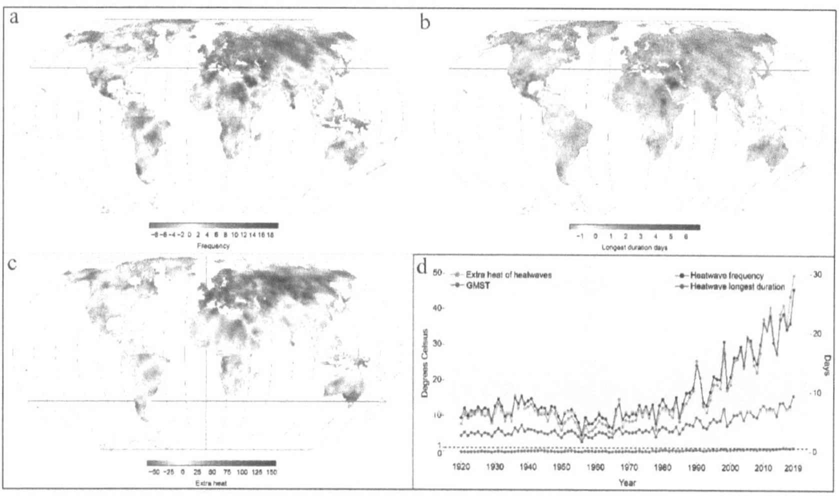

Figure 1.1 Global decadal trends of heatwaves' frequency (a), longest duration days (b) and extra heat produced (c) from 1920 to 2019. Trends were estimate using Sen's Kendal slope estimator; annual time series of global mean surface temperature anomaly (GMST) (°C) (d), heatwaves extra heat (°C), heatwave frequency (days), and longest duration (days) over the past century. Heatwave was defined as the daily maximum temperature (T_{max}) above threshold for at least three consecutive days. The threshold is the calendar day 90th percentile of T_{max} over 1920–1990, based on a 15-day window (Perkins & Alexander, 2013). *(Temperature data source: The Berkeley Earth dataset (http://berkeleyearth.org/)).*

Heat Exposure and Human Health in the Context of Climate Change
ISBN 978-0-12-819080-7
https://doi.org/10.1016/B978-0-12-819080-7.00010-0
© 2023 Elsevier Inc.
All rights reserved.

0.12 days, and 0.99°C per year, respectively (Fig. 1.1). Even a modest level of temperature warming can cause high levels of regional and seasonal heatwaves. The heatwaves are not only increasing, but also the changes are accelerating in recent years. More heat energy is trapped in the atmosphere as greenhouse gases accumulate, increasing the risk of an anticyclone (sinking air) forming, which functions as a cap or heat dome, pushing out cooler, faster-moving air currents, and squeezing clouds away.

Extreme high temperatures have been identified to cause a great health burden. Excess morbidity has been associated with high temperatures (e.g., heatwaves), particularly cardiovascular diseases, respiratory diseases, and diabetes (Moon, 2021). It has been estimated that heatwaves could directly increase mortality risks and higher temperature threshold (e.g., 95th, 97.5th, 99th) with the longer duration of heatwave days (e.g., ≥ 2, ≥ 3, ≥ 4 days) would lead to higher mortality risks. The increased mortality risk due to heatwave could last for 3 or 4 days (Guo et al., 2017). Potential mechanisms behind the heatwave—mortality/morbidity association may involve thermoregulation and dehydration. When the ambient temperature rose dramatically to an extreme level, the normal process of thermoregulation could be disrupted and the burden on the cardiovascular system increased (Guo et al., 2017). Specifically, heatwaves tend to cause a dehydrated status of the human body and thus increase the mortality and morbidity risks, especially for vulnerable groups, such as children, the elderly, pregnant women, and people with chronic diseases (Moon, 2021). Notably, the stronger urban heat island effect that accompanies climate change, as well as the potentially higher humidity in the weather system due to stronger ocean evaporation, significantly elevates the health risks of heatwaves (Matthews, 2018).

Under the enhanced climate change, the trends of temperature increase and heatwaves are projected to worsen. It is estimated that for each degree increase of global warming, the heatwave days and duration are projected to increase by between 4 and 34 days per season and 2—10 days, respectively (Perkins-Kirkpatrick & Gibson, 2017). Without immediate and efficient climate action, heatwaves will become a less rare disaster making increasingly severe health impacts. This book aims to summarize available evidence on (1) the impacts of heat exposure on vulnerable populations and its biomechanisms; (2) spatiotemporal variation in the heat effects, which can reflect human adaptation ability to local temperatures; and (3) potential strategies to mitigate the health impacts of heat exposure.

If both average temperatures and extreme heat events increase as projected (Griffiths et al., 2005), people will be at high risk of increased heat strain during summer. Vulnerable people will face profound stresses and challenges from climate change. There is great need for a better understanding of what the health effects of ambient temperatures might be, and this requires robust evidence for the health effects of heat exposure. We hope this book can inform priority setting for adaptation policies, to cope with extreme heat events. The contents of the book can be used to enhance the integrated and interdisciplinary assessment and management of health risks of extreme heat events, which have significant policy implications for health protection under climate change. Specifically, they include (1) improvement for managing or adapting to heat exposure under climate change; (2) options for responding and adapting to the impacts of heat exposure on vulnerable communities and improving the health outcomes for future climate change; and (3) providing evidence for researchers, health, social and emergency services leaders, and governments.

References

Griffiths, G. M., Chambers, L. E., Haylock, M. R., Manton, M. J., Nicholls, N., Baek, H.-J., Choi, Y., Della-Marta, P. M., Gosai, A., Iga, N., Lata, R., Laurent, V., Maitrepierre, L., Nakamigawa, H., Ouprasitwong, N., Solofa, D., Tahani, L., Thuy, D. T., Tibig, L., ..., & Zhai, P. (2005). Change in mean temperature as a predictor of extreme temperature change in the Asia–Pacific region. *International Journal of Climatology: A Journal of the Royal Meteorological Society, 25*(10), 1301–1330.

Guo, Y., Gasparrini, A., Armstrong, B. G., Tawatsupa, B., Tobias, A., Lavigne, E., Coelho, M. S. Z. S, Pan, X., Kim, H., Hashizume, M., Honda, Y., Guo, Y. L., Wu, C. F., Zanobetti, A., Schwartz, J. D., Bell, M. L., Scortichini, M., Michelozzi, P., Punnasiri, K., ..., & Tong, s. (2017). Heat wave and mortality: A multicountry, multicommunity study. *Environmental Health Perspectives, 125*(8), 087006.

Matthews, T. (2018). Humid heat and climate change. *Progress in Physical Geography: Earth and Environment, 42*(3), 391–405.

Moon, J. (2021). The effect of the heatwave on the morbidity and mortality of diabetes patients; a meta-analysis for the era of the climate crisis. *Environmental Research, 195*, 110762.

Perkins-Kirkpatrick, S., & Gibson, P. (2017). Changes in regional heatwave characteristics as a function of increasing global temperature. *Scientific Reports,* 7(1), 1–12.

Perkins, S. E., & Alexander, L. V. (2013). On the measurement of heat waves. *Journal of Climate, 26*(13), 4500–4517.

CHAPTER 2

Extreme heat and climate change

Sarah Perkins-Kirkpatrick[1] and Donna Green[2]
[1]School of Science, UNSW Canberra, Canberra, Australia; [2]Climate Change Research Centre, University of New South Wales, Sydney, NSW, Australia

1. Introduction and history of climate change

Since the industrial revolution, the global climate has warmed by approximately 1°C, with human activity being the main driver of this increase (IPCC, 2013). The release of greenhouse gases into the atmosphere by activities such as burning fossil fuels and changes in land use has disrupted the earth's energy budget such that it is no longer in equilibrium. Shortwave energy from the sun is initially absorbed by the earth's surface, and energy is then emitted as longwave radiation. Greenhouse gases work by trapping some of this longwave radiation within the lower atmosphere. While the presence of a greenhouse effect is natural and kept the earth's temperature at an average habitable 14°C during preindustrial times, enhancing it leads to increases temperatures in the lower part of the atmosphere (the troposphere), which can have severe consequences on many human and natural systems (Hoegh-Guldberg et al., 2018). In 2018, atmospheric carbon dioxide was 408 ppm (Blunden & Arndt, 2019), at its highest level in millions of years (Tripati et al., 2009; Bartoli et al., 2011). Since measurements of atmospheric carbon dioxide began in 1958, global concentrations have risen by almost 100 ppm (Blunden & Arndt 2019). Preindustrial levels were no more than 280 ppm, oscillating between 180 and 280 ppm over the past 800,000 years (Tripati et al., 2009). Human influences on atmospheric composition also extend beyond carbon dioxide. For example, aerosols can (temporarily) reduce local and regional temperatures, and other greenhouse gases such as methane, nitrous oxide, and halocarbons are released into the atmosphere (IPCC, 2013). However, the overall net effect of anthropogenic influence on global climate system is an increase of radiative forcing, thus increasing atmospheric temperature. Moreover, the greenhouse gas currently responsible for most of the increase in radiative forcing and resulting warming is carbon dioxide (Blunden & Arndt 2019; IPCC, 2013).

Heat Exposure and Human Health in the Context of Climate Change
ISBN 978-0-12-819080-7
https://doi.org/10.1016/B978-0-12-819080-7.00006-9
© 2023 Elsevier Inc.
All rights reserved.

Although most of the warming due to the anthropogenic greenhouse effect has been felt since the 1950s (IPCC, 2013), knowledge of the greenhouse properties of carbon dioxide has been known since the late 19th century. In 1856, scientist and inventor Eunice Foote was the first scientist to demonstrate the heat-trapping properties of carbon dioxide and hypothesize that increasing amounts of carbon dioxide in the atmosphere would result in warmer conditions (Foote, 1856). Using a different experimental setup, in 1861, Irish physicist John Tyndall confirmed the heat-trapping properties of carbon dioxide, as well as exploring similar properties in methane (Tyndall, 1861). Over 30 years later, Swedish physicist Svante Arrhenius calculated how increases in atmospheric carbon dioxide would enhance the greenhouse effect and thus increase atmospheric temperature (Arrhenius, 1896). Arrhenius proposed that a geometric change in carbon dioxide would result in an arithmetic change temperature (Arrhenius, 1896), the theory on which contemporary estimates of climate sensitivity are based. Guy Callendar, an English engineer, was the first to document in the late 1930s that Arrhenius' predictions were already underway, demonstrating that increases in global temperature were correlating with increases in atmospheric carbon dioxide levels (Callendar, 1938). By the 1950s, other scientists started to be convinced of Callendar's findings, which led to the establishment of the Keeling curve, which has consistently measured atmospheric carbon dioxide for the past 60 years (Keeling, 1960).

By the mid-1970s, the first numerical climate model had been built to describe how atmospheric temperature would respond to a doubling of atmospheric carbon dioxide (Manabe & Wetherald, 1975, 1980). During the 1980s, the establishment of physical climate models, along with other scientific advances and evidence, underpinned a consensus among scientists of the human influence on the climate system. The formulation of the Intergovernmental Panel on Climate Change (IPCC) occurred during the late 1980s, and the publishing of the first assessment report on climate change occurred soon after in 1990 (Houghton et al., 1990). In total, the IPCC has produced five assessment reports, which detail scientific evidence on how human activity has altered global and regional climates. More recent iterations include three working groups—the scientific basis (WGI); impacts, adaptation, and vulnerability (WGII); and mitigation of climate change (WGIII). After surveying the evidence in thousands of scientific peer-reviewed articles, the fifth assessment report (AR5) of WGI concluded that warming of the climate is unequivocal; the rate of warming since 1950 has been unprecedented; human influence on the climate system is clear; and further emissions of greenhouse gases will cause further warming and subsequent changes in the global climate system (IPCC, 2013).

In addition to increasing global average temperature, there are other physical indicators that the earth's climate is changing due to increases in atmospheric carbon dioxide. Arctic summer sea ice extent, ice sheets, and sea ice are all declining (IPCC, 2019); sea levels, ocean heat content, and acidity are rising (Rhein et al., 2013); glaciers and snow cover are retreating (IPCC, 2019); spring is starting earlier (Schwartz et al., 2006); and species are migrating (Lutz et al., 2013). Extreme temperature events are also changing, with cool extreme events occurring less frequently and heat extremes occurring more frequently (Seneviratne et al., 2012). Heatwaves, which are a particular type of heat extreme, are no exception. Significantly increasing trends in observed heatwaves have been detected over many parts of the world (Perkins-Kirkpatrick & Lewis, 2020), with these trends expected to increase as anthropogenic climate change continues (Seneviratne et al., 2016; Perkins-Kirkpatrick & Gibson, 2017).

The next section of this chapter continues with a general explanation of current mitigation and adaptation strategies that aid in reducing anthropogenic greenhouse gas emissions, as well as the impacts that climate change may cause. The last two sections then provide a specific focus on how climate change is influencing heatwaves, by providing a discussion around our climatological understanding of heatwaves: how heatwaves are measured; their physical drivers, observed changes, and future projections alongside changes in average temperature; and attributing changes in heatwaves to anthropogenic influence on the climate.

2. General mitigation and adaptation strategies to climate change

2.1 Mitigation strategies

Mitigation strategies are designed to reduce, or prevent, greenhouse gas emissions from entering the atmosphere, and they are enacted through policies at the international, national, and local levels. The majority of national-level mitigation strategies stem from the 1992 United Nations Framework Convention on Climate Change (UNFCCC), which had an objective to stabilize "greenhouse gas concentrations in the atmosphere at a level that would prevent dangerous anthropogenic interference with the climate system" (UNFCCC, 1992). Subsequent annual international agreements were made after national ratification of domestic emission reduction policies. The most recent international mitigation policy, developed from the work of the IPCC, has a goal of keeping the change in average global temperatures to less than 1.5°C degrees above preindustrial levels (IPCC, 2018). Within this overarching framework, individual nations decide themselves how they would respond to this international target. The 2016 "Paris Agreement" was considered a milestone in international negotiations over the amount of emission mitigation needed to respond effectively to scientifically informed emission scenarios (UNFCCC, 2015). To meet the Paris targets, nations submit pledges to detail their individual approaches to reduce emissions. One major concern with this approach is that the Paris Agreement and the pledges linked to it are not binding, and there is little recourse if a nation does not meet its target (Bodansky, 2016). Compounding this issue is the fact that a number of significant emission sources were not covered by these discussions (Fichert et al., 2020). And, despite a majority of nations submitting pledges to meet the Paris Agreement, as of 2020, the overwhelming majority of nations' current emissions trajectories are unlikely to meet these commitments (Watson et al., 2019).

International and national climate mitigation strategies tend to be integrated into sustainable development frameworks (such as the United Nations Sustainable Development Goals) to ensure equity is factored into international decision-making relating climate policies with environmental and social development (Marmot & Bell, 2018). This integrated approach recognizes that mitigation strategies need to respond to the collective action problem of climate change. This approach to recognize that some nations, and in many cases, some subsections of society within those nations—as well as some corporations—are disproportionally responsible for the majority of historic emissions and, for this reason, are responsible for taking a larger share of the earlier policy response to redress this inequity (Ekwurzel et al., 2017). Such international policy approaches have included the Common But Differentiated Responsibilities (CBDR) framework, which was developed in the UNFCCC process in 1992 (Pauw et al., 2014). These approaches were designed to encourage international cooperation to facilitate the development and transfer of improved renewable energy technologies and processes to encourage their swift deployment and integration into existing energy networks.

In practice, mitigation strategies to reduce greenhouse gas emissions can implement new technologies and new energy sources, engage in more efficiency practices to increase overall energy efficiency, or change consumer behavior. These strategies differ greatly at various spatial scales. Examples include international, using a markets based approach, e.g., emissions trading (Ellerman & Buchner, 2007); national, using a tax structure, e.g., carbon taxes (Baranzini et al., 2000); regional, using local resources, e.g., developing hydroelectricity; local using renewable resources, e.g., minigrids of solar electricity networks; or individual, using domestic preferences, e.g., encouraging the use of energy efficient appliances, the use of insulation, or installation of domestic solar water heating (Bruckner et al., 2014). These policies can be focused on a single sector or span multiple sectors. Mitigation strategies can be integrated with adaptation strategies or stand alone with the majority of them, focusing primarily on the reduction of carbon dioxide. The biggest reductions of carbon dioxide emissions are occurring from the transformation of our energy systems from burning fossil fuels, to the use of a range of renewable technologies to provide electricity, heat, and power (IEA, 2020a, 2020b). These energy transformations have been integrated into some national-level policy and planning approaches for decades, e.g., in Germany (Beveridge and Kern, 2013). Other examples of national-level approaches include Costa Rica's enacted decarbonization plan (Government of Costa Rica, 2019), and the proposed "Green New Deal" in the United States (Galvin & Healy, 2020), or the nongovernmental organization proposed "Million Jobs Plan for Australia" (Lenko, 2020). Due to wide variation in availability and accessibility of natural resources, expertise, domestic drivers, trade interests, and culture, nations (and states within nations) have very different approaches to reducing the carbon intensity of their electricity. The decoupling of emissions and electricity generation has been a very useful tool in understanding how effectively a nation has been in "decarbonizing" its economy (Hughes et al., 2020). Renewable energy technologies are highly modular, which means that they can be developed and installed quickly and scaled up as necessary. Not only can this decentralization reduce the inefficiencies in transmission created by centralized energy generation (Pepermans et al., 2005), but it can also reduce the potential for industry cartel behavior.

Mitigation of greenhouse gas emissions in the transport sector, especially in the domestic space, is currently undergoing rapid transition in many countries (Riesz et al., 2016). The shift from fossil-fueled cars and trucks to hybrid and full electric vehicles (EVs) is underway, with several governments' national policies mandating the phasing out of fossil fuels within two decades. Given the overwhelming impact of heatwaves in cities, and the urban heat island effect to exacerbate health concerns, reducing greenhouse gas emissions, air pollution, and localized heating can occur through the transition of the transport sector away from fossil fuels. This transition needs to take place within a strong "active" transport framework, which prioritizes walking, bicycling, and public transport to take advantage of the significant cobenefits that accrue to health and climate (Watts et al., 2015). For this to occur, urban planning and transport infrastructure need to be reconceived to encourage the use of public transport, bicycles, and walking as the preferred commuting option. There are significant health cobenefits from this approach, which include a reduction in air pollution from primary and secondary pollutants (e.g., particulate matter and ozone) caused by the predominantly fossil-fueled transport sector and which can exacerbate the impact of heatwaves on human health (Xie et al., 2018).

Different approaches to land use can contribute significantly to mitigation of greenhouse gas emissions. The UNFCCC recognizes that promoting sustainable management of forests and oceans, coastal, and marine ecosystems including management of deforestation, cropland, and grazing can contribute significantly to mitigation activities (Arnth et al., 2019). Recognition that indigenous land management practices, for example, cool season burning, mosaic clearing, and encouraging complex cropping, serves an important role in promoting sustainable activities that can sequester carbon in the ground (Rights and Resources Initiative, 2019).

Active removal of carbon dioxide through large-scale geoengineering projects is being explored and could be a useful addition to more proven mitigation techniques. However, none of these approaches are currently cost-competitive and cannot be relied on as part of an emission reduction strategy (Edenhofer et al., 2012).

2.2 Adaptation strategies

Adaptation strategies are designed to adjust systems and societies to cope with the locked-in as well as the larger, if less certain, future impacts of climate change (Mimura et al., 2014). These strategies are essential to reduce the human and ecosystem costs arising from the unavoidable impacts of climate change including sea level rise, coastal flooding, and increasing heatwaves and bushfires. Depending on how significantly we mitigate our current and future greenhouse gas emissions, ever more extreme adaptation strategies will be required (Araos et al., 2016; Brzoska & Fröhlich, 2016).

There is no clear agreement on how adaptation strategies are defined, although some key terms are useful to explain how different stakeholders generally conceive these processes more generally. Broadly, adaptation strategies can be considered as those that include planned and active intervention to adjust natural or human systems in response to actual or expected climate change (Mimura et al., 2014; Singh et al., 2020). These strategies aim to reduce existing or future harm and in some cases strengthen benefits that might accrue from their implementation.

Anticipatory adaptation strategies, as their name suggests, are developed and implemented prior to climate impacts causing actual harm, for example, the development of regional or local heatwave warning systems. Reactive adaptation occurs rapidly, for example, in response to increasing temperatures, there might be greater use of air conditioning, which can at times even serve to be maladaptive (Viguié et al., 2020).

These distinctions between different forms of adaptation are not, however, clear-cut. For example, over slightly longer time horizons, and response to increasing extreme weather, state, or national building codes might be developed to enable cooler internal temperatures to be mandated, and the requirements of stronger structures to be built or retrofitted to withstand increasingly intense weather. Policies enacted to promote these adaptation activities are reactive *and* anticipatory, and usefully, they will also mitigate emissions (i.e., they help to reduce the future impacts that we will be forced to adapt to). Many of these activities are classed as incremental adaptation (with the aim of maintaining the integrity of a system close to status quo) rather than transformational adaptation, which is what is increasingly being called for as part of the energy transition plans, as outlined in the previous section.

Adaptation strategies that build socioeconomic resilience generally develop long-term, more effective responses. Community-led activities that are shaped to promote equity and engagement and recognize the interdependent nature of health, social, and cultural systems frequently incidentally serve to build resilience (Gustfson et al., 2018; Russell-Smith et al., 2017). In this light, increasing the education of girls, the promotion of information about family planning, microloans, and hyperlocal social networks that create empowerment platforms such as http://www.kiva.org/ or http://www.atmaconnect.org/ are alternative approaches to build resilience at a local level, and in so doing, indirectly increase adaptive capacity (Ampaire et al., 2020).

Enhancing adaptive capacity can help to reduce vulnerability to climate change. Increasing intra- and international equity, reducing poverty, increasing education, designing resilient infrastructure systems, and promoting local indigenous practice and knowledge can all contribute to this process. In so doing, adaptation strategies can be maximized. There are many stakeholders in adaptation processes: to be successful, an iterative process between governments at various levels and local communities that will be affected by the activities needs to be developed.

Less industrialized nations have already felt disproportionate impacts from climate change, despite the fact that, in the main, these nations are least responsible for causing these impacts. Herold et al. (2017) use global atmospheric reanalyses to show that the world's lowest income countries are already experiencing greater increases in the occurrence of temperature extremes compared with the highest income countries and have been for over two decades. To avoid sea level rise, the issue of equity is especially significant, with the UNFCCC developing approaches to facilitate funding from countries with higher responsibility for historic emissions to provide financial resources to countries who have less responsibility. For less industrialized nations, the National Adaptation Programmes of Action (NAPA) has been created as part of the UNFCCC process. These plans allow countries to identify their most urgent and immediate adaptation projects to facilitate the flow of funding from more industrialized nations to support these self-identified priorities. It is important to recognize that only a very small amount of the billions of dollars committed to by richer nations has actually been transferred for these adaptation activities.

Traditional knowledge is being increasingly recognized as a valuable resource to draw to build resilience to change. This approach recognizes that fact that local knowledge is vital to adapt at fine scales and that prior adaptation to environmental change can provide locally relevant and tailored coping strategies with minimum outside interference and resource provision (Cochran et al., 2013). These activities will, however, not be enough to adapt to the range of conditions that are expected in coming decades, and so must be integrated into broader adaptation approaches at the regional and national levels.

In many cases, cities have proven to be the leaders in mitigation and adaptation activities. The C40 movement is a network of 94 of the world's megacities (incorporating more than 700 million people) committed to addressing climate change. The network has leveraged their ability to enact local land use planning and policies either in concert with state public health authorities, or in some cases, despite it. In so doing, these cities have shaped social and economic planning tools to incorporate adaptation and mitigation activities to respond to locally specific climate extremes—inclusive of extreme heat—as well as developing low carbon futures and energy-efficient practices (Sherlock et al., 2020; Watts, 2017).

In respect to heat exposure, public warning systems, such as those developed by Public Health England, are recent developments in adaptation strategies (Casanueva et al., 2019). By sending out text and online messaging the day(s) before forecast heatwaves, individual and local actions can be taken to prepare vulnerable individuals to respond. This might include more regular checks on elderly, keeping children inside buildings during school or reallocating patients in hospitals to prepare for an increase in admissions.

It is important to remember that climate change mitigation strategies and climate change adaptation are not mutually exclusive. Although mitigation strategies are generally aimed at tackling the causes and minimizing the possible impacts of climate change, and adaptation strategies aim to reduce the negative effects from climate change, in reality many mitigation strategies are also adaptation strategies.

3. Observed changes in heatwaves

3.1 Measuring heatwaves

Other than being described as prolonged periods of excessive heat, there is no universal definition of a heatwave (Perkins, 2015). Indeed, any heatwave definition employed or devised is likely purpose-specific. For example, when researching the impact of extreme heat on human health, a measure of humidity and/or windspeed (Steadman, 1979, 1984) and even very specific details such as an individual's metabolic rate and clothing factor may be incorporated (Fanger, 1970; Mayer & Hoppe, 1987; Pantavou et al., 2008). Climatologically, a measure of extreme temperature over a number of days is usually sufficient. Yet even within disciplines, although there are similarities across definitions, there is little agreement on the exact way to measure heatwaves. Thus, it is unlikely that a universal heatwave metric will ever be agreed upon.

Given that the rest of this chapter focuses on understanding changes in heatwaves from a climate perspective, there are some key features that climate-based heatwave definitions should encompass that allow for multiple aspects of heatwaves to be derived and studied. These include assessing *daily* extreme heat (which can be combined with humidity) via a relative, percentile-based threshold over a minimum number of consecutive days (at least 2–3; Perkins & Alexander, 2013). For example, defining a heatwave as at least 3 days in a row where average temperature exceeds the 90th percentile (e.g., Pezza et al., 2012; Perkins and Alexander. 2013; Cowan et al., 2014; Schoetter et al., 2015) allows for the interrogation of multiple characteristics, such as heatwave intensity, duration frequency, and timing (see Fischer & Schär, 2010; Perkins & Alexander, 2013; Perkins et al., 2015). Indeed, some heatwave definitions have been proposed that combine characteristics such as intensity and duration to allow a comparison of heatwave severity over space and time (Russo et al., 2014; 2015). Moreover, a relative threshold instead of an arbitrary measure (such as days over 30°C) allows for heatwaves to be derived relative to the local climate. For example, heatwaves occur over Spain and Siberia with disastrous impacts in both locations. Yet if an absolute threshold was used to define heatwaves, undoubtedly more would be measured over Spain due to the warmer underlying Mediterranean climate. Thus, given that extreme events are rare by definition, any metric defining an extreme event—inclusive of heatwaves—should be relative to local conditions, as what is rare under one climate regime may be commonplace under another.

3.2 Physical drivers of heatwaves

Generally speaking, there are three main physical drivers of heatwaves—synoptic systems, interactions with the land surface, and climate variability. All heatwaves are associated with an anticyclone, which is a persistent high-pressure system that drives heatwave conditions in a number of ways. Firstly, warm air is advected to the area affected by the heatwave from an adjacent build-up of heat (e.g., Della-Marta et al., 2007; Matsueda, 2011; Pezza et al., 2012). Secondly, clear skies allow for greater insolation, and thirdly, greater subsidence allows for adiabatic heating (i.e., descending air that heats up due to the extra pressure exerted on it; Quinting & Reeder, 2017; Loughran et al., 2019; Hochman et al., 2020). Depending on where a heatwave is occurring, these processes will have different relative roles (e.g., Hirsch et al., 2019). Moreover, the prevailing high-pressure system may last for varying lengths of time. For example, heatwave-driving persistent highs over Australia rarely last any more than a week (Pezza et al., 2012; Marshall et al., 2014), yet they can last for many weeks over Europe (Matsueda, 2011; Vautard et al., 2013). Regions where heatwaves persist for weeks or longer are generally driven by blocking highs, where upper-level atmospheric winds split due to the meandering of the jet stream, allowing a region to be "blocked" from the zonal jet stream flow, and cooler polar air cannot mix with hotter equatorial air (Cassou et al., 2005; Perkins, 2015). Yet over regions, heatwave-causing high-pressure systems are not completely blocked from cooler air flow and are usually shifted by another air mass more quickly (Marshall et al., 2014). However, regardless of how long they last, heatwave-causing high-pressure systems generally extend further aloft in the atmosphere (e.g., Boschat et al., 2015; Meehl and Tebaldi, 2004).

Land surface and atmosphere interactions are another key ingredients for heatwaves. When the land surface has plenty of moisture, latent heat is the dominant flux over sensible heat; however, these fluxes reverse when the soil is dry (Alexander, 2011). This means that during dry times, there is more energy available for heating up the atmosphere, thus increasing heatwave duration and severity (e.g., Seneviratne et al., 2006; Hirshi et al., 2011). This mechanism is more important over areas with reasonably high and consistent rainfall, as when rainfall ceases for a given reason for a relatively long time, soil moisture depletes and extreme temperatures therefore increase (Mueller & Seneviratne, 2012). It is therefore a more important heatwave driver over regions such as Europe (Fischer et al., 2007; Hirschi et al., 2011; Quesada et al., 2012) and parts of North America (Diffenbaugh et al., 2005), yet to a lesser extent over dryer regions such as Australia (Herold et al., 2016; Hirsch et al., 2019; Perkins et al., 2015).

Another important physical driver is low-frequency modes of climate variability. These modes—which include phenomena such as the El Nino/Southern Oscillation (ENSO), the Indian ocean Dipole (IOD), and the North Atlantic Oscillation—are regular or semiregular forms of natural climate variability that operate on monthly, seasonal, and yearly timescales and govern regional climate. For example, the El Nino phase of ENSO means hot and dry weather for Eastern Australia, but extreme rainfall and flooding over Western America (Trenberth, 1997). Indeed El Nino seasons also bring a higher frequency and more extreme heatwaves to Eastern Australia, whereas the Madden-Julian Oscillation and the Southern Annular Mode are more important for southern Australian heatwaves (Marshall et al., 2014). La Nina seasons—the opposite phase of El Nino—have been associated with warmer temperatures over North America (Hoerling et al., 2013; Kenyon & Hegerl, 2008; Koster et al., 2009). Over Europe and Scandinavia, the NAO and Atlantic Multidecadal Oscillation are important drivers of heatwaves (Della-Marta et al., 2007).

Climate modes are important since they influence certain weather patterns over shorter timescales. For example, a positive NAO phase over Europe is conducive to more blocking highs, which increase the duration and severity of heatwaves (Della-Marta et al., 2007). Similarly, there are clearer skies and less rainfall over Eastern Australia during El Nino, allowing for depleted soil moisture and higher incidences of heatwaves (Herold et al., 2016; Perkins et al., 2015). Indeed, all heatwaves are a manifestation of multiple physical drivers that interact with each other. For example, over Europe, summer heatwaves usually only occur after a dry winter or spring, *and* under anticyclonic conditions/blocking highs that are usually associated with a positive NAO (Quesada et al., 2012). Over some Australian regions, the combination of a desiccated land surface and persistent highs increases the frequency of heatwaves by a factor of eight (Gibson et al., 2017). Moreover, the warm air advected into a heatwave region can enhance land surface drying, which goes on to increase the duration and/or intensity of local heatwaves (Miralles et al., 2014). Thus, there are generally multiple reasons why a heatwave may occur, with varying weights of importance depending on the region affected and the interactions of the physical mechanisms.

3.3 Observed changes in temperature and heatwaves

In a stationary climate where there is no long-term change, heatwaves would occur regularly yet rarely. Since they are an extreme event, they are rare by definition but would be expected to occur sometimes due to the physical mechanisms described before. Moreover, their distribution over time would not be even: some seasons or years would experience far more heatwaves than others due to which physical drivers were present and their interactions with one another. However, over recent decades, clear and significant trends of increasing heatwave frequency, intensity, and duration have been detected over various global regions. Indeed, observed changes in heatwaves tend to closely track changes in average temperature. Since the industrial revolution, global average temperature has warmed by 1°C, with most of this warming occurring since the 1950s, and each of the past three decades has been significantly warmer than the preceding decade (IPCC, 2014, Fig. 2.1). Since the turn of the 20th century, almost everywhere in the world has experienced a significant increase in average temperature between 0.4 and 1.25°C, with higher latitude regions in the Northern Hemisphere and west Africa warming by up to 2.5°C (IPCC, 2014).

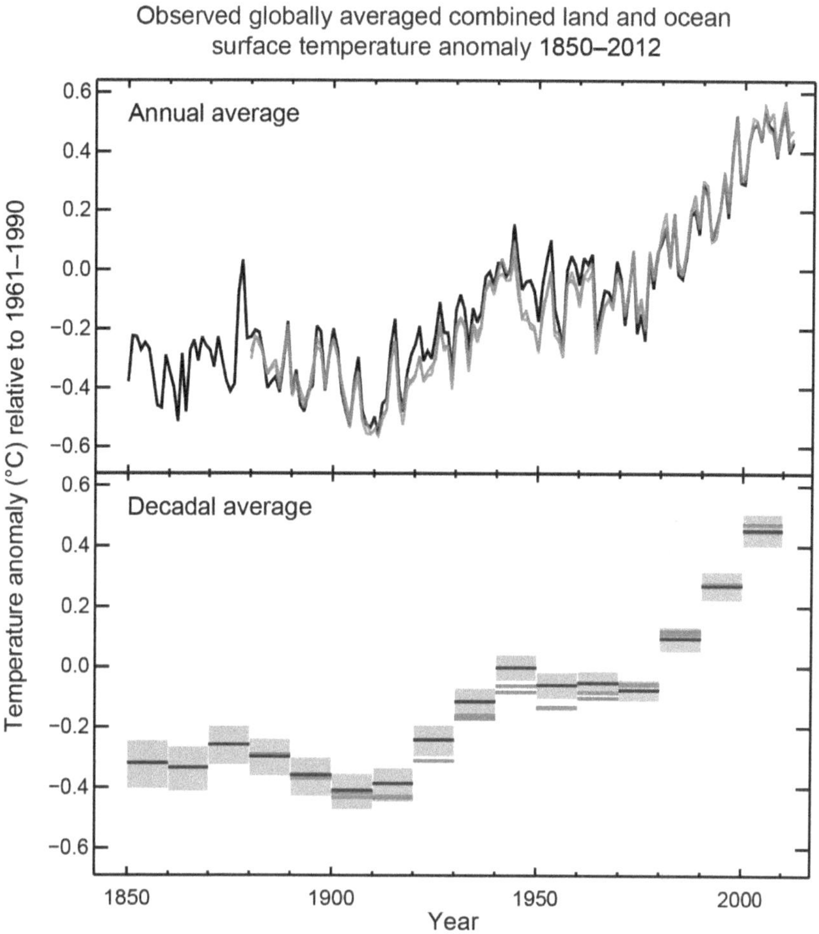

Figure 2.1 Annual average (top) and decadal average (bottom) global temperature changes from 1950 to 2012, as estimated by three observational data sets. Temperatures are anomalies relative to 1961–90. Note that decadal temperature averages for the past three decades do not overlap. *(Taken from Figure SPM 1a in IPCC. (2013). Climate change 2013: The physical science basis. Contribution of working group I to the Fifth assessment report of the intergovernmental panel on climate change. In edited by T.F. Stocker, D. Qin, G.-K. Plattner, M. Tignor, S.K. Allen, J. Boschung, A. Nauels, Y. Xia, V. Bex and P.M. Midgley, 1535. New York, NY, USA, reproduced here with permission.)*

Earlier studies focusing on observed global heatwave changes (Frich et al., 2002; Alexander et al., 2006) reported varying changes in heatwave duration, with some small areas showing a decline. However, this was in part due to the primitive indices used to detect heatwaves, and most of the western world experienced increases since the 1950s (Frich et al., 2002; Alexander et al., 2006). More recent work (Perkins et al., 2012), based on more appropriate heatwave metrics, has detected significant increases in heatwave frequency, duration, and a measure of intensity since 1950 over large parts of the world. Moreover, these trends are consistent despite the underlying definition of a heatwave (see Perkins & Alexander, 2013), and the largest and most significant trend is the increase in the number of heatwave days per season, which has risen by 1–2 days/decade across many parts of the western world. The most recent assessment of observed global heatwave trends (Perkins-Kirkpatrick & Lewis, 2020) demonstrated that these trends have continued over recent years, and also over South America and Africa, where previous information on observed trends was limited (Fig. 2.2). Moreover, while observed average heatwave intensity has changed little (see Perkins-Kirkpatrick & Lewis, 2020), there have been significant increases in cumulative heat—or heat exposure—during heatwaves (i.e., the extra heat that heatwaves generate) since at least the 1950s of 2–8°C/decade (Perkins-Kirkpatrick & Lewis, 2020). These trends are mainly driven by an increase in heatwave days and therefore more occurrences over temperatures above a heatwave threshold (see Fig. 2.2). However, caution should be exercised when interpreting heatwave trends over shorter time periods (less than a few decades), due to their high variability based on the physical mechanisms described before, which may swamp detectable signals over shorter temporal lengths (Perkins-Kirkpatrick et al., 2017; Perkins-Kirkpatrick & Lewis, 2020).

While all heatwave characteristics are broadly increasing across the world, there are regional variations in trends. Over Europe, heatwave duration has significantly increased at many locations since the 1880s (Della-Marta et al., 2007). Yet over Australia, heatwave frequency demonstrates the largest and most significant trends (Perkins & Alexander, 2013). Indeed, the size and significance of heatwave trends fluctuate at local and city scales, where individual cities can demonstrate very different rates of change (e.g., Steffen et al., 2014). Over the central United States, large numbers of heatwaves occurred during the 1930s, concurrent with the dust-bowl drought (Hoerling et al., 2013), with increasing frequency trends over parts of the United States since the 1960s (Kunkel et al., 2010). Fifty US cities have also reported increasing frequency, duration, and intensity trends over a similar timeframe, though exact trends differ depending on the city of interest (Habeeb et al., 2015). Alaska has also experienced an alarming and dramatic increase in heatwaves since 2000 (Peterson et al.,

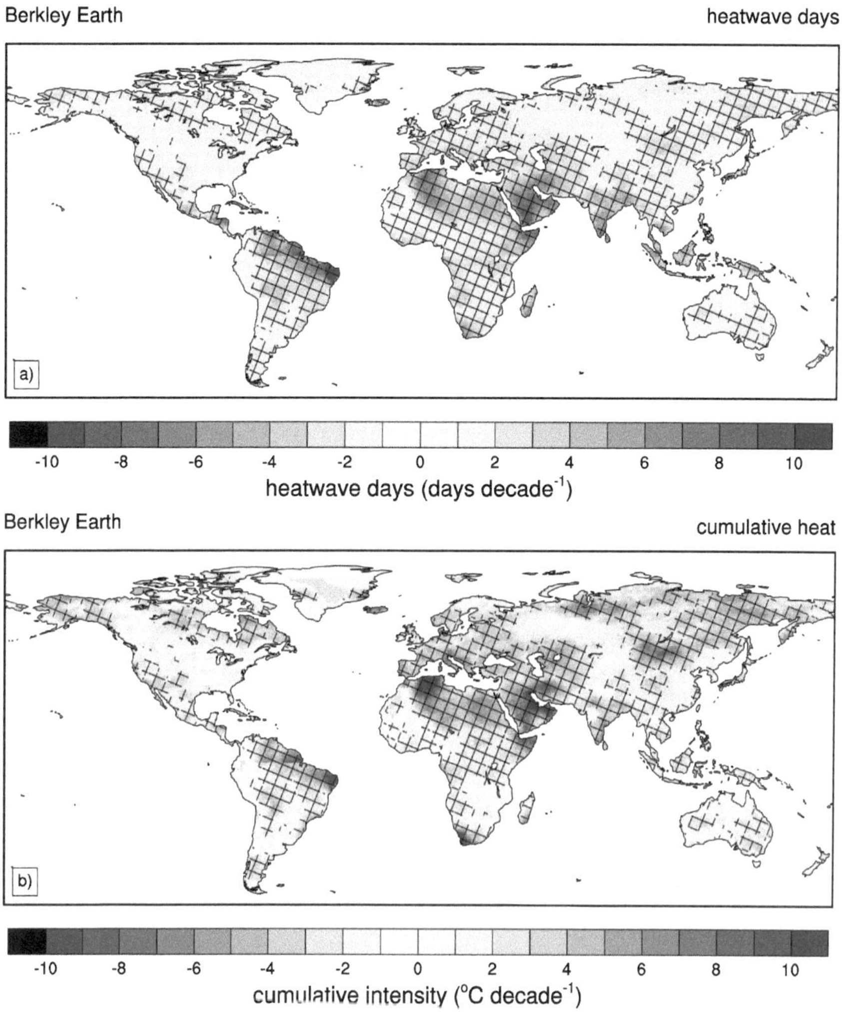

Figure 2.2 Changes in (A) heatwave days and (B) cumulative heat (heat exposure) between 1950 and 2014 using the Berkley Earth observational data set. Hatching indicates areas where trends are statistically significant. Trends in cumulative heat are largely (though not completely) driven by heatwave days. *(Adapted from Figure 1 by Perkins-Kirkpatrick, S. E., & Lewis, S. C. (2020). Increasing trends in regional heatwaves.* Nature Communications, 11, *3357.)*

2013). However, little change has been detected over southern Central United States likely due to the "warming hole" phenomena and its connection to the hydrological cycle (Pan et al., 2004; Portmann et al., 2009). Over China, significant increases in heatwave duration have been detected since 1960 (Ding et al., 2011) with more rapid changes since 1990 (You et al., 2017), and heatwaves have significantly increased in frequency, intensity, and duration over northwest India (Rohini et al., 2016), and in frequency for many other stations across the country (Pai et al., 2013). Increasing duration trends since the 1960s using a number of heatwave definitions have also been reported for Argentina, with the most severe heatwaves occurring during recent decades (Rusticucci et al., 2016).

4. Future changes in temperature and extreme heat

4.1 Physical climate models and emission scenarios

To understand future changes in climate, physical climate models are employed. These are mathematical representations of the global climate system underpinned by the laws of physics. While fundamental climate physics such as the conservation of energy, mass, and momentum are identical across climate models, how they represent other processes can differ. For example, interactions between the land surface and atmosphere and how this influences local latent and sensible heat fluxes, which contributes to local heating or cooling, may be represented differently across models (e.g., Hirsch et al., 2019; Ukkola et al., 2018, 2020). Thus, when investigating changes in future climate, an ensemble rather than a single model is employed, such that a spread of plausible outcomes is sampled.

The most commonly used and state-of-the-art ensemble is the Coupled Model Intercomparison Project (CMIP) archive, which is the suite of global climate models (GCMs) that underpin our understanding of future climatic changes in each iteration of the IPCC assessment reports. AR5 focused on four key scenarios (known as Representative Concentration Pathways or RCPs), which included socioeconomic changes such as population growth, adoption of mitigation policies, and reliance on fossil fuels, which then infer different estimates in the radiative forcing on the climate system. Climate models then include these estimates of radiative forcing over time (for example, decades to centuries) to assess how the state of the global climate changes dependent on which scenario transpires. This is useful in helping us interrogate future changes in many climate-related entities, from global average temperature to regional changes in heatwaves.

The overall increase in global average temperature by 2100 is dependent on which scenario plays out (Fig. 2.3) scaling with the radiative forcing of the respective scenario. For example, under a midrange scenario (RCP-4.5, Fig. 2.3), global warming by 2100 will likely be in the range of 1.1–2.6°C,

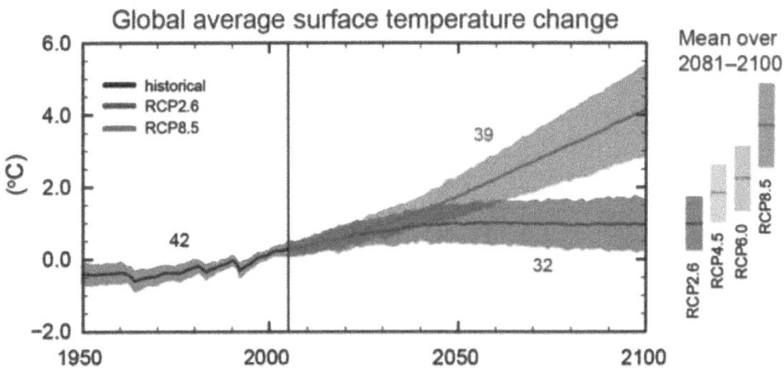

Figure 2.3 Global average surface temperature changes as projected by the CMIP5 model ensemble. Changes are anomalies relative to 1961–90 for each model. Projections are for the historical period (1950–2005, black), and the RCO 2.6 and 8.5 scenarios (2006–2100; blue [dark gray in print version] and red [light gray in print version], respectively). Numbers indicate the amount of models participating in each scenario. The ensemble range of average projected temperature range for four emission scenarios is shown on the far right. *(Taken from Figure SPM 7a in IPCC. (2013). Climate change 2013: The physical science basis. Contribution of working group I to the Fifth assessment report of the Intergovernmental Panel on climate change. In edited by T.F. Stocker, D. Qin, G.-K. Plattner, M. Tignor, S.K. Allen, J. Boschung, A. Nauels, Y. Xia, V. Bex and P.M. Midgley, 1535. New York, NY, USA, and reproduced with permission.)*

compared with the end of the 20th century (IPCC, 2018). Under the high-range scenario (RCP-8.5, Fig. 2.3), this range increases to 2.6–4.8°C (IPCC, 2018), and under the lowest-range scenario (RCP-2.6, Fig. 2.3), the range is 0.3–1.7°C (IPCC, 2018). However, it is worth noting that global warming under all scenarios is substantially larger when compared with earlier periods, due to less anthropogenic influence on the climate. When compared with 1850–1900, it is likely that global warming will exceed 1.5°C for all scenarios except RCP-2.6 (IPCC, 2018), which is an important threshold regarding the extent of impacts, as well as adaptive and mitigative measures to climate change by the global community (IPCC, 2018).

4.2 Projected changes in heatwaves

While the aforementioned changes in global average temperature may not initially seem large, there is a tight and sensitive coupling between average temperature and extreme events, such as heatwaves. A small shift in average temperature toward warmer conditions—even 1°C or less—is enough to see a large increase in the *frequency* of extremely hot temperatures, as well as higher occurrences of record-hot *intensities* (Fig. 2.4). It is important to

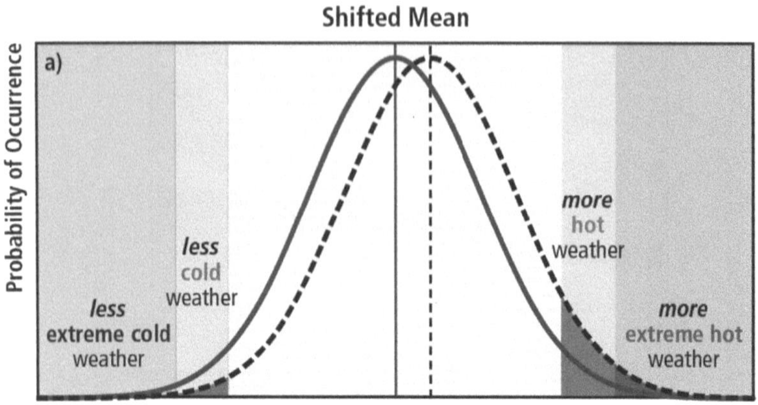

Figure 2.4 Changes in extreme temperatures when the mean of the distribution is shifted. When average temperature warms, there is an increase in the occurrence of hot weather, with extreme hot weather also occurring more often. Similarly, there is a reduction in the frequency of cold extremes. *(Taken from Figure SPM 3a from IPCC. (2012). Managing the risks of extreme events and disasters to advance climate change adaptation. A special report of working groups I and II of the intergovernmental Panel on climate change. In edited by C.B. Field, V. Barros, T.F. Stocker, and D.J. Dokken D. Qin, K.L. Ebi, M.D. Mastrandrea, K.J. Mach, G.-K. Plattner, S.K. Allen, M. Tignor, and P.M. Midgley, 582. Cambridge, UK, and New York, NY, USA: Cambridge University Press, reproduced with permission.)*

remember that extreme events—inclusive of heatwaves—are rare by definition, and although an increase in frequency technically makes them less rare, associated impacts inclusive of those on human health will still be large and adverse due to factors such as physical limitations of the human body to regulate temperature (e.g., Sherwood & Huber, 2010), and specific changes in the respective local climate. It follows that a warming of average temperature generally means a decrease in the frequency and intensity of cold extremes (Fig. 2.4), and over some regions, this is clearly evidenced by the occurrence of record-breaking warm extremes well outnumbering record-breaking cold extremes (e.g., Lewis & King, 2015).

Given the scaling relationship between average temperature and extreme heat, increases in the intensity, frequency, and duration of heatwaves as a function of global average temperature are guaranteed (e.g., Guo et al., 2017; Perkins-Kirkpatrick et al., 2017; Seneviratne et al., 2016). The exact change depends on the heatwave characteristic and region of interest; however, almost all characteristics and region combinations display a robust increase relative to global warming (Fig. 2.5; Perkins-Kirkpatrick & Gibson, 2017). The exception is a number of tropical regions where the number of

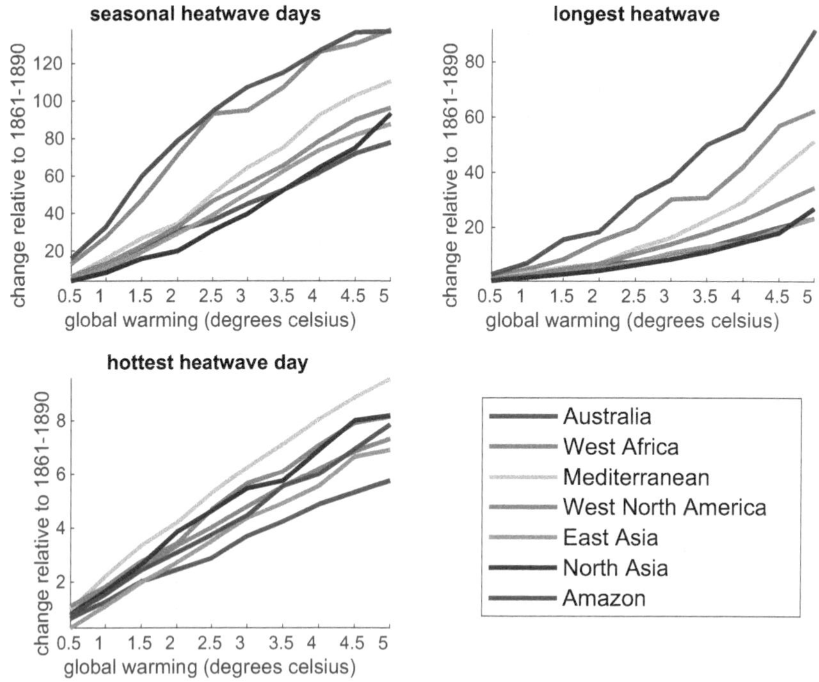

Figure 2.5 Projections of three heatwave metrics relative to global warming for seven regions. Projections are calculated as the median of the CMIP5 global climate model ensemble, with both heatwaves and average temperature anomalies calculated relative to 1861–90 from each models' historical simulation. Projections are calculated by splicing the historical and RCP8.5 simulations together for each model. *(Adapted from Figure 2 in Perkins-Kirkpatrick, S. E., & Gibson, P. B. (2017). Changes in regional heatwave characteristics as a function of increasing global temperature.* Scientific Reports, 7, *12256.)*

heatwave events begins to decline after reaching 2–3°C global warming, as there are less discrete events that instead turn into larger, longer events (Perkins-Kirkpatrick & Gibson, 2017). Similar to the observed changes described before, the biggest changes in heatwaves relative to global average temperature are seen for heatwave frequency (Fig. 2.5; Perkins-Kirkpatrick & Gibson, 2017). Over some regions, the intensity of heatwaves warms at the same pace as global warming (e.g., Australia), while over others they warm almost two times faster (e.g., Mediterranean; Seneviratne et al., 2016; Perkins-Kirkpatrick et al., 2017). There is also evidence that heatwaves become longer more quickly beyond 3–3.5°C, concurrent with the overall number of events decreasing (Fig. 2.5). Due to the differences in climate models described before, there is a spread in the severity of the projected relationship between global average temperature and heatwaves; however, all show an increase (e.g., Perkins-Kirkpatrick & Gibson, 2017). The strength of the relationship is also irrespective of emission scenario (Seneviratne et al., 2016; Perkins-Kirkpatrick & Gibson, 2017); however, given that less global warming will occur by 2100 under lower emission scenarios, it follows that increases in heatwaves will also be less.

Future time-based changes in regional heatwaves are also well documented. Over Australia, increases in heatwave frequency are higher in the tropical north (up to 50 extra heatwave days) compared with the temperate south (up to 15 extra heatwave days) by 2100 under RCP8.5, whereas projections of heatwave intensity are larger in the south (up to 3°C) than the north (up to 1°C; Cowan et al., 2014). Indeed, other tropical regions such as Pacific Islands and Central and South America are projected to see sharp increases in heatwave frequency with relatively modest changes in intensity (Angeles-Malaspina et al., 2018; Feron et al., 2019; Perkins, 2011). This is because there is little annual variance in the overall temperature distribution of these climates, making the average-extreme relationship described before more tightly coupled. Therefore a small shift in average temperature exacerbates the increase in the frequency of warm extremes. Moreover, because of the small overall temperature range of tropical regions, increases in heatwave frequency and intensity alone may matter little, since such extreme temperatures are not that much larger than the average. However, coupling these increases increasing humidity would be problematic for human health and heat stress in tropical regions (Kjellstrom et al., 2016).

Over Europe, increases in the frequency, duration, intensity, and spatial extent of heatwaves are well documented (e.g., Schoetter et al., 2015; Russo et al., 2014, 2015; Guerreiro et al., 2018). These increases scale with the respective emissions scenario, with longer, larger, and hotter events more common under a high emission future (e.g., Schoetter et al., 2015). Very intense events such as the 2003 European and 2010 Russian heatwaves will become more common by 2100 under a high emission scenario and will be regularly exceeded by larger record-breaking events; however, they would still be considered reasonably rare under a midrange emission scenario (Russo et al., 2014). By 2100, the number of heatwave days is projected to increase 20-fold over Spain and the Mediterranean (Fischer et al., 2010; Fischer and Shar, 2009), with a 200%−500% increase in hot temperatures and heat extremes also over this period (Diffenbaugh and

Ashaq, 2010). Over China, regions that currently experience more severe heatwaves are projected to experience events of larger intensities and greater frequency under enhanced global warming (Guo et al., 2017), with significant increases detectable by 2050. Detectable increases in heatwave intensity and frequency are projected over Northern America by middle of the 21st century (Diffanbaugh and Asfaq, 2010; Jeong et al., 2016). There is high subregional variation in projections over Canada (Jeong et al., 2016), larger increases in frequency over the southwest United States, and larger increases in intensity in the northeast (Diffenbaugh and Ashfaq, 2010; Lau and Nath, 2012; Wuebbles et al., 2014). Under a medium emission scenario, record-breaking warm extremes in the United States are projected outnumber cold events by 50-fold by 2100 (Meehl et al., 2009). Increases in heatwave intensity and frequency are more prominent under a high emission compared with a medium emission scenario over all of South America, which are concurrent with larger changes in extremely warm summers (Feron et al., 2019). Indian heatwaves are also projected to intensify by 2100 with some events reaching over 45°C under a high emission scenario, and increased frequencies in heatwaves over Southern India where heatwaves have rarely occurred historically (Murari et al., 2015).

It is very clear that regardless of whether changes are presented temporally or relative to global average temperature, heatwaves will get worse—both regionally and globally—under intensified anthropogenic climate change. However, the impacts of heatwaves on human health can be amplified by humidity, since higher amounts of atmospheric moisture inhibit the effectiveness of perspiration. A theoretical limit of human heat stress is a sustained wet-bulb (T_W, which measures the saturation level of the air) temperature of 35°C, which would induce hyperthermia (Sherwood & Huber, 2010). This is because skin temperature is regulated to be 35°C or below such that core temperature can be sustained at around 37°C (Sherwood & Huber, 2010). Thus, a T_W value this high, where means the air is fully saturated at 35°C, means that humans cannot regulate their temperature via perspiration, as there is no moisture gradient between the skin and the air. Projections over the Middle East show T_W reaching 30–33°C by 2100 under a high emission scenario, calling into question the habitability of the region without drastic adaptation measures (Pal and Eltahir, 2015). Heat stress projections for the North China plain are also concerning, with T_W projected to reach 34°C under a high emission scenario by 2100 (Kang and Eltahir, 2018). Concerningly, a T_W of 35°C is projected by 2100 over some low-latitude regions under a high emission scenario; however, this theoretical heat stress threshold is not reached under lower emission scenarios (Coffel et al., 2017). Another concern is the disproportional impact of increased heat stress on disadvantaged communities. Many areas where heat stress is projected to increase are highly populated and/or have little adaptive capacity (e.g., Coffel et al., 2017), bringing to light the inequality of impacts that are exacerbated by climate change (King et al., 2017).

4.3 Finding the anthropogenic signal behind extreme heat events

As discussed before, it is clear that the intensity, frequency, and duration of heatwaves will worsen under climate change. However, climate scientists can also estimate how anthropogenic influence on the climate has increased the frequency and severity of extreme events such as heatwaves. This field—known as event detection and attribution—employs two sets of climate model experiments, one where historical climate change has occurred, and another where climate change is not included (e.g., Allen, 2003; Stott et al., 2004). The change in the frequency or intensity of an observed event between the two experiments is determined, providing evidence of how climate change has altered the occurrence of that specific event.

For example, the extreme European summer of 2003 had increased in frequency by a factor of two due to climate change (Stott et al., 2004). Similarly, Australia's "Angry Summer" of 2012/2013 increased in frequency by a factor of five due to climate change (Lewis & Karoly, 2013), whereas the frequency of heatwaves within the summer increased by a factor of three. The record-breaking 2017 heatwave over the Yangtze River delta in China has increased in occurrence by 23% due to climate change (Zhou et al., 2019), whereas climate change has increased the frequency of the 2013 Korean heatwave by at least fivefold (Kim et al., 2018). It is important to remember that the role of anthropogenic climate change behind specific events will increase over time. For example, extremely warm summers over China have increased 60-fold since the 1950s and will increase further to occur 50% of the time by 2050 (Sun et al., 2014). Similarly, the likelihood of a European summer similar to 2003 has increased another 10-fold since the early 2000s (Christidis et al., 2015).

It is also important to remember that the anthropogenic signal reported for a given heatwave is specific to that event only. The overall signal depends on the event and underpinning characteristics, such as the spatial scale of the event, how long it lasted for and where exactly it occurred. Also, the exact climate model methods employed may result in a spread of signal estimates (e.g., Kim et al., 2018). However, given our understanding of how heatwaves are changing under enhanced climate change, it is extremely likely that all recently observed heatwaves have a measurable anthropogenic signal.

References

Alexander, L. (2011). Climate science: Extreme heat rooted in dry soils. *Nature Geoscience, 4*, 12–13.

Alexander, L. V., Zhang, X., Peterson, T. C., Caesar, J., Gleason, B., Klein Tank, A. M. G., Haylock, M., Collins, D., Trewin, B., Rahimzadeh, F., Tagipour, A., Rupa Kumar, K., Revadekar, J., Griffiths, G., Vincent, L., Stephenson, D. B., Burn, J., Aguilar, E., Brunet, M., … Vazquez-Aguirre, J. L. (2006). Global observed changes in daily climate extremes of temperature and precipitation. *Journal of Geophysical Research, 111*, D05109.

Allen, M. (2003). Liability for climate change. *Nature, 421*, 891–892.

Ampaire, E. L., Acosta, M., Huyer, S., Kigonya, R., Muchunguzi, P., Muna, R., & Jassogne, L. (2020). Gender in climate change, agriculture, and natural resource policies: Insights from East Africa. *Climatic Change, 158*(1), 43–60. https://doi.org/10.1007/s10584-019-02447-0

Angeles-Malaspina, M., González-Cruz, J. E., & Ramírez-Beltran, N. (2018). Projections of heat waves events in the intra-Americas region using multimodel ensemble. *Advances in Meteorology, 2018*, 16. https://doi.org/10.1155/2018/7827984, 7827984 2018.

Araos, M., Berrang-Ford, L., Ford, J. D., Austin, S. E., Biesbroek, R., & Lesnikowski, A. (2016). Climate change adaptation planning in large cities: A systematic global assessment. *Environmental Science and Policy, 66*, 375–382. https://doi.org/10.1016/j.envsci.2016.06.009

Arneth, A., et al. (2019). *Climate Change and Land: An IPCC Special Report on climate change, desertification, land degradation, sustainable land management, food security, and greenhouse gas fluxes in terrestrial ecosystems - summary for Policymakers*. Geneva, Switzerland: IPCC.

Arrhenius, S. (1896). XXXI. On the influence of carbonic acid in the air upon the temperature of the ground. *The London, Edinburgh, and Dublin Philosophical Magazine and Journal of Science, 41*, 237–276.

Baranzini, A., Goldemberg, J., & Speck, S. (2000). A future for carbon taxes. *Ecological Economics, 32*(3), 395–412. https://doi.org/10.1016/S0921-8009(99)00122-6

Bartoli, G., Hönisch, B., & Zeebe, R. E. (2011). Atmospheric CO2 decline during the Pliocene intensification of Northern Hemisphere glaciations. *Paleoceanography, 26*, PA4213.

Beveridge, R., & Kern, K. (2013). The Energiewende in Germany: background, developments and future challenges. *Renewable Energy Law and Policy Review, 4*, 3.

Blunden, J., & Derek, S. A. (2019). State of the climate in 2018. *Bulletin of the American Meteorological Society, 100*, Si–S306.

Bodansky, D. (2016). The legal character of the Paris Agreement. *Review of European, Comparative & International Environmental Law, 25*(2), 142–150.

Boschat, G., Alexandre, P., Simmonds, I., Perkins, S., Cowan, T., & Purich, A. (2015). Large scale and sub-regional connections in the lead up to summer heat wave and extreme rainfall events in Eastern Australia. *Climate Dynamics, 44*, 1823–1840.

Bruckner, T., Bashmakov, I.A., Mulugetta, Y., Chum, H., De la Vega Navarro, A., Edmonds, J., Faaij, A., Fungtammasan, B., Garg, A., Hertwich, E. & Honnery, D., (2014). Energy systems.

Brzoska, M., & Fröhlich, C. (2016). Climate change, migration and violent conflict: Vulnerabilities, pathways and adaptation strategies. *Migration and Development, 5*(2), 190–210. https://doi.org/10.1080/21632324.2015.1022973

Callendar, G. S. (1938). The artificial production of carbon dioxide and its influence on temperature. *Quarterly Journal of the Royal Meteorological Society, 64*, 223–240.

Casanueva, A., Burgstall, A., Kotlarski, S., Messeri, A., Morabito, M., Flouris, A. D., Nybo, L., Spirig, C., & Schwierz, C. (2019). Overview of existing heat-health warning systems in Europe. *International Journal of Environmental Research and Public Health, 16*(15). https://doi.org/10.3390/ijerph16152657

Cassou, C., Laurent, T., & Phillips, A. S. (2005). Tropical Atlantic influence on European heat waves. *Journal of Climate, 18*, 2805–2811.

Christidis, N., Jones, G. S., & Stott, P. A. (2015). Dramatically increasing chance of extremely hot summers since the 2003 European heatwave. *Nature Climate Change, 5*, 46–50.

Cochran, P., Huntington, O. H., Pungowiyi, C., Tom, S., Chapin, F. S., III, Huntington, H. P., Maynard, N. G., & Trainor, S. F. (2013). Indigenous frameworks for observing and responding to climate change in Alaska. *Climatic Change, 120*(3), 557–567. https://doi.org/10.1007/s10584-013-0735-2

Coffel, E. D., Horton, R. M., & Sherbinin, A. (2017). Temperature and humidity based projections of a rapid rise in global heat stress exposure during the 21st century. *Environmental Research Letters, 13*, 014001.

Cowan, T., Purich, A., Perkins, S., Alexandre, P., Boschat, G., & Sadler, K. (2014). More frequent, longer, and hotter heat waves for Australia in the twenty-first century. *Journal of Climate, 27*, 5851–5871.

Della-Marta, P. M., Luterbacher, J., von Weissenfluh, H., Elena, X., Brunet, M., & Wanner, H. (2007). Summer heat waves over western Europe 1880–2003, their relationship to large-scale forcings and predictability. *Climate Dynamics, 29*, 251–275.

Diffenbaugh, N. S. (2005). Sensitivity of extreme climate events to CO2-induced biophysical atmosphere-vegetation feedbacks in the Western United States. *Geophysical Research Letters, 32*, L07702.

Diffenbaugh, N. S., & Ashfaq, M. (2010). Intensification of hot extremes in the United States. *Geophysical Research Letters, 37*(15).

Ding, Z., Chen, L., & Sun, F. (2011). Modeling and performance analysis of energy selective electron (ESE) engine with heat leakage and transmission probability. *Science China Physics, Mechanics and Astronomy, 54*, 1925.

Edenhofer, O., Wallacher, J., Lotze-Campen, H., Reder, M., Knopf, B., & Müller, J. (Eds.). (2012). *Climate change, justice and sustainability: Linking climate and development policy*. Springer Science & Business Media.

Ekwurzel, B., Boneham, J., Dalton, M. W., Heede, R., Mera, R. J., Allen, M. R., & Frumhoff, P. C. (2017). The rise in global atmospheric CO2, surface temperature, and sea level from emissions traced to major carbon producers. *Climatic Change, 144*(4), 579–590. https://doi.org/10.1007/s10584-017-1978-0

Ellerman, A. D., & Buchner, B. K. (2007). The European union emissions trading scheme: Origins, allocation, and early results. *Review of Environmental Economics and Policy, 1*(1), 66–87. https://doi.org/10.1093/reep/rem003

Fanger, P. O. (1970). *Thermal comfort. Analysis and applications in environmental engineering*. Copenhagen: Danish Technical Press.

Feron, S., Cordero, R. R., Damiani, A., Llanillo, P. J., Jorquera, J., Sepulveda, E., Asencio, V., Laroze, D., Labbe, F., Carrasco, J., & Torres, G. (2019). Observations and projections of heat waves in South America. *Scientific Reports, 9.*

Fichert, F., Forsyth, P., & Niemeier, H. M. (2020). *Aviation and climate change: Economic perspectives on greenhouse gas reduction policies.* London: Routledge.

Fischer, E. M., & Schär, C. (2009). Future changes in daily summer temperature variability: Driving processes and role for temperature extremes. *Climate Dynamics, 33*, 917.

Fischer, E. M., & Schär, C. (2010). Consistent geographical patterns of changes in high-impact European heatwaves. *Nature Geoscience, 3*, 398–403.

Fischer, E. M., Sonia, I. S., Luigi Vidale, P., Lüthi, D., & Schär, C. (2007). Soil moisture–atmosphere interactions during the 2003 European summer heat wave. *Journal of Climate, 20*, 5081–5099.

Foote, E. (1856). ART. XXXI.–Circumstances affecting the heat of the sun's rays. *American Journal of Science and Arts, 22*, 382, 1820–1879.

Frich, P., Alexander, L. V., Della-Marta, P. M., Gleason, B., Amg Klein Tank, M., & Peterson, T. (2002). Observed coherent changes in climatic extremes during the second half of the twentieth century. *Climate Research, 19*, 193–212.

Galvin, R., & Healy, N. (2020). *The green new deal in the United States: What it is and how to pay for it energy research and social science* (Vol. 67). https://doi.org/10.1016/j.erss.2020.101529

Gibson, P. B., Pitman, A. J., Lorenz, R., & Perkins-Kirkpatrick, S. E. (2017). The role of circulation and land surface conditions in current and future Australian heat waves. *Journal of Climate, 30*, 9933–9948.

Government of Costa Rica. (2019). *National decarbonization plan Costa Rica.* San José: Government of Costa Rica).

Guerreiro, S. B., Dawson, R. J., Kilsby, C., Lewis, E., & Ford, A. (2018). Future heat-waves, droughts and floods in 571 European cities. *Environmental Research Letters, 13*, 034009.

Guo, Y., Gasparrini, A., Armstrong, B. G., Tawatsupa, B., Tobias, A., Lavigne, E., De Sousa, M., Staglior Coelho, Z., Pan, X., Kim, H., Hashizume, M., Honda, Y., Liang, Y., Guo, L., Wu, C.-F., Zanobetti, A., Schwartz, J. D., Bell, M. L., Scortichini, M., … Tong, S. (2017). Heat wave and mortality: A multicountry, multicommunity study. *Environmental Health Perspectives, 125*, 087006.

Gustafson, S., Joehl Cadena, A., Ngo, C. C., Kawash, A., Saenghkaew, I., & Hartman, P. (2018). Merging science into community adaptation planning processes: A cross-site comparison of four distinct areas of the lower mekong basin. *Climatic Change, 149*(1), 91–106. https://doi.org/10.1007/s10584-016-1887-7

Habeeb, D., Vargo, J., & Stone, B. (2015). Rising heat wave trends in large US cities. *Natural Hazards, 76*, 1651–1665.

Herold, N., Alexander, L., Green, D., & Donat, M. (2017). Greater increases in temperature extremes in low versus high income countries. *Environmental Research Letters, 12*(3). https://doi.org/10.1088/1748-9326/aa5c43

Herold, N., Kala, J., & Alexander, L. V. (2016). The influence of soil moisture deficits on Australian heatwaves. *Environmental Research Letters, 11*, 064003.

Hirsch, A. L., Evans, J. P., Virgilio, G. D., Perkins-Kirkpatrick, S. E., Argüeso, D., Pitman, A. J., Carouge, C. C., Kala, J., Andrys, J., Petrelli, P., & Rockel, B. (2019). Amplification of Australian heatwaves via local land-atmosphere coupling. *Journal of Geophysical Research: Atmospheres, 124*, 13625–13647.

Hirschi, M., Seneviratne, S. I., Alexandrov, V., Boberg, F., Boroneant, C., Christensen, O. B., Formayer, H., Orlowsky, B., & Stepanek, P. (2011). Observational evidence for soil-moisture impact on hot extremes in southeastern Europe. *Nature Geoscience, 4*, 17–21.

Hochman, A., Scher, S., Julian, Q., Pinto, J. G., & Messori, G. (2020). A new view of heat wave dynamics and predictability over the Eastern Mediterranean. In *Copernicus GmbH.*
Hoerling, M., Kumar, A., Randall, D., Nielsen-Gammon, J. W., Eischeid, J., Perlwitz, J., Quan, X.-W., Zhang, T., Pegion, P., & Chen, M. (2013). Anatomy of an extreme event. *Journal of Climate, 26,* 2811−2832.
Hoegh-Guldberg, O., Jacob, D., Bindi, M., Brown, S., Camilloni, I., Diedhiou, A., Djalante, R., Ebi, K., Engelbrecht, F., Guiot, J. & Hijioka, Y. (2018). Impacts of 1.5 C global warming on natural and human systems. Global warming of 1.5° C.
Houghton, J. T., Jenkins, G. J., & Ephraums, J. J. (1990). In *Climate change: The IPCC scientific assessment* (p. 414). New York, NY, USA.
Hughes, L., de Jong, M., & Thorne, Z. (2020). (De)coupling and (De)carbonizing in the economies and energy systems of the G20. *Environment, Development and Sustainability.* https://doi.org/10.1007/s10668-020-00834-7
IEA. (2020a). *Global energy review 2020 − analysis.* Paris: IEA.
IEA. (2020b). *Buildings − a source of enormous untapped efficiency potential.* https://www.iea.org/topics/buildings
Imran, M. A., Ali, A., Ashfaq, M., Hassan, S., Culas, R., & Ma, C. (2018). Impact of climate smart agriculture (CSA) practices on cotton production and livelihood of farmers in Punjab, Pakistan. *Sustainability, 10*(6), 2101. https://doi.org/10.3390/su10062101
IPCC. (2012). A special report of working groups I and II of the intergovernmental Panel on climate change. In C. B. Field, V. Barros, T. F. Stocker, D. J. Dokken D. Qin, K. L. Ebi, M. D. Mastrandrea, K. J. Mach, G.-K. Plattner, S. K. Allen, M. Tignor, & P. M. Midgley (Eds.), *Managing the risks of extreme events and disasters to advance climate change adaptation* (p. 582). Cambridge, UK, and New York, NY, USA: Cambridge University Press.
IPCC. (2013). In T. F. Stocker, D. Qin, G.-K. Plattner, M. Tignor, S. K. Allen, J. Boschung, A. Nauels, Y. Xia, V. Bex, & P. M. Midgley (Eds.), *Climate change 2013: The physical science basis. Contribution of working group I to the fifth assessment report of the intergovernmental Panel on climate change* (p. 1535). New York, NY, USA.
IPCC. (2014). *Climate change 2014: Synthesis report. Contribution of working groups I, II and III to the fifth assessment report of the intergovernmental Panel on climate change.* In R. K. Pachauri, M. R. Allen, R. Vicente, J. Broome, W. Cramer, R. Christ, J. A. Church, L. Clarke, Q. Dahe, & P. Dasgupta (Eds.). *151.* Geneva, Switzerland.
IPCC. (2018). Summary for Policymakers. In: Global Warming of 1.5°C. An IPCC Special Report on the impacts of global warming of 1.5°C above pre-industrial levels and related global greenhouse gas emission pathways. In P. Zhai, H.-O. Pörtner, D. Roberts, J. Skea, P. R. Shukla, A. Pirani, W. Moufouma-Okia, C. Péan, R. Pidcock, S. Connors, J. B. R. Matthews, Y. Chen, X. Zhou, M. I. Gomis, E. Lonnoy, T. Maycock, M. Tignor, & T. Waterfield (Eds.), *The context of strengthening the global response to the threat of climate change, sustainable development, and efforts to eradicate poverty [Masson-Delmotte, V]* (p. 32). Geneva, Switzerland: World Meteorological Organization.
IPCC. (2019). In H.-O. Pörtner, V. Masson-Delmotte, D. C. Roberts, P. Zhai, M. Tignor, E. Poloczanska, K. Mintenbeck, A. Alegría, A. O. M. Nicolai, J. Petzold, & B. Rama (Eds.), *IPCC special report on the ocean and cryosphere in a changing climate.* Weyer. (in press).
Jeong, H., Yim, H. W., Song, Y. J., Ki, M., Min, J. A., Cho, J., & Chae, J. H. (2016). Mental health status of people isolated due to Middle East Respiratory Syndrome. *Epidemiology and Health, 38.*
Kang, S. & Eltahir, E.A (2018). North China Plain threatened by deadly heatwaves due to climate change and irrigation. *Nature communications*, 9(1), 1-9
Keeling, C. D. (1960). The concentration and isotopic abundances of carbon dioxide in the atmosphere. *Tellus, 12,* 200−203.

Kenyon, J., & Hegerl, G. C. (2008). Influence of modes of climate variability on global temperature extremes. *Journal of Climate, 21*, 3872–3889.
Kim, Y.-H., Min, S.-K., Stone, D. A., Shiogama, H., & Wolski, P. (2018). Multi-model event attribution of the summer 2013 heat wave in Korea. *Weather and Climate Extremes, 20*, 33–44.
King, A. D., Karoly, D. J., & Henley, B. J. (2017). Australian climate extremes at 1.5 C and 2 C of global warming. *Nature Climate Change*, 7(6), 412–416.
Kjellstrom, T., Briggs, D., Freyberg, C., Lemke, B., Otto, M., & Hyatt, O. (2016). Heat, human performance, and occupational health: a key issue for the assessment of global climate change impacts. *Annual review of public health, 37*, 97–112.
Koster, R. D., Schubert, S. D., & Suarez, M. J. (2009). Analyzing the concurrence of meteorological droughts and warm periods, with implications for the determination of evaporative regime. *Journal of Climate, 22*, 3331–3341.
Kunkel, K. E., Liang, X.-Z., & Zhu, J. (2010). Regional climate model projections and uncertainties of US summer heat waves. *Journal of Climate, 23*, 4447–4458.
Lau, N. C., & Nath, M. J. (2012). A model study of heat waves over North America: Meteorological aspects and projections for the twenty-first century. *Journal of Climate, 25*(14), 4761–4784.
Lenko, E., (2020). Beyond Zero Emissions (2020) The Million Jobs Plan: A unique opportunity to demonstrate the growth and employment potential of investing in a low-carbon economy should be attributed to Beyond Zero Emissions. available for download from http://bze.org.au
Lewis, S. C., & Karoly, D. J. (2013). Anthropogenic contributions to Australia's record summer temperatures of 2013. *Geophysical Research Letters, 40*, 3705–3709.
Lewis, S. C., & King, A. D. (2015). Dramatically increased rate of observed hot record breaking in recent Australian temperatures. *42*, 7776–7784.
Loughran, T. F., Pitman, A. J., & Perkins-Kirkpatrick, S. E. (2019). The El Niño–Southern Oscillation's effect on summer heatwave development mechanisms in Australia. *Climate Dynamics, 52*, 6279–6300.
Lutz, D. A., Powell, R. L., & Silman, M. R. (2013). Four decades of andean timberline migration and implications for biodiversity. *Loss with Climate Change, 8*, e74496.
Manabe, S., & Wetherald, R. T. (1975). The effects of doubling the CO2 concentration on the climate of a general circulation model. *Journal of the Atmospheric Sciences, 32*, 3–15.
Manabe, S., & Wetherald, R. T. (1980). On the distribution of climate change resulting from an increase in CO2 content of the atmosphere. *Journal of the Atmospheric Sciences, 37*(1), 99–118.
Marmot, M., & Bell, R. (2018). The sustainable development goals and health equity. *Epidemiology, 29*(1), 5–7. https://doi.org/10.1097/EDE.0000000000000773
Marshall, A. G., Hudson, D., Wheeler, M. C., Alves, O., Hendon, H. H., Pook, M. J., & Risbey, J. S. (2014). Intra-seasonal drivers of extreme heat over Australia in observations and POAMA-2. *Climate Dynamics, 43*, 1915–1937.
Matsueda, M. (2011). Predictability of Euro-Russian blocking in summer of 2010. *Geophysical Research Letters, 38*, L06801.
Mayer, H., & Hoppe, P. (1987). 'Thermal comfort of man in different urban environments. *Theoretical and Applied Climatology, 38*, 43–49.
Meehl, G. A. (2004). More intense, more frequent, and longer lasting heat waves in the 21st century. *Science, 305*, 994–997.
Meehl, G. A., Tebaldi, C., Walton, G., Easterling, D., & McDaniel, L. (2009). Relative increase of record high maximum temperatures compared to record low minimum temperatures in the US. *Geophysical Research Letters, 36*(23).
Mimura, N., Pulwarty, R. S., Duc, D. M., Elshinnawy, I., Redsteer, M. H., Huang, H. Q., Nkem, J. N., & Sanchez Rodriguez, R. A. (2014). Adaptation planning and

implementation. In C. B. Field, V. R. Barros, D. J. Dokken, K. J. Mach, M. D. Mastrandrea, T. E. Bilir, M. Chatterjee, K. L. Ebi, Y. O. Estrada, R. C. Genova, B. Girma, E. S. Kissel, A. N. Levy, S. MacCracken, P. R. Mastrandrea, & L. L. White (Eds.), *Climate change 2014: Impacts, adaptation, and vulnerability. Part A: Global and sectoral aspects. Contribution of working group II to the fifth assessment report of the intergovernmental Panel on climate change* (pp. 869–898). Cambridge, United Kingdom and New York, NY, USA: Cambridge University Press.

Miralles, D. G., Adriaan, J. T., Van Heerwaarden, C. C., & Jordi Vilà-Guerau, D. A. (2014). Mega-heatwave temperatures due to combined soil desiccation and atmospheric heat accumulation. *Nature Geoscience, 7*, 345–349.

Mueller, B., & Seneviratne, S. I. (2012). Hot days induced by precipitation deficits at the global scale. *Proceedings of the National Academy of Sciences, 109*, 12398–12403.

Murari, K. K., Ghosh, S., Patwardhan, A., Daly, E., & Salvi, K. (2015). Intensification of future severe heat waves in India and their effect on heat stress and mortality. *Regional Environmental Change, 15*(4), 569–579.

Pai, D. S., Anil Nair, S., & Ramanathan, A. N. (2013). Long term climatology and trends of heat waves over India during the recent 50 years (1961–2010. *Mausam, 64*, 585–604.

Pal, J. S., & Eltahir, E. A. (2016). Future temperature in southwest Asia projected to exceed a threshold for human adaptability. *Nature climate change, 6*(2), 197–200.

Pan, Z., Arritt, R. W., Takle, E. S., Gutowski, W. J., Anderson, C. J., & Segal, M. (2004). *Altered hydrologic feedback in a warming climate introduces a warming hole* (Vol. 31, p. L17109).

Pantavou, K., Theoharatos, G., Nikolopoulos, G., Katavoutas, G., & Asimakopoulos, D. (2008). Evaluation of thermal discomfort in Athens territory and its effect on the daily number of recorded patients at hospitals' emergency rooms. *International Journal of Biometeorology, 52*, 773–778.

Pauw, P., Mbeva, K., & Van Asselt, H. (2019). Subtle differentiation of countries' responsibilities under the Paris Agreement. *Palgrave communications, 5*(1), 1–7.

Pepermans, G., Driesen, J., Haeseldonckx, D., Belmans, R., & D'Haeseleer, W. (2005). Distributed generation: Definition, benefits and issues. *Energy Policy, 33*(6), 787–798. https://doi.org/10.1016/j.enpol.2003.10.004

Perkins, S. E. (2011). Biases and model agreement in projections of climate extremes over the tropical Pacific. *Earth Interactions, 15*, 1–36.

Perkins, S. E. (2015). A review on the scientific understanding of heatwaves—their measurement, driving mechanisms, and changes at the global scale. *Atmospheric Research, 164–165*, 242–267.

Perkins-Kirkpatrick, S. E., Fischer, E. M., Oliver, A., & Gibson, P. B. (2017). The influence of internal climate variability on heatwave frequency trends. *Environmental Research Letters, 12*, 044005.

Perkins-Kirkpatrick, S. E., & Gibson, P. B. (2017). Changes in regional heatwave characteristics as a function of increasing global temperature. *Scientific Reports, 7*, 12256.

Perkins-Kirkpatrick, S. E., & Lewis, S. C. (2020). Increasing trends in regional heatwaves. *Nature Communications, 11*, 3357.

Perkins, S. E., & Alexander, L. V. (2013). On the measurement of heat waves. *Journal of Climate, 26*, 4500–4517.

Perkins, S. E., Alexander, L. V., & Nairn, J. R. (2012). Increasing frequency, intensity and duration of observed global heatwaves and warm spells. *Geophysical Research Letters, 39*, L20714.

Perkins, S. E., Argüeso, D., & White, C. J. (2015). Relationships between climate variability, soil moisture, and Australian heatwaves. *Journal of Geophysical Research: Atmospheres, 120*, 8144–8164.

Peterson, T. C., Heim, R. R., Jr., Hirsch, R., Kaiser, D. P., Brooks, H., Diffenbaugh, N. S., Dole, R. M., Giovannettone, J. P., Guirguis, K., & Karl, T. R. (2013). Monitoring and understanding changes in heat waves, cold waves, floods, and droughts in the United States: State of knowledge. *Bulletin of the American Meteorological Society, 94*, 821–834.

Pezza, A. B., Van Rensch, P., & Cai, W. (2012). Severe heat waves in southern Australia: Synoptic climatology and large scale connections. *Climate Dynamics, 38*, 209–224.

Portmann, R. W., Solomon, S., & Hegerl, G. C. (2009). Spatial and seasonal patterns in climate change, temperatures, and precipitation across the United States. *Proceedings of the National Academy of Sciences, 106*, 7324–7329.

Quesada, B., Vautard, R., Pascal, Y., Martin, H., & Seneviratne, S. I. (2012). Asymmetric European summer heat predictability from wet and dry southern winters and springs. *Nature Climate Change, 2*, 736–741.

Quinting, J. F., & Reeder, M. J. (2017). Southeastern Australian heat waves from a trajectory viewpoint. *Monthly Weather Review, 145*, 4109–4125.

Rhein, M., Rintoul, S. R., Aoki, S., Campos, E., Chambers, D., Feely, R. A., Gulev, S., Johnson, G. C., Josey, S. A., & Kostianoy, A. (2013). Observations: ocean.

Riesz, J., Sotiriadis, C., Ambach, D., & Donovan, S. (2016). Quantifying the costs of a rapid transition to electric vehicles. *Applied Energy, 180*, 287–300. https://doi.org/10.1016/j.apenergy.2016.07.131

Rohini, P., Rajeevan, M., & Srivastava, A. K. (2016). On the variability and increasing trends of heat waves over India. *Scientific Reports, 6*, 26153.

RRI. (2019). *IPCC Agrees with indigenous Peoples and local Communities on climate change. Rights and Resources Institute (RRI).* https://rightsandresources.org/blog/ipcc-agrees-with-indigenous-peoples-and-local-communities-on-climate-change/files/3879/ipcc-agrees-with-indigenous-peoples-and-local-communities-on-climate-change.html

Russell-Smith, J., Monagle, C., Jacobsohn, M., Beatty, R. L., Bilbao, B., Millán, A., Vessuri, H., & Sánchez-Rose, I. (2017). Can savanna burning projects deliver measurable greenhouse emissions reductions and sustainable livelihood opportunities in fire-prone settings? *Climatic Change, 140*(1). https://doi.org/10.1007/s10584-013-0910-5, 47–6.

Russo, S., Dosio, A., Graversen, R. G., Jana, S., Carrao, H., Dunbar, M. B., Singleton, A., Montagna, P., Barbola, P., & Vogt, J. V. (2014). Magnitude of extreme heat waves in present climate and their projection in a warming world. *Journal of Geophysical Research: Atmospheres, 119*(12), 500–512, 12.

Russo, S., Sillmann, J., & Fischer, E. M. (2015). Top ten European heatwaves since 1950 and their occurrence in the coming decades. *Environmental Research Letters, 10*(12), 124003.

Rusticucci, M., Jan, K., Almeira, G., & Lhotka, O. (2016). Long-term variability of heat waves in Argentina and recurrence probability of the severe 2008 heat wave in Buenos Aires. *Theoretical and Applied Climatology, 124*, 679–689.

Schwartz, M. D., Ahas, R., & Aasa, A. (2006). Onset of spring starting earlier across the Northern Hemisphere. *Global Change Biology, 12*, 343–351.

Seneviratne, S. I., Donat, M. G., Pitman, A. J., Knutti, R., & Wilby, R. L. (2016). Allowable CO2 emissions based on regional and impact-related climate targets. *Nature, 529*, 477–483.

Sherlock, L., et al. (2020). *Constructing a new low-carbon future: How Chinese cities are scaling ambitious building energy-efficiency solutions. New York: C40 and China.* Energy Group - Lawrence Berkeley National Laboratory.

Sherwood, S. C., & Huber, M. (2010). An adaptability limit to climate change due to heat stress. *Proceedings of the National Academy of Sciences, 107*, 9552–9555.

Singh, C., Ford, J., Ley, D., Bazaz, A., & Revi, A. (2020). Assessing the feasibility of adaptation options: Methodological advancements and directions for climate adaptation

research and practice. *Climatic Change, 162*, 255–277. https://doi.org/10.1007/s10584-020-02762-x
Steadman, R. G. (1979). The assessment of sultriness. Part I: A temperature-humidity index based on human physiology and clothing science. *Journal of Applied Meteorology, 18*, 861–873.
Steadman, R. G. (1984). A universal scale of apparent temperature. *Journal of Climate and Applied Meteorology, 23*, 1674–1687.
Steffen, W., Hughes, L., & Perkins, S. (2014). In *Heatwaves: Hotter, longer, more often* (p. 70). Climate Council of Australia.
Schoetter, R., Cattiaux, J., & Douville, H. (2015). Changes of western European heat wave characteristics projected by the CMIP5 ensemble. *Climate Dynamics, 45*, 1601–1616.
Stott, P. A., Stone, D. A., & Allen, M. R. (2004). Human contribution to the European heatwave of 2003. *Nature, 432*, 610–614.
Sun, Y., Zhang, X., Zwiers, F. W., Song, L., Wan, H., Hu, T., Yin, H., & Ren, G. (2014). Rapid increase in the risk of extreme summer heat in Eastern China. *Nature Climate Change, 4*, 1082–1085.
Trenberth, K. E. (1997). The definition of el nino. *Bulletin of the American Meteorological Society, 78*, 2771–2778.
Tripati, A. K., Roberts, C. D., & Eagle, R. A. (2009). Coupling of CO2 and ice sheet stability over major climate transitions of the last 20 million years. *Science, 326*, 1394–1397.
Tyndall, J. (1861). XXIII. On the absorption and radiation of heat by gases and vapours, and on the physical connexion of radiation, absorption, and conduction.—the bakerian lecture. *The London, Edinburgh, and Dublin Philosophical Magazine and Journal of Science, 22*, 169–194.
Ukkola, A. M., De Kauwe, M. G., Roderick, M. L., Abramowitz, G., & Pitman, A. J. (2020). Robust future changes in meteorological drought in CMIP6 projections despite uncertainty in precipitation. *Geophysical Research Letters, 47*, GL087820.
Ukkola, A. M., Pitman, A. J., De Kauwe, M. G., Abramowitz, G., Herger, N., Evans, J. P., & Decker, M. (2018). Evaluating CMIP5 model agreement for multiple drought metrics. *Journal of Hydrometeorology, 19*, 969–988.
UNFCCC. (1992). *United nations framework convention on climate change.* Bonn, Germany: UNFCCC.
UNFCCC. (2015). What is the Paris agreement? https://unfccc.int/process-and-meetings/the-paris-agreement/what-is-the-paris-agreementfiles/3852/what-is-the-paris-agreement.html
Vautard, R., Gobiet, A., Jacob, D., Belda, M., Colette, A., Déqué, M., Fernández, J., García-Díez, M., Goergen, K., & Güttler, I. (2013). The simulation of European heat waves from an ensemble of regional climate models within the EURO-CORDEX project. *Climate Dynamics, 41*, 2555–2575.
Viguié, V., Lemonsu, A., Hallegatte, S., Beaulant, A. L., Marchadier, C., Masson, V., Pigeon, G., & Salagnac, J. L. (2020). Early adaptation to heat waves and future reduction of air-conditioning energy use in Paris. *Environmental Research Letters, 15*(7). https://doi.org/10.1088/1748-9326/ab6a24
Watson, R., McCarthy, J. J., Canziani, P., Nakicenovic, N., & Hisas, L. (2019). *The truth Behind the climate pledges – FEU-US (District of Columbia, US: FEU-US).*
Watts, M. (2017). Commentary: Cities spearhead climate action. *Nature Climate Change,* 7(8), 537–538. https://doi.org/10.1038/nclimate3358
Watts, N., et al. (2015). Health and climate change: Policy responses to protect public health. *The Lancet, 386*(10006), 1861–1914. https://doi.org/10.1016/S0140-6736(15)60854-6

Wuebbles, D., Meehl, G., Hayhoe, K., Karl, T. R., Kunkel, K., Santer, B., Wehner, M., Colle, B., Fischer, E. M., Fu, R., & Goodman, A. (2014). CMIP5 climate model analyses: climate extremes in the United States. *Bulletin of the American Meteorological Society, 95*(4), 571–583.

Xie, Y., Dai, H., Xu, X., Fujimori, S., Hasegawa, T., Yi, K., Masui, T., & Kurata, G. (2018). Co-benefits of climate mitigation on air quality and human health in Asian countries. *Environment international, 119*, 309–318.

You, Q., Jiang, Z., Kong, L., Wu, Z., Bao, Y., Kang, S., & Pepin, N. (2017). A comparison of heat wave climatologies and trends in China based on multiple definitions. *Climate Dynamics, 48*, 3975–3989.

Zhou, C., Wang, K., Qi, D., & Tan, J. (2019). Attribution of a record-breaking heatwave event in summer 2017 over the Yangtze River Delta. *Bulletin of the American Meteorological Society, 100*, S97–S103.

Seneviratne, S., Nicholls, N., Easterling, D., Goodess, C., Kanae, S., Kossin, J., Luo, Y., Marengo, J., McInnes, K., Rahimi, M. & Reichstein, M. (2012). Changes in climate extremes and their impacts on the natural physical environment.

Further reading

Barnett, J., & O'Neill, S. (2010). Maladaptation. *Global Environmental Change, 20*(2), 211–213. https://doi.org/10.1016/j.gloenvcha.2009.11.004

Boucher, O., Gruber, N., & Blackstock, J. (2011). In O. Edenhofer, C. Field, R. Pichs-Madruga, Y. Sokona, T. Stocker, V. Barros, Q. Dahe, J. Minx, K. Mach, G.-K. Plattner, S. Schlömer, G. Hansen, & M. Mastrandrea (Eds.), *IPCC working group III technical support unitSummary of the synthesis session in: IPCC expert meeting report on geoengineering* (p. 7). Potsdam, Germany: Potsdam Institute for Climate Impact Research.

Bruckner, T., Bashmakov, I. A., Mulugetta, Y., Chum, H., De la Vega Navarro, A., Edmonds, J., Faaij, A., Fungtammasan, B., Garg, A., Hertwich, E., Honnery, D., Infield, D., Kainuma, M., Khennas, S., Kim, S., Nimir, H. B., Riahi, K., Strachan, N., Wiser, R., & Zhang, X. (2014). Energy systems. In O. Edenhofer, R. Pichs-Madruga, Y. Sokona, E. Farahani, S. Kadner, K. Seyboth, A. Adler, I. Baum, S. Brunner, P. Eickemeier, B. Kriemann, J. Savolainen, S. Schlömer, C. Von Stechow, T. Zwickel, & J. C. Minx (Eds.), *Climate change 2014: Mitigation of climate change. Contribution of working group III to the fifth assessment report of the intergovernmental Panel on climate change*. Cambridge, United Kingdom and New York, NY, USA: Cambridge University Press.

BZE. (2020). *The million Jobs plan*. Melbourne, Australia: Beyond Zero Emissions.

Dannenberg, A. L., Frumkin, H., Hess, J. J., & Ebi, K. L. (2019). Managed retreat as a strategy for climate change adaptation in small communities: Public health implications. *Climatic Change, 153*, 1–2. https://doi.org/10.1007/s10584-019-02382-0

Gambhir, A., Green, F., & Pearson, P. J. G. (2018). *Towards a just and equitable low-carbon energy transition: Grantham Institute Briefing paper No 26* (London: Imperial college).

Harrington, L. J., Frame, D., King, A. D., & EL Otto, F. (2018). How uneven are changes to impact-relevant climate hazards in a 1.5°C world and beyond? *Geophysical Research Letters, 45*, 6672–6680.

Markandya, A., Sampedro, J., Smith, S. J., Van Dingenen, R., Pizarro-Irizar, C., Arto, I., & González-Eguino, M. (2018). Health co-benefits from air pollution and mitigation costs of the Paris agreement: A modelling study. *The Lancet Planetary Health, 2*(3), e126–e133. https://doi.org/10.1016/S2542-5196(18)30029-9

Nichols, A. (2019). Climate change, natural hazards, and relocation: Insights from Nabukadra and Navuniivi villages in Fiji. *Climatic Change, 156*(1–2), 255–271. https://doi.org/10.1007/s10584-019-02531-5

Nurse, A., & Dunning, R. (2020). Is COVID-19 a turning point for active travel in cities? *Cities & Health*, 1–3. https://doi.org/10.1080/23748834.2020.1788769

Ochieng, J., Kirimi, L., Ochieng, D. O., Njagi, T., Mathenge, M., Gitau, R., & Ayieko, M. (2020). Managing climate risk through crop diversification in rural Kenya. *Climatic Change*. https://doi.org/10.1007/s10584-020-02727-0

Perkins, S. E., & Gibson, P. B. (2015). Increased risk of the 2014 Australian May heatwave due to anthropogenic activity. *Bulletin of the American Meteorological Society, 96*, S154–S157.

Schröder, A., & Traber, T. (2012). *The economics of fast charging infrastructure for electric vehicles energy policy* (Vol. 43). https://doi.org/10.1016/j.enpol.2011.12.041

Stanley, J., Ellison, R., Loader, C., & Hensher, D. (2018). Reducing Australian motor vehicle greenhouse gas emissions. *Transportation Research A: Policy and Practice, 109*, 76–88. https://doi.org/10.1016/j.tra.2018.01.002

CHAPTER 3

The biologic mechanism for heat exposure and human health

Micheline de Sousa Zanotti Stagliorio Coelho, Mariana Matera Veras and Paulo Hilario Nascimento Saldiva
Department of Pathology, University of Sao Paulo, School of Medicine, Sao Paulo, Brazil

1. Introduction

Temperature is more than a basic parameter for comfort. Extreme temperature or its variability has direct and indirect effects on human health, and changes in temperature patterns occasioned by climate change raised the concern about its global health impacts. By 2100, an increase between 1.8 and 4 degrees in average temperature is expected (IPCC, 2018), and the recent understanding of adaptation has moved away from biophysiological factors, to wider social and economic dimensions of vulnerability and capacity of people to respond to challenges (Watts et al., 2018a). Consequences of climate change can be analyzed from different perspectives. Briefly, it can be addressed by its economic impacts, social consequences, global food security, changes in the build environment, mitigation strategies, health impacts, susceptibilities of minorities, changes in the distribution of vector-borne diseases, and many others (IPCC, 2018). No matter from which perspective the analysis is conducted, there is no doubt that human health will be affected, directly or indirectly (Watts et al., 2018b).

In the past years, climate changed faster, and despite acclimatization capacity of humans and the technological development, we are seeing a growing number of studies showing that extreme and even moderate changes in temperature are related to negative health outcomes and in the worst scenario associated to premature death (Heng, 2017). However, findings are diverse and in some cases opposite depending on the location where the studies were conducted, even after controlling for cofounding factors. As observed by Liu and Ma (2019) in a recent comment, we need more information about the factors that modify the effects of temperature on health and social/anthropological particularities associated with susceptibility to better understand its expected consequences. It is well known that there are differences in thermal comfort, physiological responses to

Heat Exposure and Human Health in the Context of Climate Change
ISBN 978-0-12-819080-7
https://doi.org/10.1016/B978-0-12-819080-7.00001-X
© 2023 Elsevier Inc.
All rights reserved.

temperature changes, and performance between young and elderly people, and this is also true for individuals with preexisting diseases (e.g., diabetes, hypertension) and for different ethnical groups, e.g., Bantu are tolerant to very hot and humid environments (Kjellstrom et al., 2010; Wyndham et al., 1968).

Therefore, we have to understand how humans physiologically respond to both extreme and moderate-, short-, or long-term temperature variability to better develop strategies to cope with these climate changes, protecting communities as a whole.

2. Thermoregulation general concepts

2.1 Anthropologic and evolutionary aspects

Adaptation to diverse environmental conditions made possible the consolidation of the human species. The search for new habitats was associated with both genetic and epigenetic modifications of human physiology and behavioural adaptation, as well as a driving force to boost human creativity, allowing opportunities to ingenuity to adapt housing, clothing, and social habits (Fan et al., 2016). There is a considerable body of evidences pointing out that the first *Homo sapiens* was born around latitude 10, most probably in Africa (hot and dry climates). Since then, the migration to other areas of our planet encompassed a series of needs to adapt to different light cycle, temperature, and production of food (Mcbrearty and Brooks, 2000).

Indeed, the process of adaptation to hard conditions was an evolutionary pressure, favoring more adequate gene mutations to a given environment, as well as to force the large brain, a bet of this particular evolution pathway chosen by the *H. sapiens*, to make more functional the enormous repertoire of neuronal connections, which, ultimately, resulted in human creativity (Potts and Faith, 2015).

H. sapiens emerged about 300,000 years ago in the African grasslands (Mcbrearty & Brooks, 2000). Changes in body shape, partially induced by epigenetic[1] and environmental factors, played a major role in the process of adaptation to temperature. Humans have adapted to many different environmental conditions, with temperatures ranging from −68°C (Arctic

[1] Epigenetic changes by definition are heritable changes in gene expression without DNA sequence modification; these changes are normal and can be induced by several factors including age, the environmental conditions, lifestyle, nutrition, and disease state.

region) to 45°C (deserts). Adaptation was possible to physiological plasticity, which in the long term changed body size and shape the physiological mechanisms involved in thermoregulation, for example, an African cattle herder has an inherited phenotype, with a high proportion of muscular tissue in body composition. The high surface-to-mass ratio helps in the process of heat dissipation during intense outdoor physical activities, and the high muscular body composition helps to produce heat during the night due to high metabolic rate. On the other hand, the Arctic hunter has a brevilineous body shape, with shorter legs and high proportion of fat tissue working as an efficient thermal insulator (McElroy & Towsend, 2018). The shape of the external nose seems not to be simply the result of genetic variation but may be adaptations to climate. A recent study conducted by Zaidi et al. (2017) showed that the width of the nares is correlated with temperature and absolute humidity, but not with relative humidity. The function of the nose is to condition the air (humidify and warm) before it enters the lungs, which is necessary for the optimal function of the mucociliary apparatus (one of the defense systems of the respiratory tract) and for a more efficient gas exchange.

The circulatory system is an crucial component of the control system of body temperature; changes in heart rate, plasma volume, and vasomotor responses are basic mechanisms inherent to this system (Sanin et al., 2016). Thus, thermal stress (hot or cold) acts as an adaptive force for the response of peripheral vessels located in the body extremities. As a rule, cold induces constriction, and the reverse situation, vasodilation, occurs during exposure to hot conditions. These changes are transitory, lasting for few minutes (Sanin et al. 2016). An example of vascular response adapted to climate condition can be seen in Arctic hunters; they have a more rapid dilation after exposure to cold, thus preventing ischemic (absence of blood flow) damage to extremities, the "frost bites"[2] (Brown et al., 1963)

The skin, as the circulatory system, also played a major role in thermal adaptation. Reduction of dermal hair to accommodate sudoriparous glands was a necessary event for adaptation to the excess of sunlight for the first humans. However, having an efficient sudomotor mechanism and consequently a hairless skin, early humans had to adapt to the absence of the shield played by epidermal hair to UV radiation. The excess of solar radiation compromises cell division by the degradation of folate, whereas too low solar radiation impedes the transformation of dermal tococalciferol in

[2] Freezing of the skin and subjacent tissues due to exposure to low temperatures.

vitamin D, a fundamental element for adequate bone homeostasis (Borradale & Kim, 2012; Saraff & Shaw, 2016). Moreover, adaptation to different solar radiation was possible by variations in skin color. A particular polymorphism of the melanocortin receptor gene that promotes a higher production of melanin and eumelanin (determine the color of the skin), became prevalent in African (McElroy & Townsend, 2018).

Along history, the process of conquering more living spaces, the migratory movements, and the capacity of adaption to cold and hot conditions occurred within a time window of thousands of years. More recently, the migratory events produced by wars, famine, and other factors, including climate change, represent a new challenge to human adaptation to temperature, since people move globally in much faster speeds (Gemenne at al., 2015). In addition, global climate change not only will promote more intense migratory events but also will change the regional climate in an unprecedented degree (IPCC, 2018).

2.2 Pathophysiology of thermal regulation

Humans control body core temperature (Tcore) within a narrow range (35−37°C) for survival to cold and warm variations by means of a series of physiologic and behavioral adaptations, being classified as homeothermic. We may produce heat internally through shivering by muscles and through hormones produced by the thyroid causing increased metabolism when exposed to cold. Endothermic and homeothermic physiology implies that the organisms have a system composed by thermal sensors, a central processing unit, and effector organs and structures (Tansey & Johnson, 2015). The sensor system, composed by neural receptors, is stimulated by the temperature variation (body or external environment) (Kumar, 2018). Most of the cold impulses are originated from cutaneous receptors, whereas hot signals are derived with more intensity from peripheral and central receptors. Such pattern reflects the relative distribution frequency of cold and warm receptors, as well as the relative higher risk of damage to internal organs of increases in body temperature.

The signals generated reach the central control in the hypothalamus and are processed resulting in a response through the effector signals that are used to restore the core temperature to its set point around 37°C (Fig. 3.1). Changes in Tcore can alter the permeability of their cell membranes affecting the exchange diffusion of substances between intra- and extracellular fluid disturbing body homeostasis (Hall, 2015). Thermoregulation

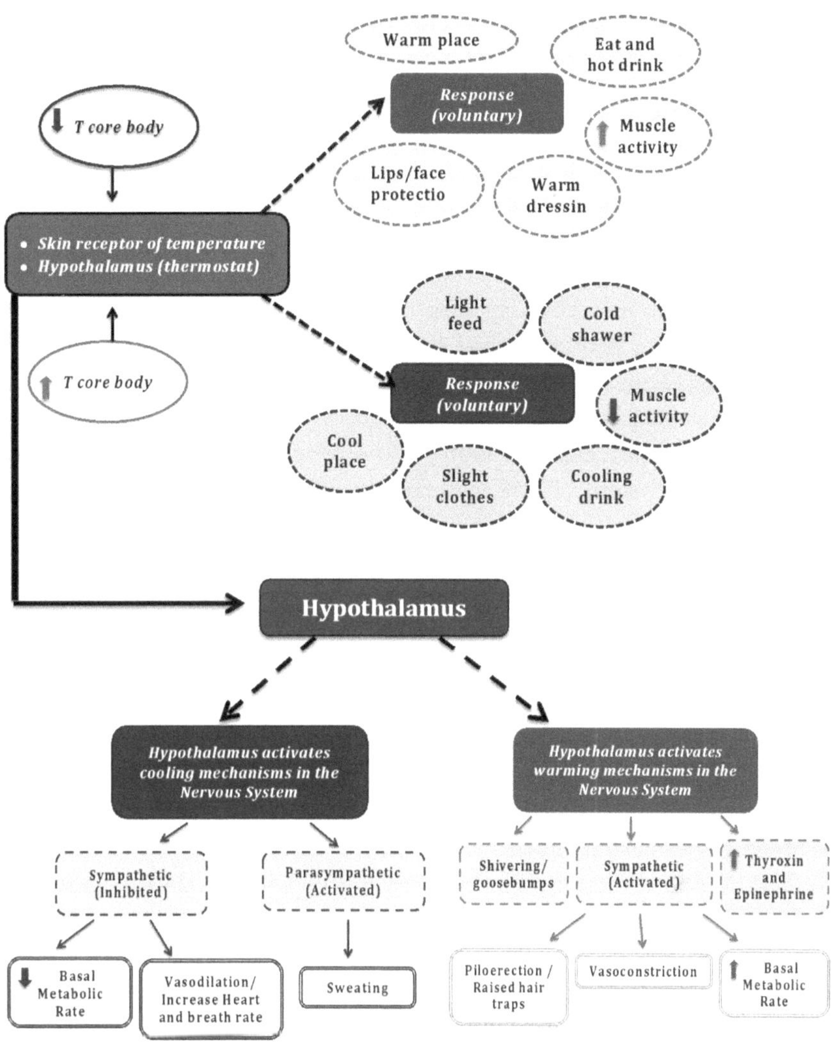

Figure 3.1 Negative feedback loop for the temperature homeostasis of body from hypothalamus, and voluntary responses. The direction of thick arrows into circles and rectangles red (light gray in print version) up and blue (dark gray in print version) down means increased and decrease rates, respectively.

represents a perfect integrative role of the hypothalamus in generating patterns of autonomic, endocrine, motor, and behavioral responses to adapt to extrinsic temperature challenges and intrinsic changes in the Tcore during the inflammatory response (fever) and behavioral states and in response to declining energy homeostasis (Morrinson, 2011).

The thermal receptors in the skin and other tissues are free nerve endings, and the sensations of heat and cold also travel by delta A and C fiber neurons (6–30 m/s conduction velocity, 2–5 micrometer of diameter) (Marieb et al., 2018). First and second order axons are responsible for the neurological transmission of pain and temperature. The axons first-order neuron entering the spinal cord from the posterior root ganglion in the gray column and divide into ascending and descending branches. These branches travel for spinal cord and form the posterolateral tract of Lissauer. The second-order neurons have their axons cross obliquely to the opposite side in the anterior gray and white commissures within one spinal segment of the cord, ascending in the contralateral white column as the lateral spinothalamic tract. The second-order neuron in the substantia gelatinosa gives rise to an axon that decussates and ascends to a higher level of the central nervous system, where it synapses with the third-order neuron in the ventral posterolateral nucleus of thalamus and gives rise to a projection fiber that passes to a sensory region of the cerebral cortex. It is believed that the inputs of the sensations of temperature are modified by excitatory or inhibitory information from other sensory inputs and by information from the posterior central gyrus in the cerebral cortex (Kumar, 2018). The hypothalamus has a controlling influence on the autonomic nervous system and appears to integrate the temperature regulation. The anterior hypothalamic nuclei control those mechanisms that dissipate heat loss through dilatation of skin blood vessels and sweating. On the other hand, the posterior hypothalamic nuclei control cold temperature with cutaneous vasoconstriction regulating heat loss (Zhao, 2017). The skeletal muscles can also receive messages from hypothalamus to cause shivering to produce heat. Brown adipose tissue is also capable of producing heat via non-shivering thermogenesis (burned triglycerides produced heat increasing body temperature) (Jendritzky, 2000).

As briefly described before, the response system is composed by the action of different organs that are recruited depending on if it is hot or cold and if changes in temperature are moderate or extreme. More detailed description of the response system is presented in the following paragraphs.

Skin: Changes in the dermal blood flow are an important mechanism of adaptation to temperature alterations. The vascular network located in the dermis is able to modify substantially its flow regimen in response to cold (constriction) and hot (dilation). Between these two extremes, dermal blood flow may vary up to 60 times. In response to the indications of the hypothalamic processor, the vasodilator or vasoconstrictor stimuli will modify the lumen and recruit or exclude vessels in the dermal territory, accordingly. The same stimuli regulate the activity of sweat glands, causing inhibition or hypersecretion in response to cold or hot conditions (Tansey & Johnson, 2015). The sweat glands excrete a saline solution in other to evaporate rapidly improving heat transfer and maintain Tcore adequate temperature (Bates et al., 2008).

Three types of sensory receptors distinguish thermal sensations from the skin. The first and second sensory receptors are cold and heat, respectively. They are stimulated in the interval from 15 to 44°C. The last type is a pain sensory receptor being stimulated by extreme heat (above 44°C) or extreme cold (below 15°C) (Fig. 3.2). In most areas of the body, there are more points of cold than points of heat, and these points differ in the areas of the body, being more numerous on the lips (Hall, 2015). Climbers are commonly affected by cold injuries (frostnip). Frostnip, which occurs due to vasoconstriction, can progress to ulceration and permanently damage nerve endings. Initial symptoms are numbness or pain, which can lead to tissue destruction, leading to loss of fingers, toes, nose tip, and lips.

- *Metabolism and function of vital organs:* Hot temperatures promote a parasympathetic stimulus, which decreases heart rate, cardiac output, blood pressure, and the secretion of cortisol and thyroid hormones. The decrease in cardiac output may be amplified by the excessive loss of circulating blood volume, due to abundant transpiration, and may play a significant role in determining acute cardiovascular death, mainly

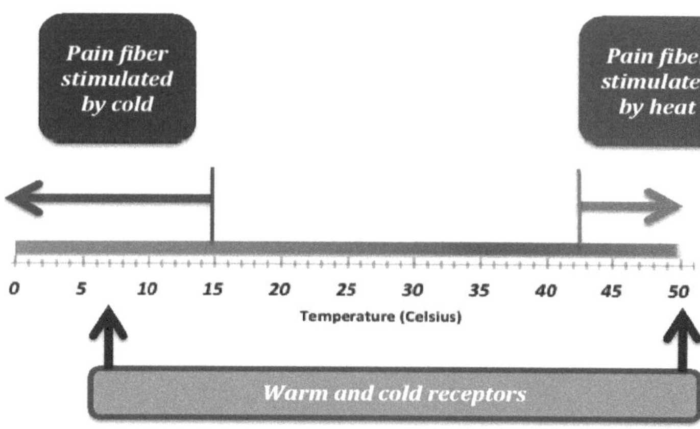

Figure 3.2 Illustrative scheme about three types of thermal sensory receptors. The blue (dark gray in print version) (cold-frostnip) and red (light gray in print version) (warm-burning) arrows show thermal and pain simultaneously.

in individuals who already have impaired circulation in important target organs such as brain and heart. In addition, parasympathetic stimulation decreases the activity of sinoatrial node, the natural pacemaker of the heart, increasing the possibility that an erratic heart beat escapes from the control enhancing the chance of developing arrhythmias (Giorgini et al. 2017). Decreases in temperature elicit opposite alterations by increasing the sympathetic tonus. Peripheral vasoconstriction increases metabolic rate; elevations of frequency poses an increased workload to the myocardium and blood vessels, predisposing to infarctions in the brain and heart (Giorgini et al., 2017). Low temperatures also represent a significant challenge to the respiratory system, to perform an adequate gas exchange; the entire cardiac output flows through the lung microcirculation, which irrigates more than 70 square meters of gas exchange surface in an adult. In this setting—pulmonary alveolar microcirculation—blood and the inspired air are separated by the alveolar epithelia metabolism, which decreases the motility of epithelial cilia and the production of antimicrobial capillary barrier, which average thickness is about 1 micrometer (Hogan et al., 1986). And to avoid an excessive heat loss in the lungs, the inspired air must be conditioned before it reaches the alveoli. Airways are responsible to a considerable heat exchange with the inhaled air, a situation that decreases proteins and immunoglobulins locally secreted and hardens the mucous gel that lines airway mucosa, predisposing to respiratory infections (Franchini et al., 2016).

Heat exhaustion and heatstroke are the most severe forms of heat-related illness. In heat exhaustion, core body temperature reaches no more than 40°C where mental status remains preserved (Beachy et al. 2011). On the other hand, core body temperature above 40°C leads to heatstroke and causes serious cell damage and organ dysfunction. Heat stroke often causes hypercoagulation simultaneously with enhanced fibrinolysis leading to disseminated intravascular coagulation (Bouchama et al. 2002). Heatstroke effects of hyperthermia that are harmful to the immune system, intercellular adhesion molecule-1 (ICAM-1), and von Willebrand factor are significantly elevated in the plasma (Hammami et al., 1997)

- *Brown adipose tissue*: The so-called brown adipose tissue (BAT) is a type of fat tissue clearly identified in neonates, which has a high oxidative (aerobic) metabolic rate. The particular color—brown—at macroscopy is due to a large number of mitochondria in the brown fat cells. Briefly, under cold stimulus, BAT increases its metabolic rate, which is uncoupled to the generation of ATP, thus promoting the liberation of heat. BAT is probably the main mechanism that permits mammals to overcome the challenge of leaving the intrauterine environment (cavity with a very precise and narrow temperature range) to enter in a new environment where ambient temperature varies constantly. When BAT is not enough, such as in prematurity, sometimes it is necessary to mimic the uterus by housing babies in incubators in the immediate neonatal period. Present evidence clearly demonstrated that the normal adipose tissue may exhibit a "browning" transformation as a consequence of several conditions, such as chronic infections, stress, and, more importantly, adaptation to cold climate. Thus, it is interesting to notice that evolution has brought to us a greater physiological capacity to adapt to cold than to hot conditions (Fenzi & Kiefer, 2014; Tansey & Johnson, 2015).

- *Behavioral responses:* Changes in behavior are one of the strategies that animals use to cope with changes in external temperature and maintain internal temperature (McCafferty et al., 2017; Schlader et al., 2011). Our physiological thermoregulatory system has a limited capacity to deal with temperature extremes; however, our behavioral responses are tremendous. Behavioral responses start first to avoid extra energy costs and wastewater. Among behavioral responses, shelter search, adding or removing clothes, increasing or decreasing physical activity, posture changes, and the use of technological resources (e.g., air-conditioners) and in an extreme weather scenario, migration is more common (Terrien et al., 2011). It is easy to perceive the risks and the direct impacts on health of extreme weather events (e.g., droughts, storms, floods); however, changes in temperature variability, average temperature, local weather patterns, and their consequence on daily life activities and human behavior are not easy to perceive, and consequently, it is underestimated. Behavioral response to changes in temperature should also be evaluated from a psychosocial perspective. Studies have shown that subtle mental processes may also be affected by climate. Recent evidence pointed out that mood behavior, as expressed in terms of suicides, is associated with hot temperatures (Kim et al., 2016; Page et al., 2007). Moreover, there are evidence that elevations in temperature are associated with increased aggressive behaviors and violent crimes (Tiihonen et al. 2017; Younan et al., 2018). Disorders of thermoregulation may be attributable to focal hypothalamic, brainstem, or spinal lesions, or autonomic failure could alter the sensitivity to heat and cold (Benarroch, 2007). In this way, studies of the impacts of temperature on health should be taken into account patients with chronic and debilitating diseases such as Raynaud's disease, endocrine system disorders, neurological lesions of anterior and posterior nerve roots, lamination of the ascending tracts, multiple sclerosis, Brown-Séquard syndrome, and vertebrobasilar artery occlusion (Benjamin et al. 2015).

In general, human beings seem physiologically well adapted to hot dry weather. In arid climate (dry and hot weather), the human beings feel fresher than hot and wet environment because sensible sweat evaporates from the skin more rapidly than in humid climates, where the air is already saturated with water. This carries heat away from the skin rapidly, leading to a cool sensation. In humid regions, the sweat does not evaporate as readily, and thus the body's cooling mechanism is hampered. However, effective thermoregulation is reduced. In hot-humid conditions, clothing can impede efficient evaporation because high water vapor (satured) in the atmosphe (Jendritzky, 2009).

2.3 Individual and pathological modifiers of thermal regulation

In face of climate changes, the number of publications investigating the effects of temperature on health is increasing, and most of them focus on mortality and cardiovascular effects. Besides the growing number of studies, there are still many open questions concerning effect modifiers, such as sex, age, and previous health conditions as mentioned before (Hossain et al., 2019; Lee et al., 2019). Taking into account the pathophysiological knowledge, the intrinsic vulnerability of the exposed population should be considered as it influences the individual thermoregulatory processes (Kalkstein & Greene, 1997). Age and the presence of chronic health conditions, such as obesity and diabetes, modulates individual resilience to temperature changes, as depicted in the following:

- *Age:* Thermoregulation control is also markedly affected by age, and its efficiency decreases with aging, predisposing the elderly to hypothermia or hyperthermia. Besides altered thermal perception, age affects the primary response mechanisms that maintain body temperature, which are the vascular response (constriction and dilation), sweating, shivering, and metabolic heat production. World population is aging; in 2050, it is expected 2.1 billion people will be aged 60 or older and 79% of this population will be living in developing regions (UN, 2017). Several studies have showed that in the face of drift changes in thermal comfort, physiological responses, and performance between young and elderly people experience health disturbance, especially if we consider cities adaptation and preparedness and adequate public services are extremely important (World Bank, 2010). Babies and children are also less effective in controlling body temperature and are more prone to the negative effects of great variability in environmental temperature and extreme conditions (Díaz et al., 2004; 2015). Leaving the intrauterine environment to outer world is a thermoregulatory challenge for neonates, mainly for those delivered in premature births. This is the reason why newborn babies are kept after birth in heating cradles. When facing ordinary infections, children may develop fever with higher temperatures than those experienced by adults. Conversely, elderly people may not have fever in the course of severe infection, such as pneumonia and urinary infections. Thus, we may postulate that thermoregulatory mechanisms maturate, stabilize, and deteriorate along our life span (Székely & Garai, 2018). For heat tolerance, age also influences the effectiveness of the "cooling system"; sweat gland is fully developed only in the youth (Cui & Schlesinger, 2015). It has long been known from animal studies that ambient temperature can harm fetal development (Gwazdauskas, 1985; Sawyer & Narayan, 2019). However, in humans' studies, the temperature was most frequently approached as a modifier or confounding factor in studies and not as a risk factor. Although deviations in the normal development of the fetus resulting in low birth weight (LBW) and preterm birth (PTB) are in most cases the result of the interaction of individual biopsychosociological and abiotic factors, recent studies indicate that variations in temperature may directly contribute (Wu et al., 2019). Evidence to date indicates that temperature viability as well as high and low temperatures may affect pregnancy, and the gestational period during which exposures occur determines the outcome (Son, 2019; Li et al., 2018; Wu et al., 2019; Basu et al., 2018). The aging process per se may interfere in our capacity to adapt to changes in temperature; however, studies show that the biggest problem is diseases that come with advancing age and physical inactivity. This means that a healthy elderly person is less at risk when facing critical changes in temperature. Systemic hypertension, a common condition in the elderly, induces structural alterations in the walls of small arteries and arterioles, characterized by smooth muscle hyperplasia and fibrosis. These changes impair the capacity the vessels responsible for the fine distribution of blood flow through body to respond to changes in external temperature, increasing the risks for mortality (Liu et al., 2015; Sun et al., 2018).

- *Health condition:* Obesity, a global epidemic, can also influence thermal regulation. Subcutaneous adipose tissue acts as a thermal insulator, since it has relative lower-temperature conductivity in comparison with other tissues. In fact, subcutaneous fat plays a significant role in the adaptation of homeothermic animals to cold environments. Experimental studies in humans indicate that the amount of subcutaneous fat influences the changes in core temperature during immersions in cold water (Speakman 2018), suggesting that body composition may affect the transference of ambient temperature to cutaneous thermoreceptors, as well as the heating dissipation mechanisms. Diabetes may also influence thermoregulation. One of the most frequent complications of diabetes is the peripheral neuropathy, which is, in brief, the result of obliteration of peripheral afferent and efferent innervation. As consequence, diabetes with peripheral neuropathy may compromise the communication between the sensory system (cutaneous thermoreceptors) and the response system, which means impairing the adaptive physiological responses to changes in external temperature, such as sudoresis and changes of cutaneous blood flow (Fealey, 2019). It is important to notice that risk for type II diabetes is influenced by obesity, due to the higher resistance to insulin of adipose tissue. Thus, only for didactic purposes, these two conditions, diabetes and obesity, are presented separately in the present manuscript (Al- Goblan et al., 2014). Congestive heart failure (CHF) is characterized by the incapacity of our circulatory system to move efficiently the circulating blood. In response to such condition, human body launches a series of adaptive mechanisms, including a significant increase in sympathetic tonus, which affects the compensatory vasodilator response to elevated temperatures (Balmain et al., 2017). Considering that CHF prevalence is increasing globally, and temperature and air pollutants have synergic effects for this health outcome, it should be also considered as an important determinant of human resilience to climate (Morris & Naumova, 1988).

Diabetic patients can have compromised thermoregulatory capacity as impaired functioning of the sweat glands and low blood flow reduction, leading to inability to disperse body heat effectively (Petrofsky et al., 2005). In general, medications used to treat patients with type 1 diabetes mellitus show pharmacodynamic effects in the fluid homeostasis getting worse dehydration during high-temperature exposure (Yardley et al., 2013). Heat exhaustion can lead to heatstroke, resulting in a hypoglycemic episode due to insulin peak and worse glucose homeostasis (Al-Qaissi et al., 2019; Dumke et al., 2015).

Dissolved in the blood stream, the electrolytes (potassium, magnesium, sodium, calcium, chloride) play a role to maintain the proper functioning of human body. For some reason, there is a change in the levels of these electrolytes, and some symptoms begin to occur ranging from a light weakness to life-threatening. There are many causes associated to electrolyte imbalances associated with hot weather such as diarrhea, vomiting, excess sweating or fever, undernutrition, malabsorption, hormonal or endocrine disorders, kidney disease, and tumor lysis syndrome (Lorenzo & Liaño, 2017).

Some studies have showed a correlation between heat exhaustion/heatstroke and kidney failure. The dehydration and rhabdomyolysis are mechanisms that may contribute in the development of acute kidney injury (Tan et al., 1995). Dehydration from exposure to high temperature may lead to hypovolemia or a reduction in the effective circulating volume, such as decreased cardiac output, systemic vasodilation, or intrarenal vasoconstriction results in impaired renal autoregulation diminishing glomerular filtration rate and in turn decreasing the kidney hypoperfusion (Lameire et al., 2005). The sequence of injury in renal autoregulation is related to preglomerular arteriolar vasodilation (by prostaglandin I2 and nitric oxide) and postglomerular arteriolar vasoconstriction (by angiotensin II), resulting in prerenal acute kidney injury (Varghese et al., 2005).

2.4 Influence of climate on human health

Several studies have showed strong association between chronic and acute diseases with polluting agents and extreme weather conditions. Heat-related illness can vary range from mild conditions such as a discomfort, rash or cramps to very serious conditions such as heatstroke and death. (Atha, 2013). Acclimation to heat is a process that takes some time to be achieved (Wenger, 1988), and physical fitness, health status, and age are determining factors (Hanna & Brown, 1983). It is estimated that first adjustments in thermal regulation occur in 7—30 days, and to be acclimatized/resistant, our body needs 8 weeks (Périard et al., 2015). In the near future, heatwaves, urban heat islands, and increased diurnal temperature range will become more frequent, challenging human health and technological adaptation capacity.

Heatwaves are considered the most dangerous climate extreme to human health, because it is not easily predicted, and the human body, even for young healthy individuals, has no sufficient time to adapt and respond to the sudden increase in temperature. Moreover, other factors related to societal vulnerability and city preparedness to these events may aggravate the impacts on health (Coumou & Robinson, 2013; Heaviside et al., 2017; Perkins et al., 2012). In the scientific literature, many reports concerning heatwaves and excess deaths in different countries can be found (Nitschke et al., 2011; Poumadère et al., 2005; Tan et al., 2010).

The sudden rise in temperature requires the activation of the physiological response to allow thermal balance. In addition to the individual factors that increase the negative health impacts of higher temperatures (age, health status), we have to consider if the challenge is dry or humid heat. This is important because in both situations our physiological mechanisms can be depleted. In the case of dry heat, water loss through sweating intensifies, and without fluid replacement, the maintenance of homeostasis is impaired. On the other hand, in the case of humid heat, the difficulty of heat dissipation by evaporation due to the high humidity compromises the cooling of our body (Maughan et al., 2012). Thus, it is clear that access to supporting technologies and the preparation and adaptation of cities to assist their inhabitants in these episodes is imperative.

Nevertheless, there are also diseases caused by vectors, which designates a group of microorganisms that share the common characteristic of being transmitted to vertebrate hosts by arthropods. It is estimated at 100, the number of arboviruses that cause disease in humans, and at 40, the number of arboviruses that cause disease in domestic animals. Most arboviruses are zoonoses, diseases caused by viruses whose natural host is some vertebrate, in particular birds and mammals. The most important vectors are mosquitoes of various genera, followed by ticks, phlebotomids, and some types of blood-sucking flies. Diseases caused by vectors are very prominent in the tropical part of the planet, where the climatic conditions of the tropics favor the development of insects, which are the hosts of various diseases. In this part of the globe, diseases caused by vectors gain prominence, since they usually occur causing epidemics, causing great human and financial losses (Watts, 2018a; Patz, 2002).

2.4.1 Global climate aspects

The perception that human action can cause warming of the atmosphere is not new. Late 19th century, the Swedish scientist Arrhenius claimed that carbon dioxide production by human activities could affect the climate by inducing warming of the atmosphere. However, in 1954, the chemist Charles David Keeling started working at the California Institute of Technology in the United States, and as part of his research, Keeling began storing carbon dioxide concentration media, daily and in increasingly remote locations (Keeling, 1960). The results have shown that there is a daily fluctuation concentration of carbon dioxide throughout planet, with the minimum value shortly after the half-day and maximum value at dawn. This variation can be explained due to the behavior of green plants. However, it is observed that the equilibrium between plants and carbon dioxide shows disrupted due to the inability of atmosphere to absorb all the carbon or carbon dioxide used by human actions. As part of the effort to verify this hypothesis, a research base was set up on the volcanic island of Mauna Loa, Hawaii, in the middle of the tropical Pacific and Keeling started as measurements in 1958 (Marx et al., 2017). As a result, the scientist left 47 years of measurements of these gases and a surprising discovery: an anthropic action increased the levels of this gas in the planet excessively, and a speed of this increase demonstrated that it is the first time that a human action can cause climate alterations without geological forcing (IPCC, 2018). The legacy left by Keeling encouraged the scientific community awoke to serious problems of global development and its possible consequences for humanity.

In 1979, the first World Climate Conference, organized by the World Meteorological Organization (WMO), expressed that "the continued expansion of human activities on Earth can cause significant changes in the climate, with regional and even global extension." Such an alert pointed to the need for global cooperation to consider climate change in planning the future development of human society. Although many researchers observe this evidence, there is still no coordinated work to study this phenomenon. This only came about in 1987, at the 10th WMO Congress, where the demand for an internationally coordinated scientific assessment was recognized, which was also objective and balanced, to understand the effects of increasing concentrations of greenhouse gases (GHGs) in the earth's climate and how these changes could affect socioeconomic patterns. In this way, the WMO Executive Council together with the United Nations Environment Program (UNEP) established the Intergovernmental Panel on Climate Change, and in November 1988 (https://library.wmo.int), the IPCC held its first Plenary Session establishing three working groups that could prepare evaluation reports (https://www.ipcc.ch/assessment-report/ar6/), which would be as follows:

- Working Group I—focused on the scientific evaluation of a wide range of topics, including GHG and aerosols, radiative forcing, processes and modeling, observation of climatic variations and their changes, and detection of GHG through observations.

- Working Group II—focused on a summary assessment of scientific understanding about the impacts of climate change on agriculture and forests, natural terrestrial ecosystems, hydrology and water resources, human settlements, coastal and ocean areas, and also areas covered seasonally by snow, ice, and "permafrost."
- Working Group III—focused on strategic responses. Subgroups were established for options for mitigation and adaptation responses in the areas of energy and industry, agriculture and forestry, and other human activities, including impacts on human health.

Unlike from medical research studies whereupon experiments can be controlled into laboratory, climate studies, it is impossible once the Chaos theory governs atmosphere, that is, the interactions between the various temporal and spatial scales are uncontrollable in reality. Thereby all atmosphere studies comprise to call upon of mathematical models (Hoskins, 1983). These models are computer programs compiled from numerical mathematical equations representing the physical laws that govern the air movements of the atmosphere; thus, the outputs from models are probabilities and certainly not deterministic. These models simulate scenarios of atmosphera compartments and their interactions with soil, oceans, cryosphere, vegetation, biogeochemical cycles, chemistry of the atmosphere, etc. Each day, new studies are carried out to improve these compartments and their interrelations. Based on these models, it is possible to simulate future climate evolution scenarios based on various GHG emission scenarios. However, despite scientific advances in the area of climate modeling, there are two major sources of uncertainties: (1) unpredictable trajectory of GHG and aerosol emissions and (2) such emissions depend on human decisions on mitigation effectively implemented. In summary, the mathematical models are imperfect representations of nature. Besides these limitations of the MCG, several projections have been confirmed with increasing of frequency and severity of weather extremes such as droughts, summer, windstorms, severe storms, and floods. Like this, we are already experiencing serious environmental catastrophes in recent years around the planet causing deaths and illnesses mainly in the more vulnerable population.

2.5 Local climate aspects

The 2014 revision of the World Urbanization Prospects relates that, we reached a historically significant trademark with more than 54% of the world's population residing in urban areas, and this percentage can reach 66% in 2050 (United Nations, 2018). Considering the accelerated urban expansion due to migration of population from rural areas to cities has caused changes in the landscape and soil coverage, so the replacing of land and vegetation by buildings, roads, and other infrastructure, it leads to phenomenon of heat islands (Imran et al., 2018) (urban areas experience elevated temperatures compared with their outlying rural surroundings). In addition, the modifications of the urban thermal profile affect sensible heat, latent heat and momentum fluxes, surface convergence, optical properties of the atmosphere, boundary layer height, and other boundary layer features. Such changes impact upon pollution dispersion, thunderstorm initiation, and many other physical and chemical processes in the urban environment making large cities more vulnerable to weather extremes event associated (or not) with the climate change (Freitas et al., 2007).

The magnitude of surface urban heat islands can also exacerbate the impact of heatwaves, which are periods of abnormally hot. Sensitive people, such as children, older adults, and those with existing health conditions, are at particular risk from these events such as general discomfort, respiratory difficulties, heat cramps and exhaustion, nonfatal heatstroke, and heat-related mortality. The World Health Organization (WHO) estimates that from 1998 to 2017, excessive heat exposure contributed to more than 160,000 early deaths on the world (WHO, 2018). On the other hand, when analyzing the meteorological variables of large cities, one has to take into consideration the emission of pollutant gases (Coelho et al., 2010). Urban growth causes increase of toxic substances emitted by anthropogenic sources such as vehicles; industries, mining, and others have contaminated the atmosphere. These pollutants are unsafe to human health and contribute to complex atmospheric processes (Andrade et al., 2004).

When studying large urban centers, one should not separate the influence of meteorological variables and air pollution, as there is a synergy between them and, therefore, we refer to such variables as "atmospheric variables," as the human being is exposed to the set of atmospheric variables and not only through a meteorological element, or a single pollutant (Coelho et al., 2010). According to Jendritzky (2000), three factors act synergically on the people: the exchange of heat between humans and the atmosphere, to maintain thermal balance; radiation flows in short and long waves; and air pollution. The main stressors for humans related to the atmospheric environment are excessive heat and cold; air pollution; intense UV-B radiation; atmospheric pressure with little oxygen; sudden changes in weather conditions; and travel to places of very different climate. According to Jendritzky, healthy people have a great capacity to adapt to extreme changes in atmospheric variations. However, there are also people who find it more difficult to adapt, especially more sensitive individuals, such as the elderly, chronically ill, pregnant women, and children (Wang et al., 2019). Faced with this scenario, a question still unanswered: The raise extreme meteorological events in recent years are due to climate change? or local alteration? or the synergy of both? How to quantify each one? Whatever the answer, the truth is that those who live in this city suffer the impacts of these atmospheric and environmental changes on their own health, especially the most vulnerable people.

2.6 Scientific methods used on climate and human health studies

Human health studies comprise a wide range of designs and statistic approaches. We will briefly describe the types of methodologies (research designs) used in this type of research. Note that, from this point on, some words could be unfamiliar to some readers.

In general, medical studies are divided into two approaches: clinical trials and epidemiological studies. Clinical trials are *experimental studies*, which involve an intervention (treatment) on disease in a controlled experimental setting. On the other hand, epidemiological researchers are *observational studies*, and there is no intervention. An observational study differs from experimental studies for being multifactorial, and the causal factor can only be controlled through statistics; thus the effects of confounding may be reduced through appropriate design or through special multivariate statistical methods to control causal factors other than the principle explanatory variable (Table 3.1). In this way, studies about impacts of environmental on human health are analyzed from the perspective of epidemiology.

Within this context, some definitions and explanations about epidemiology must be considered, as it is the science that studies the health–disease process in human communities. In a simplified way, epidemiology can be conceptualized as the *"Epidemiology is the study of the distribution and determinants of health-related states or events in specified populations, and the application of this study to the control of health problems"*.

Epidemiology studies are conducted using human populations to estimate whether there is an association between exposure to a substance and adverse health effects. Epidemiological studies quantity the probability of illness or death in an exposed population compared with that probability in

Table 3.1 Types of observational studies.

Types of study temporality	Measure of risk	Advantages	Disadvantages
Ecological Retrospective	**Prevalence ratio**	Inexpensive, fast easy collected data	Inaccuracy of data. Inability to control confounders
Cross-sectional Retrospective	**Odds ratio**	Inexpensive, individualized data, ability to control confounders	Not good for rare diseases. Cannot examine change over time
Longitudinal Retrospective	**Odds ratio**	Examines changes within individuals over time, power cause-and-effect relationships	Expensive. Hard logistically. Losing respondents
Case −control Retrospective	**Odds ratio**	Inexpensive, good for rare diseases, ability to control confounders	Cannot calculate prevalence. Difficult to get good sample
Cohort Retrospective and prospective	**Relative risk**	Ability to control confounders, individualized data	Expensive. Time-intensive. Losing respondents

an identical unexposed population. The epidemiological approach that compares disease coefficients (or rates) in population subgroups became common practice in the late 19th and early 20th centuries. The epidemiology is also concerned with the evolution and outcome (natural history) of diseases in individuals and population groups. There are four principal types of epidemiological studies: cohort, case−control, cross-sectional, and ecological studies (Table 3.1). In order to analyze how the effects of the pollutants and meteorological variables impact on human health, we should use the *ecological time series studies* type.

Ecological studies sometimes offer fragile confirmation on causality due to extrapolations to individuals from geographic data leading to fallacies. On the other hand, for certain exposures that cannot normally be measured individually (such as air pollution, pesticide residues in food, water), evidence from ecological studies is very important. When the causal relationship has been well demonstrated (well-designed, well-controlled measures of variables), ecological studies can be very useful. However, on rare occasions, an ecological study will provide good evidence for establishing causality (for more information on advantages and disadvantages of methodological designs, we suggest reading the bibliography (Woodward, 2014) indicated at the end of the chapter).

2.6.1 Development of statistical methods from 1960s to 2020

Since the great smog of London of 1952, studies on the effects of air pollution on human health have multiplied, and with each study, the statistical methods have been improved to facilitate the understanding of these effects. The evolution of the statistical techniques of time series analysis has allowed to capture the effects of air pollution on health, even when the pollution levels are not so marked, when compared with the pioneer studies of the 1950s in the United States and England. The following is a brief description of how the statistical models were improved from the 1960s to the present day.

1960s—In the early 1960s, the researchers Martin and Bradley (1960) published a study on the fog incident in the London winter of 1958–59, assessing the health effect of pollution on the population. They used correlation analysis and simple linear regression between daily deaths and pollutant concentrations. In this study, meteorological factors were discussed but were not considered in the analysis. In the same decade, Sterling et al. (1969) investigated the possible effects of air pollution on the rate of hospital admissions in Los Angeles (United States) for different diseases, and the influence of the days of the week on the increase in the number of hospital admissions was observed; this way for the first time the calendar effect, as well as the seasonal effects due to weather conditions, was demonstrated. These two effects were identified as potential confounding factors in the association of interest.

In the same year, Greenburg et al. (1962) examined the pattern of mortality from all diseases during an episode of high pollution that occurred in New York between January 29 and February 12, 1963. During this period, it was assumed that deaths followed a Poisson distribution. The association between the total numbers of deaths from all diseases from 1961 to 1965 was performed using a chi-square trend test.

1970s—In the 1970s, new statistical techniques were used, favored by computational improvements and database from environmental policies done by developed countries. from then on multiple regression models started to be used, as well as control of meteorological factors and long-term fluctuation. This decade was also marked by constant methodological discussions and several studies that evaluated the absence and inadequate treatment of meteorological variables and the multicollinearity between predictive variables and spatiotemporal patterns as alternatives for regression analysis (Farrar, 1967).

1980s—In the 1980s, studies continued in the direction of multiple regression, including control for potential confounding factors and the search for alternative nonlinear regression models to fit nonlinear variables. In this decade, publications became clearer in representing the models used, the adjustments, and the control of confounding factors. In California, Shumway et al. (1988) sought to adjust linear and nonlinear models to explain the possible associations between mortality, pollution, and climatic variables in Los Angeles, in the period between 1970 and 1979. They used time series analysis techniques for 11 series of daily data: 2 series of mortality (total, from respiratory and cardiovascular diseases), 2 from climate (temperature and relative humidity), in addition to 6 types of pollutants. After smoothing the original data, they are transformed into weekly averages. The independent variables are the temperature and the pollution. The errors, additive and correlated, supposedly satisfy an autoregressive model of an unspecified order. The adjusted multiple regression models show a positive and significant association between SO_2 and daily mortality, regardless of the effects of temperature, humidity, secular variations, seasonality, monthly, and annual variations, in addition to possible interactions between meteorological factors and the seasons of the year.

1990–2000—In the second half of the 1990s (1995–99), the regression model for time series proposed by Zeger (1988) stood out, which allowed to adequately adjust the data when they presented overdispersion and/or autocorrelation. This model became known as the autoregressive Poisson model (log-linear model with autoregressive Poisson errors or Poisson time series regression model). The series analyzed (outcome) were daily counts of mortality or hospital admissions for a specific cause, mainly in European cities associated with the project Air Pollution and Health: an European Approach (APHEA). Several studies on the effect of air pollution used this model (Saldiva et al., 1994; Schwartz et al., 1995). It should also be noted that, in the second half of the 1990s, Poisson's generalized additive models (MAG), such as those described by Hastie and Tibshirani (1995), were adopted in some studies, allowing nonparametric adjustment of meteorological variables, trends and seasonality. This model provided greater flexibility in describing the relationship between the outcome and nonlinear covariates (Hastie, 2017).

2000s—Over the years 2000–02, the MLG (generalized linear model) and MAG, applied to Poisson regression, were widely used for these types of ecological studies. In particular, MAG provides greater flexibility in describing complex association patterns from smoothed functions, such as the weighted moving average, locally weighted scatterplot smoother, or cubic smoothing splines (splines), to control meteorological factors and the trend and seasonality of the outcome series. However, the two models showed associations between pollution and hospitalizations (Braga, 2002). In these studies, we observed that hospitalizations, deaths, or even outpatient visits occur hours or days after a peak of atmospheric pollution, or an extreme weather event (cold/heatwave, low relative humidity, etc.). Thus, statistical models need to address this time lag (known as "lags"). To simplify the statistical analysis, until then, lag effects were considered as a linear function (over time). However, these associations were often not linear, and statistical models tend to increase the error in the estimate. How to take into account the effect of each day individually? And is the effect of each day equal (linear) or does the effect of one day influence the following days? To address these deficiencies, a new methodology has been developed to address this problem, further improving the statistical models.

From the 2010s—Developed by Gasparrini (2010), the methodology represents a modeling structure capable of flexibly capturing associations that show nonlinear lag effects in time series. This method was called "distributed lag nonlinear models" (distribution of nonlinear lags). This is a powerful approach, in which several researchers have developed innovative statistical techniques and disseminated this information. The most recent studies have applied this methodology to assess the impacts of climate change on human health through a network of researchers who have shared data and knowledge from their respective countries (http://mccstudy.lshtm.ac.uk)

In this decade, new approaches, both methodological and statistical, have been developed to have a clearer picture as environmental variables affect human health. The thermal comfort indices deal with two or more meteorological variables (temperature, humidity, and wind, for instance), given that the human body reacts to all environmental variables at the same time. Thermal comfort indices become a useful tool to measure thermal stress, mainly in vulnerable patients (Jendritzky, 2000; Sharafkhani et al., 2020). In this way, Coelho et al. (2010) used the thermal comfort index, hospitalizations data, and output from meteorological forecasting to build a hybrid model to predict respiratory hospital admissions. The Brazilian Weather Service used this model to warning public health service in weather extremes situation.

Human activities in large cities promote changes in the cities creating heat islands (from modifications of the urban thermal profile) and increasing air pollution. In this way, it was necessary to come up with a methodology that can deal with the influence of both weather variables and air pollution, and therefore to analyze its impacts on human health (Passos et al. 2014). The impact of weather extremes, as well as the effects of air pollution, is already known. In this way, both effects must be taken into account when analyzing large cities. Thus, principal components analysis technique was used to create an urban air index. This method embeds latent variables instead of the original variables, achieving synergy between pollutants and meteorological variables (Coelho and Saldiva, 2011).

The advancement of computational tools became available environmental variables from satellites. Numerous variables have been measured on a global scale and with better spatial resolutions. Measurements of atmospheric aerosols, vegetation indices, temperature of the land surface, and water could be obtained readily. The obtaining of variables through terrestrial measurements (fixed stations) is still very limited, mainly in poorer countries, since the cost to keep these measuring stations in operation is not viable. Even so, they still have limited measurements in both time and space. In this way, the use of variables made available by satellites becomes an important tool to expand epidemiological studies to regions where there are no terrestrial measurements of atmospheric pollution and meteorological variables (Xu et al., 2014).

2.6.2 Final considerations

Humans are sensitive to many environmental aspects, and recently, climate change brought the focus to the impacts of temperature on health. As explained before, human has the capacity to cope with changes in temperature; however, based on mortality data of recent episodes of heat waves in Europe, it is clear that this physiological adaptation is not so simple and will require behavioral and technological support (Fouda & Santamaouris, 2017; Giorgini, 2017). Humans will face different scenarios of temperature increase, heatwaves, higher mean annual temperature, increased diurnal temperature range, and urban heat islands, and each of them is characterized by different exposure duration and thermic stress.

This chapter is written within the scope of producing a brief summary of the physiology of human thermoregulation, highlighting the importance of some conditions such as age and chronic noncommunicable diseases that may interfere with thermoregulatory mechanisms. At the same time, the methods used to study the impacts of climate change on health are described. We exclude from this work the health effects of other consequences of global warming, such as natural disasters, food and water scarcity, migratory flows, and vector- and waterborne infectious diseases. We also exclude from this text aspects of city adaptation to climate changes. The focus on human physiology and pathophysiology is deliberately chosen in the context of an evolutionary perspective. Along evolution, humans were capable to adapt their physiology and habits to populate the globe across a wide range of climate profiles. This achievement was possible due to a combination of ingenuity with genetic/epigenetic events. The question is if the velocity of such adaptations will move in the same pace of changes in global climate. Up to now, the evidence that we, human population, may promote short-term physiologic adaptations to global warming are inconclusive. In fact, humans are involved in a natural experiment, induced by their own behavior and consumption patterns, and at this moment, the consequences of it cannot be adequately anticipated.

References

Al-Goblan, A. S., Al-Alfi, M. A., & Khan, M. Z. (2014). Mechanism linking diabetes mellitus and obesity. *Diabetes, Metabolic Syndrome and Obesity, 7*, 587–591. https://doi.org/10.2147/DMSO.S67400

Al-Qaissi, A., Papageorgiou, M., Javed, Z., Heise, T., Rigby, A. S., Garrett, A. T., Hepburn, D., Kilpatrick, E. S., Atkin, S. L., & Sathyapalan, T. (2019). Environmental Effects of Ambient Temperature and Relative Humidity on Insulin Pharmacodynamics in Adults with Type 1 Diabetes Mellitus. *Diabetes, Obesity & Metabolism, 21*(3), 569–574.

Andrade, M. F., Ynoue, R. Y., Harley, R., & Miguel, A. H. (2004). Air Quality Model Simulating Photochemical Formation of Pollutants: The Sao Paulo Metropolitan Area, Brazil. *International Journal of Environment and Pollution, 22*(4), 460.

Atha, W. F. (2013). Heat-related illness. *Emergency Medicine Clinics of North America, 31*(4), 1097–1108. https://doi.org/10.1016/j.emc.2013.07.012

Balmain, B. N., Sabapathy, S., Jay, O., Adsett, J., Stewart, G. M., Jayasinghe, R., & Morris, N. R. (2017). Heart failure and thermoregulatory control: can patients with heart failure handle the heat? *Journal of Cardiac Failure, 23*(8), 621–627.

Basu, R., et al. (2018). Temperature and term low birth weight in California. *American Journal of Epidemiology, 1 87*(11), 2306–2314. https://doi.org/10.1093/aje/kwy116

Bates, G. P., & Miller., V. S. (2008). Sweat Rate and Sodium Loss during Work in the Heat. *Journal of Occupational Medicine and Toxicology (London, England), 3*, 4.

Beachy, S. H., & Repasky, E. A. (2011). Toward Establishment of Temperature Thresholds for Immunological Impact of Heat Exposure in Humans. *International Journal of Hyperthermia: The Official Journal of European Society for Hyperthermic Oncology, North American Hyperthermia Group, 27*(4), 344–352.

Benarroch, E. E. (2007). Thermoregulation: Recent concepts and remaining questions. *Neurology, 69*(12), 1293–1297. https://doi.org/10.1212/01.wnl.0000275537.71623.8e

Benjamin, I., et al. (2015). *Andreoli and carpenter's Cecil essentials of medicine elsevieron vitalsource* (9th ed.) Saunders.
Borradale, D. C., & Kimlin, M. G. (2012). Folate degradation due to ultraviolet radiation: possible implications for human health and nutrition. *Nutrition Reviews, 70*(7), 414–422. https://doi.org/10.1111/j.1753-4887.2012.00485.x
Bouchama, A., & Knochel, J. P. (2002). Heat stroke. *The New England Journal of Medicine, 346*(25), 1978–1988.
Braga, A. L. F., Zanobetti, A., & Schwartz, J. (2002). The effect of weather on respiratory and cardiovascular deaths in 12 U.S. cities. *Environmental Health Perspectives, 110*(9), 859–863.
Brown, G. M. (1963). Response to cold of eskimos of the easter Canadia Arctic. *Journal of Applied Physiology, 18*, 970–974. https://doi.org/10.1152/jappl.1963.18.5.970
Coelho, M.S.Z.S., Alves, F.L.T.G., Latorre, M.R.D.O. (2010). Statistical analysis aiming at predicting respiratory tract disease hospital admissions from environmental variables in the city of Sao Paulo. Journal of Environmental and Public Health, 2010, Article ID: 209270. 10.1155/2010/209270
Coelho, M. S. Z. S., & Saldiva, P. H. N. (2011 Dec 5–9). *Use of the "Urban Air Index" to estimate morbidity and mortality in large cities: case study São Paulo, Brazil. 19th Congress International of the Biometeorology*. Auckland, New Zealand.
Coumou, D., & Robinson, A. (2013). Historic and future increase in the global land area affected by monthly heat extremes. *Environmental Research Letters, 8*(3). https://doi.org/10.1088/1748-9326/8/3/034018, 34018-6.
Cui, C. Y., & Schlessinger, D. (2015). Eccrine sweat gland development and sweat secretion. *Experimental Dermatology, 24*(9), 644–650. https://doi.org/10.1111/exd.12773
Díaz, J. (2015). Comparison of the effects of extreme temperatures on daily mortality in Madrid (Spain), by age group: The need for a cold wave prevention plan. *Environmental Research, 143*(Pt A), 186–191. https://doi.org/10.1016/j.envres.2015.10.018
Díaz, J., et al. (2004). Impact of temperature and air pollution on the mortality of children in Madrid. *Journal of Occupational and Environmental Medicine, 46*(8), 768–774. https://doi.org/10.1097/01.jom.0000135542.12974.49
Dumke, C. L., Slivka, D. R., Cuddy, J. S., Hailes, W. S., Rose, S. M., & Ruby, B. C. (2015). The Effect of Environmental Temperature on Glucose and Insulin after an Oral Glucose Tolerance Test in Healthy Young Men. *Wilderness & Environmental Medicine, 26*(3), 335–342.
Fan, S., et al. (2016). Going global by adapting local: A review of recent human adaptation. *Science, 354*(6308), 54–59. https://doi.org/10.1126/science.aaf5098
Farrar, D. E., & Glauber, R. R. (1967). Multicollinearity in regression analysis: The problem revisited. *The Review of Economics and Statistics, 49*(1), 92.
Fealey, R. D. (2019). Thermoregulation in neuropathics. *Handbook of Clinical Neurology, 157*, 777–787. https://doi.org/10.1016/B978-0-444-64074-1.00048-3
Fenzi, A., & Kiefer, F. W. (2014). Brown adipose tissue and thermogenesis. *Hormone Molecular Biology and Clinical Investigation, 19*(1), 25–37. https://doi.org/10.1515/hmbci-2014-0022
Founda, D., & Santamouris, M. (2017). Synergies between urban heat island and heat waves in Athens (Greece), during an extremely hot summer (2012). *Science Reports,* 7(1), 10973. https://doi.org/10.1038/s41598-017-11407-6, 2017 Sep 8.
Franchini, M. L., et al. (2016). Oxygen with cold bubble humidification is no better than dry oxygen in preventing mucus dehydration, decreased mucociliary clearance, and decline in pulmonary function. *Chest, 150*(2), 407–414. https://doi.org/10.1016/j.chest.2016.03.035

Freitas, E. D., Rozoff, C. M., Cotton, W. R., & Silva Dias, P. L. (2007). Interactions of an urban heat island and sea-breeze circulations during winter over the metropolitan area of São Paulo, Brazil. *Boundary - Layer Meteorology, 122*(1), 43–65.

Gasparrini, A., Armstrong, B., & Kenward, M. G. (2010). Distributed lag non-linear models. *Statistics in Medicine, 29*(21), 2224–2234.

Gemenne, F., Zickgraf, C., & Ionesco, D. (Eds.). (2015). *International Organization on Migration (IOM) The State of Environmental Migration 2015- A review of 2014.* Geneva, Switzerland: IOM Publications. Available at: https://environmentalmigration.iom.int/state-environmental-migration-2015-%E2%80%93-review-2014.

Giorgini, P. (2017). Climate changes and human health: A review of the effect of environmental stressors on cardiovascular diseases across epidemiology and biological mechanisms. *Current Pharmaceutical Design, 23*(22), 3247–3261. https://doi.org/10.2174/1381612823666170317143248

Greenburg, L., Jacobs, M. B., Drolette, B. M., Field, F., & Braverman, M. M. (1962). Report of an air pollution incident in New York City, November 1953. *Public Health Reports,* 77(1), 7.

Gwazdauskas, F. C. (1985). Effects of climate on reproduction in cattle. *Journal of Dairy Science, 68*(6), 1568–1578. https://doi.org/10.3168/jds.S0022-0302(85)80995-4

Hall, J. E. (2015). *Guyton and Hall textbook of medical physiology* (13th ed.). London, England: W B Saunders.

Hammami, M. M., Bouchama, A., Al-Sedairy, S., Shail, E., AlOhaly, Y., & Mohamed, G. E. (1997). Concentrations of Soluble Tumor Necrosis Factor and Interleukin-6 Receptors in Heatstroke and Heatstress. *Critical Care Medicine, 25*(8), 1314–1319.

Hanna, J. M., & Brown, D. E. (1983). Human heat tolerance: an anthropological perspective. *Annual Review of Anthropology, 12*, 259–284. https://doi.org/10.1146/annurev.an.12.100183.001355

Hastie, T. J. (2017). Generalized Additive Models. In *Statistical Models in S* (pp. 249–307). Routledge.

Hastie, T., & Tibshirani, R. (1995). Generalized additive models for medical research. *Statistical methods in medical research, 4*(3), 187–196.

Heaviside, C., Macintyre, H., & Vardoulakis, S. (2017). The urban heat island: Implications for health in a changing environment. *Current Environmental Health Reports, 4*(3), 296–305. https://doi.org/10.1007/s40572-017-0150-3

Heng, N. (2017). Tackling the health impacts of climate change in the twenty-first century. *Medicine, Conflict and Surviva, 33*(4), 306–318. https://doi.org/10.1080/13623699.2017.1420409

Hogan, J., et al. (1986). The thickness of the alveolar capillary wall in the human lung at high and low altitude. *British Journal of Diseases of the Chest, 80*(1), 13–18. https://doi.org/10.1016/0007-0971(86)90004-5

Hoskins, B. J. (1983). Dynamical Processes in the Atmosphere and the Use of Models. *Quarterly Journal of the Royal Meteorological Society, 109*(459), 1–21. Royal Meteorological Society (Great Britain).

Hossain, M. Z., et al. (2019). Sociodemographic, climatic variability and lower respiratory tract infections: a systematic literature review. *International Journal of Biometeorology, 63*(2), 209–219. https://doi.org/10.1007/s00484-018-01654-1. Epub 2019 Jan 24.

IPCC. (2018). Global Warming of 1.5°C. An IPCC Special Report on the impacts of global warming of 1.5°C above pre-industrial levels and related global greenhouse gas emission pathways. In P. Zhai, H. O. Pörtner, D. Roberts, J. Skea, P. R. Shukla, A. Pirani, W. Moufouma-Okia, C. Péan, R. Pidcock, S. Connors, J. B. R. Matthews, Y. Chen, X. Zhou, M. I. Gomis, E. Lonnoy, T. Maycock, M. Tignor, & T. Waterfield (Eds.), *the context of strengthening the global response to the threat of climate change, sustainable development, and efforts to eradicate poverty [V. Masson-Delmotte.*

Jendritzky, G., Bucher, K., Laschewski, G., & Walther, H. (2000). Atmospheric Heat Exchange of the Human Being, Bioclimate Assessments, Mortality and Thermal Stress. *International Journal of Circumpolar Health, 59*(3–4), 222–227.

Jendritzky, G., & de Dear, R. (2009). Adaptation and Thermal Environment. In *Biometeorology for Adaptation to Climate Variability and Change* (pp. 9–32). Dordrecht: Springer Netherlands.

Kalkstein, L. S., & Greene, J. S. (1997). An evaluation of climate/mortality relationships in large U.S. cities and the possible impacts of a climate change. *Environmental Health Perspectives, 105*(1), 84–93. https://doi.org/10.1289/ehp.9710584

Keeling, C. D. (1960). The concentration and isotopic abundances of carbon dioxide in the atmosphere. *Tellus, 12*(2), 200–203.

Kim, Y., et al. (2016). Suicide and ambient temperature in east 576 Asian countries: A time-stratified case-crossover analysis. *Environmental Health Perspectives, 124*, 75–80. https://doi.org/10.1289/ehp.1409392

Kjellstrom, T., et al. (2010). Public health impact of global heating due to climate change: potential effects on chronic non-communicable diseases. *International Journal of Public Health, 55*(2), 97–103. https://doi.org/10.1007/s00038-009-0090-2

Kumar, V., Abbas, A. K., & Aster, J. C. (2018). *Robbins basic pathology Elsevier on vital source.* Elsevier.

Lameire, N., Vanbiesen, W., & Vanholder, R. (2005). Acute renal failure. *Lancet, 365*(9457), 417–430.

Lee, J. Y., et al. (2019). Predicted temperature-increase-induced global health burden and ist regional variability. *Environment International, 131*, 105027. https://doi.org/10.1016/j.envint.2019.105027

Li, S., et al. (2018). Exploring associations of maternal exposure to ambient temperature with duration of gestation and birth weight: a prospective study. *BMC Pregnancy Childbirth, 18*(1), 513. https://doi.org/10.1186/s12884-018-2100-y

Liu, T., & Ma, W. (2019). Climate change and health: more research on adaptation is needed. *Lancet Planet Health, 3*(7), e281–e282. https://doi.org/10.1016/S2542-5196(19)30112-3

Liu, C., Yavar, Z., & Sun, Q. (2015). Cardiovascular response to thermoregulatory challenges. *American Journal of Physiology-Heart and Circulatory Physiology, 309*(11), H1793–H1812. https://doi.org/10.1152/ajpheart.00199.2015

Lorenzo, A., & Liaño, F. (2017a). Altas temperaturas y nefrología: a propósito del cambio climático. *Nefrologia: publicacion oficial de la Sociedad Espanola Nefrologia, 37*(5), 492–500.

Marieb, E. N., & Katja, H. (2018). *Study guide for human anatomy & physiology* (11th ed.). Upper Saddle River, NJ: Pearson.

Martin, A. E., & Bradley, W. H. (1960). Mortality, Fog and Atmospheric Pollution: An Investigation during the Winter of 1958–59. *Monthly Bulletin of the Ministry of Health and the Public Health Laboratory Service, 19*, 56–72.

Marx, W., Haunschild, R., French, B., & Lutz, B. (2017). Slow Reception and Under-Citedness in Climate Change Research: A Case Study of Charles David Keeling, Discoverer of the Risk of Global Warming. *Scientometrics, 112*(2), 1079–1092.

Maughan, R. J., Otani, H., & Watson, P. (2012). Influence of relative humidity on prolonged exercise capacity in a warm environment. *European Journal of Applied Physiology, 112*(6), 2313–2321. https://doi.org/10.1007/s00421-011-2206-7. Epub 2011 Oct 20.

Mcbrearty, S., & Brooks, A. S. (2000). The revolution that wasn't: a new interpretation of the origin of modern human behaviour. *Journal of Human Evolution, 39*(5), 453–563.

McCafferty, D. J., et al. (2017). Animal thermoregulation: A review of insulation, physiology and behaviour relevant to temperature control in buildings. *Bioinspiration & Biomimetics, 13*(1), 011001. https://doi.org/10.1088/1748-3190/aa9a12

McElroy, A., & Towsend, P. K. (2018). *Medical anthropology in ecological perspective* (sixth edition). Westview Press.

Morris, R. D., & Naumova, E. N. (1998). Carbon monoxide and hospital admissions for congestive heart failure: Evidence of an increased effect at low temperatures. *Environmental Health Perspectives, 106*(10), 649–653. https://doi.org/10.1289/ehp.98106649

Morrison, S. F., & Blessing, W. W. (2011). Central nervous system regulation of body temperature. In *Central Regulation of Autonomic Functions* (pp. 324–344). Oxford University Press.

Nitschke, M., et al. (2011). Impact of two recent extreme heat episodes on morbidity and mortality in Adelaide, South Australia: a case-series analysis. *Environmental Health, 10*, 42. https://doi.org/10.1186/1476-069X-10-42

Page, L. A., Hajat, S., & Kovats, R. S. (2007). Relationship between daily suicide counts and temperature in England 594 and Wales. *British Journal of Psychiatry, 191*, 106–112. https://doi.org/10.1192/bjp.bp.106.031948

Passos, S. D., Gazeta, R. E., Felgueiras, A. P., Beneli, P. C., & Coelho, M. D. S. (2014). Do pollution and climate influence respiratory tract infections in children? *Revista da Associação Médica Brasileira, 60*(3), 276–282. https://doi.org/10.1590/1806-9282.60.03.018

Patz, J. A. (2002). Global Climate Change and Health: Challenges for Future Practitioners. *JAMA: The Journal of the American Medical Association, 287*(17), 2283–2284.

Périard, J. D., Racinais, S., & Sawka, M. N. (2015). Adaptations and mechanisms of human heat acclimation: Applications for competitive athletes and sports. *Scandinavian Journal of Medicine & Science in Sports, 25*(Suppl 1), 20–38. https://doi.org/10.1111/sms.12408

Perkins, S. E., Alexander, L. V., & Nairn, J. R. (2012). Increasing frequency, intensity and duration of observed global heat waves and warm spells. *Geophysical Research Letters, 39*(20), L20714. https://doi.org/10.1029/2012GL053361

Petrofsky, J. S., Scott, L., & Cuneo-Libarona, M. (2005). The Impact of Rosiglitazone on Heat Tolerance in Patients with Type 2 Diabetes. *Medical Science Monitor: International Medical Journal of Experimental and Clinical Research, 11*(12), CR562–C569.

Potts, R., & Faith, J. T. (2015). Alternating high and low climate variability: The context of natural selection and speciation in Plio-Pleistocene hominin evolution. *Journal of Human Evolution, 87*, 5–20. https://doi.org/10.1016/j.jhevol.2015.06.014

Poumadère, M., et al. (2005). The heat wave in France: Dangerous climate change here and now. *Risk Analysis, 25*, 1483–1494. https://doi.org/10.1111/j.1539-6924.2005.00694.x

Saldiva, P. H., Lichtenfels, A. J., Paiva, P. S., Barone, I. A., Martins, M. A., Massad, E., Pereira, J. C., Xavier, V. P., Singer, J. M., & Böhm, G. M. (1994). Association between air pollution and mortality due to respiratory diseases in children in São Paulo, Brazil: A preliminary report. *Environmental Research, 65*(2), 218–225.

Sanin, L., Cabrera, A. M. Z., & Morales, A. M. T. (2016). Adaptive responses to thermal stress in mammals. *Revista de Medicina Veterinaria, 31*, 121–135. http://www.scielo.org.co/pdf/rmv/n31/n31a12.pdf.

Saraff, V., & Shaw, N. (2016). Sunshine and vitamin D. *Archives of Disease in Childhood, 101*(2), 190–192. https://doi.org/10.1136/archdischild-2014-307214. Epub 2015 Aug 31.

Sawyer, G. M., & Narayan, E. J. (2019). A review on the influence of climate change on sheep reproduction. In E. J. Narayan (Ed.), *Comparative Endocrinology of Animals* (pp. 1–21). https://doi.org/10.5772/intechopen.86799

Schlader, Z. J., Simmons, S. E., Stannard, S. R., & Mündel, T. (2011). Skin temperature as a thermal controller of exercise intensity. *European journal of applied physiology, 111*(8), 1631–1639.
Schwartz, J., & Morris, R. (1995). Air pollution and hospital admissions for cardiovascular disease in detroit, Michigan. *American Journal of Epidemiology, 142*(1), 23–35.
Sharafkhani, R., et al. (2020). The effect of physiological equivalent temperature index variations on mortality in Urmia (The Northwest of Iran). *Urban Climate, 32*(100595), 100595.
Shumway, R. H., Azari, A. S., & Pawitan, Y. (1988). Modeling mortality fluctuations in Los Angeles as functions of pollution and weather effects. *Environmental Research, 45*(2), 224–241.
Son, J. Y. (2019). Impacts of high temperature on adverse birth outcomes in Seoul, Korea: Disparities by individual- and community-level characteristics. *Environmental Research, 168*, 460–466. https://doi.org/10.1016/j.envres.2018.10.032
Speakman, J. R. (2018). Obesity and thermoregulation. *Handbook of Clinical Neurology, 156*, 431–443. https://doi.org/10.1016/B978-0-444-63912-7.00026-6
Sterling, T. D., Pollack, S. V., & Weinkam, J. (1969). Measuring the effect of air pollution on urban morbidity. *Archives of Environmental Health, 18*(4), 485–494.
Sun, Z., et al. (2018). Effects of ambient temperature on myocardial infarction: A systematic review and meta-analysis. *Environmental Pollution, 241*, 1106–1114. https://doi.org/10.1016/j.envpol.2018.06.045, 2018 Oct.
Székely, M., & Garai, J. (2018). Thermoregulation and age. *Handbook of Clinical Neurology, 156*, 377–395. https://doi.org/10.1016/B978-0-444-63912-7.00023-0
Tan, J., et al. (2010). The urban heat island and its impact on heat waves and human health in Shanghai. *International Journal of Biometeorology, 54*, 75–84. https://doi.org/10.1007/s00484-009-0256-x
Tan, W., Herzlich, B. C., Funaro, R., Koutelos, K., Pagala, M., Amaladevi, B., & Grob, D. (1995). Rhabdomyolysis and Myoglobinuric Acute Renal Failure Associated with Classic Heat Stroke. *Southern Medical Journal, 88*(10), 1065–1068.
Tansey, E. A., & Johnson, C. D. (2015). Recent advances in thermoregulation. *Advances in Physiology Education, 39*(3), 139–148. https://doi.org/10.1152/advan.00126.2014
Terrien, J., Perret, M., & Aujard, F. (2011). Behavioral thermoregulation in mammals: a review. *Frontiers in Bioscience (Landmark Ed)., 16*, 1428–1444. https://doi.org/10.2741/3797
Tiihonen, J., et al. (2017). The association of ambient temperature and violent crime. *Science Reports,* 7, 6543. https://doi.org/10.1038/s41598-017-06720-z
UN. (2017). *United nations, department of economic and social affairs, population division.* World Population Ageing 2017 - Highlights (ST/ESA/SER.A/397). Available at: https://www.un.org/en/development/desa/population/publications/pdf/ageing/WPA2017_Highlights.pdf.
United Nations, Department of Economic and Social Affairs, Population Division (2018). The World's Cities in 2018 – Data Booklet (ST/ESA/ SER.A/417). Available from: https://www.un.org/en/events/citiesday/assets/pdf/the_worlds_cities_in_2018_data_booklet.pdf
Varghese, G. M., John, G., Thomas, K., Abraham, O. C., & Mathai, D. (2005). Predictors of Multi-Organ Dysfunction in Heatstroke. *Emergency Medicine Journal: EMJ, 22*(3), 185–187.
Wang, Jiajia, Tong, Shilu, Williams, Gail, & Pan, Xiaochuan (2019). Exposure to heat wave during pregnancy and adverse birth outcomes: An exploration of susceptible windows. *Epidemiology (Cambridge, Mass, 30*(Suppl 1), S115–21.

Watts, N., et al. (2018a). The Lancet Countdown on health and climate change: from 25 years of inaction to a global transformation for public health. *Lancet, 391*(10120), 581–630. https://doi.org/10.1016/S0140-6736(17)32464-9

Watts, N., et al. (2018b). The 2018 report of the Lancet Countdown on health and climate change: shaping the health of nations for centuries to come. *Lancet, 392*(10163), 2479–2514. https://doi.org/10.1016/S0140-6736(18)32594-7

Wenger, C. B. (1988). Human heat acclimatization. In K. B. Pandolf, M. N. Sawka, & R. R. Gonzalez (Eds.), *Human performance physiology and environmental medicine at terrestrial extremes. Indianapolis* (pp. 153–197). Benchmark Press.

https://www.who.int/health-topics/heatwaves#tab=tab_1.

Woodward, M. (2014). *Epidemiology: Study design and data analysis* (3rd ed.). Philadelphia, PA: Chapman & Hall/CRC.

World Bank. (2010). *Cities and climate change : an urgent agenda (English). Urban development series knowledge papers; no. 10.* Washington, DC: World Bank. Available at: http://documents.worldbank.org/curated/en/194831468325262572/Cities-and-climate-change-an-urgent-agenda.

Wu, M., et al. (2019). Prenatal exposure of diurnal temperature range and preterm birth: Findings from a birth cohort study in China. *Science of The Total Environment, 656*, 1102–1107. https://doi.org/10.1016/j.scitotenv.2018.11.305

Wyndham, C. H. (1968). Tolerance of very hot humid environments by highly acclimatized Bantu at rest. *British Journal of Industrial Medicine, 25*(1), 22–39. https://doi.org/10.1136/oem.25.1.22

Xu, Z., Liu, Y., Ma, Z., Li, S., Hu, W., & Tong, S. (2014). Impact of temperature on childhood pneumonia estimated from satellite remote sensing. *Environmental Research, 132*, 334–341.

Yardley, J. E., Stapleton, J. M., Sigal, R. J., & Kenny., G. P. (2013). Do heat events pose a greater health risk for individuals with type 2 diabetes? *Diabetes Technology & Therapeutics, 15*(6), 520–529.

Younan, D., et al. (2018). Long term ambient temperature and externalizing behaviors in adolescents. *American Journal of Epidemiology, 187*(9), 1931–1941. https://doi.org/10.1093/aje/kwy104

Zaidi, A. A., et al. (2017). Investigating the case of human nose shape and climate adaptation. *PLoS Genetics, 13*(3), e1006616. https://doi.org/10.1371/journal.pgen.1006616

Zeger, S. L. (1988). A regression model for time series of counts. *Biometrika, 75*(4), 621.

Zhao, Zheng-Dong, Yang, Wen Z., Gao, Cuicui, Fu, Xin, Zhang, Wen, Zhou, Qian, Chen, Wanpeng, et al. (2017). A Hypothalamic Circuit That Controls Body Temperature. *Proceedings of the National Academy of Sciences of the United States of America, 114*(8), 2042–2047.

Further reading

Goldman, L., & Schafer, A. I. (2011). *Goldman's cecil medicine* (24th ed.). London, England: Grune & Stratton.

Grecequet, M., et al. (2017). Climate Vulnerability and human migration in global perspective. *Sustainability, 9*(5), 720. https://doi.org/10.3390/su9050720. pii.

Guo, P., Liu, T., Zhang, Q., Wang, L., Xiao, J., Zhang, Q., Luo, G., Li, Z., He, J., Zhang, Y., & Ma, W. (2017 Oct 16). Developing a dengue forecast model using machine learning: A case study in China. *PLOS Neglected Tropical Diseases, 11*(10), e0005973. https://doi.org/10.1371/journal.pntd.0005973. PMID: 29036169; PMCID: PMC5658193.

Porta, M., & Last, J. M. (2008). *Dictionary of epidemiology* (5th ed., p. 81). New York, NY: Oxford University Press.
Sixth Assessment Report. International Panel Climate Change. IPCC. Ch. Accessed October 25, 2020. https://www.ipcc.ch/assessment-report/ar6/.
World Meteorological Organization (WMO). WMO.int. Accessed October 25, 2020b. https://library.wmo.int/doc_num.php?explnum_id=4858.
World Meteorological Organization (WMO). Heatwaves. Wmo.int. Accessed October 30.

CHAPTER 4

Heat exposure and child health

Naoto Fujii[1] and Yasushi Honda[1,2]
[1]Faculty of Health and Sport Sciences, University of Tsukuba, Tsukuba, Japan; [2]Center for Climate Change Adaptation, National Institute for Environmental Studies, Tsukuba, Japan

1. Background

In the WHO report on climate change impact on health, the chapter for heat-related mortality only dealt with elderly (65+ years of age) (Honda et al., 2014). This is because other causes of deaths (malnutrition, diarrhea, malaria, and dengue fever) in the report were concentrated on children, and including all the age groups for heat-related mortality had a possible threat of double counting. However, this does not mean we can ignore the direct heat impact on children. As discussed in the following chapters, children are more susceptible to heat than adults, and substantial proportion of ambulance car visits for heat stroke are occupied by children, and the burden on this public health infrastructure can be critical under the COVID-19 turmoil, by which the medical systems of many countries have already been severely damaged.

In this chapter, we first introduce the thermophysiological property of children, and a wide range of epidemiological findings follow. We do not discuss specific adaptation measures here, but we believe that the readers can find several hints in the following sections.

2. Thermophysiology of children

Studies showed that human temperature regulation can be modulated by individual factors such as age (Holowatz & Kenney, 2010; Inoue, Havenith, et al., 1999; Inoue, Shibasaki et al., 19999; Kenney et al., 1990; Larose et al., 2013), sex (Charkoudian & Johnson, 2000; Gagnon & Kenny, 2012; Inoue et al., 2005; Kuwahara et al., 2005; Stephenson & Kolka, 1993), chronic disease (Fujii et al., 2016; Holowatz & Kenney, 2007; Kellogg et al., 1998; Sokolnicki et al., 2009; Wick et al., 2006), fitness level (Ho et al., 1997; Inoue, Havenith, et al., 1999; Lamarche et al., 2017; Notley et al., 2019), and body morphology (Notley et al., 2019). Here, we discuss thermoregulation in prepubertal children by comparing that in young

Heat Exposure and Human Health in the Context of Climate Change
ISBN 978-0-12-819080-7
https://doi.org/10.1016/B978-0-12-819080-7.00011-2
© 2023 Elsevier Inc.
All rights reserved.

adults with an emphasis on (1) heat dissipation, (2) heat production, (3) body size, (4) body core temperature, and (5) heat acclimation (adaptation to heat).

2.1 Heat dissipation

Human body temperature is determined by the balance between heat production and heat dissipation; the latter is achieved through the heat loss responses of cutaneous vasodilation (dry heat loss) and sweating (evaporative heat loss). Previous work consistently showed that the heat loss response of sweating is smaller in prepubertal children relative to young adults, though this attenuation is compensated by augmented heat loss response of cutaneous vasodilation (Inoue et al., 2009; Shibasaki et al., 1999; Wagner et al., 1974; Woloschuk et al., 2020). Since the number of sweat gland remains unchanged throughout the lifespan (Kawabata, 1950), sweat gland density per given skin area is higher in prepubertal children in comparison with young adults. However, prepubertal children exhibit lower sweat gland output during heat stress, resulting in the lower sweat rate observed in this population (Shibasaki et al., 1997). This lower sweat output can be due to the smaller sweat gland size in prepubertal children compared with young adults (Landing et al., 1970). This is supported by a previous in vitro study that demonstrated sweat gland size is highly correlated with sweat gland output capacity (Sato & Sato, 1983).

2.2 Heat production

Since body temperature can be influenced by both heat production and heat dissipation, the former should be considered in addition to heat dissipation when evaluating children's thermoregulation. During exercise, only a portion of energy produced in the active muscle is utilized for external work. The remaining large portion of energy is converted to heat, causing body temperature to increase. Thus, the extent to which energy is used for external work (i.e., mechanical efficiency) is critical in determining body core temperature. It has been shown that prepubertal children show higher oxygen uptake and therefore lower mechanical efficiency in comparison with young adults when walking or running at the same speed (Cureton et al., 1997; Rowland et al., 1987). In other words, prepubertal children yield greater heat production relative to young adults under these conditions. However, mechanical efficiency during cycling (weight-supported exercise) is not different between prepubertal children and young adults (Rowland et al., 1990).

2.3 Body size

Prepubertal children and young adults largely differ in body size, which can greatly differentiate body temperature regulation. As described in the recent review by Notley et al. (2020), assuming a similar average tissue-specific heat capacity in prepubertal children and young adults (3.47 kJ/kg/°C), a 139 kJ increase in body heat content would increase mean body temperature by 1.0°C in 40 kg prepubertal children, whereas the same heat content would increase only 0.5°C in 80 kg young adults.

2.4 Body core temperature

Body core temperature is a critical factor causing heat stroke. As mentioned before, differences between prepubertal children and young adults exist in heat dissipation, heat production, and body size, all of which can modulate body core temperature regulation. However, prepubertal children can maintain their body core temperature during exercise as do young adults under nonhumid cool or temperate conditions (Delamarche et al., 1990; Drinkwater et al., 1977; Leites et al., 2016; Rowland et al., 2008; Smolander et al., 1992; Wagner et al., 1972). This may not be surprising as elevated dry heat loss associated with cutaneous vasodilation can play a major role in heat dissipation when ambient temperature is low (Nielsen, 1938). However, when ambient temperature is high (35°C), sweating plays a pivotal role in facilitating heat loss (Nielsen, 1938). Therefore, lower sweating response thus smaller evaporative heat loss in prepubertal children relative to young adults could lead to augmented heat strain during exercise under these conditions (Bar-Or, 1980, 1998; Falk, 1998; Inoue et al., 2004). Surprisingly, several studies reported that prepubertal children can well maintain body core temperature during exercise compared with young adults even under heat-stressed conditions (≥30°C) (Drinkwater et al., 1977; Inbar et al., 2004; Leites et al., 2016; Rivera-Brown et al., 2006; Rowland et al., 2008; Shibasaki et al., 1997). This may be as a result of more efficient sweat evaporation (Inbar et al., 2004). Based on recent findings, the revision of the American Academy of Pediatrics (AAP) policy in 2011 states that physiological burden associated with exercise-induced heat stress is not different between children and adults (Council on Sports et al., 2011).

It should be, however, highlighted that the aforementioned experiments did not consider the effect of solar radiation. Solar radiation can increase radiative heat gain as reflected by elevations in skin temperature

(Otani et al., 2016). This elevation in skin temperature would be further augmented with high ambient temperature, which may ultimately increase body core temperature in prepubertal children in a greater manner in comparison with young adults. Also, as Notley et al. (2020) suggest, differences in thermoregulation between prepubertal children versus young adults may be more pronounced as exercise-induced heat load (i.e., exercise intensity) increases. Along these lines, high-intensity exercise can attenuate the heat loss response of cutaneous vasodilation (Fujii et al., 2014), a major avenue for heat loss in prepubertal children, ultimately causing greater body core temperature elevation in this population. Clearly, future studies are needed to delineate these possibilities.

2.5 Heat acclimation

Heat acclimation is a human adaptation to heat stress mediated by artificial means such as under controlled environment. Heat acclimation in young adults can be achieved by 6—14 consecutive heat exposures, mediating several adaptations such as improved exercise capacity in the heat, plasma volume expansion, reduced heart rate and body core temperature, enhanced heat loss responses of cutaneous vasodilation and sweating, diluted sweat, and reduced metabolic rate during exercise (i.e., better exercise economy) (Daanen et al., 2018; Fujii et al. 2012, 2015; Gibson et al., 2020; Periard et al., 2015; Sawka et al., 2011). Exercise in the heat is the most effective intervention that mediates heat acclimation; however, passive heat exposure such as hot water immersion or sauna can also help achieve heat acclimation (Heathcote et al., 2018). Given the aforementioned adaptations, heat acclimation is effective to prevent extreme elevations in body core temperature and thus heat stroke. Prepubertal children can acclimatize to heat, but the magnitude may be smaller relative to postpubertal boys as demonstrated in the study by Wager et al. (1972). In addition, Bar-Or (1980) demonstrated that prepubertal children (8—10 years) acclimated at a slower rate relative to young adults during repeated heat exposure. The slower heat adaptation, if at all, could increase the risk of heat stroke in prepubertal children especially during seasonal changes such as transition from spring to summer wherein natural heat adaptation (i.e., heat acclimatization) occurs.

3. Epidemiological findings of heat—health relations

As seen in the previous sections, children are more vulnerable to heat exposure compared with adults. In this section, we review epidemiological studies to evaluate to what extent heat exposure affects children's health. First, we explore the breadth of impacts by finding the various health outcomes among the available literature. Then, we will address the most serious impact, heat-related mortality.

3.1 Wide range of the health outcome due to heat exposure

The effect of heat begins even before birth. Dreiling et al. (1991) reported that heat stress reduces uterine blood flow by 20%—30%, and this reduced circulation can be considered as one of the reasons for the heat impacts on fetuses. Actual impacts from epidemiologic studies are given in the following.

3.1.1 Placental abruption

He et al. (2018) reported the impact on placental abruption in Quebec, Canada. Their main exposure was the maximum temperature reached during the week before abruption. Using the case-crossover design, they computed the odds ratio (=OR). The reference temperature was 15°C, and the overall OR for 30 degrees was 1.07, with the 95% confidence interval (=CI) 0.99—1.16. At term, OR was 1.12 (95%CI: 1.02—1.24), but preterm abruption showed no elevated risk.

3.1.2 Premature rupture of membrane

Ha et al. (2018) used a nationwide US obstetrics cohort (2002—08) and conducted a case-crossover study on the relation between outdoor temperature and premature rupture of membranes. One degree celsius increase in temperature during the week before delivery was associated with 5% (95%CI: 3—6) increased risk of preterm premature rupture of membranes and 4% (95%CI: 3—5) increased risk of term premature rupture of membranes for the warm season (May—September). On the other hand, 1 degree increase was associated with 2% (95%CI: 1—3) risk reduction in both preterm and term premature rupture of membranes during the cold season (October—April).

3.1.3 Preterm delivery

In their case-crossover study, Avalos et al. (2017) showed the relation between the high apparent temperature and spontaneous preterm delivery in California. With the 5.6°C increase (average of the same day and the previous 6 days) during warm season (May—October), they found 11.6% (95%CI: 4.1—19.7) increase in spontaneous preterm delivery. During cold season (November—April), on the other hand, they did not find a significant relation. Ilango et al. examined whether the extreme heat events affect the risk of preterm birth in California. They created a population-based cohort of mothers who had the live, singleton births for the period of 2005—13 during the warm season (May—September). The exposure was the extreme heat during the last week of gestation; for extreme heat, they tried several definitions based on the previous reports. For all definitions of extreme heat, the risk of preterm birth was higher among the mothers with extreme heat during their last week of gestation. The hazard ratios were from 1.008 (95%CI: 0.997—1.021) to 1.128 (95%CI: 1.052, 1.210) depending on the definition of extreme heat; the association was stronger when the exposure was more extreme and longer.

3.1.4 Stillbirth

Basu et al. (2016) reported elevated apparent temperatures were related to stillbirth in California. Using the time-stratified case-crossover design, they found that every 5.6°C increase in the cumulative average of lags 2—6 days of apparent temperature resulted in 10.4% (95%CI: 4.4—16.8) increased risk of stillbirth during warm season (May—October). Similar to the preterm delivery, they found no association during the cold season (November—April). Using the data of 12 US multicenter retrospective cohort study, Ha et al. (2017) also showed the relation between the temperature and stillbirth. For acute effect, they used case-crossover design, and for chronic effect, they compared the risk of hot (higher than 90th percentile) temperatures with mild (10th—90th percentile) temperatures. Whole-pregnancy exposure to hot showed higher risk compared with mild; adjusted odds ratio (OR) was 3.71 (95%CI: 3.07—4.47). For short-term effect, they found that 1°C increase during the week preceding delivery was associated with a 6% (95%CI: 3—9) increase in stillbirth risk during warm season (May—September).

3.1.5 Low birth weight

Kloog et al. (2015) used the satellite-based spatiotemporal resolved air temperature and examined its relation to birth outcomes in Massachusetts. They used linear and logistic mixed models and accelerated failure time models. An interquartile range increase (8.4°C) during the last trimester was associated with 16.7g (95%CI: 29.7−3.7) lower birth weight. Basu et al. reported the relation between temperature and term low birth weight in California. In their retrospective cohort study, they identified 43,629 full-term low-birth-weight infants and 2,032,601 normal-weight infants during the period 1999−2013. With the logistic regression models, they found that 5.6°C increase in apparent temperature during the full gestation was associated 13.0% (95% CI: 4.1−22.7) increase in odds of term low birth weight. If the exposure period was restricted, the third trimester showed the highest risk of 15.8% (95%CI: 5.0−27.6).

3.1.6 Congenital cataracts

A population-based case−control study found that multiple heat indicators, such as universal apparent temperature (= UAT), heat wave, the number of heat waves, and the number of days with UAT above 90th percentile, were associated with higher risk of congenital cataract (Van Zutphen et al., 2012). In case of UAT, adjusted odds ratio for a 5-degree Fahrenheit increase in 1 week mean of daily minimum UATs was 1.51 (95%CI: 1.14−1.99) in summer.

Heat impact can affect children shortly after birth; Basagana et al. (2011) reported that the higher infant mortality on hot days was detected only for conditions originating in the perinatal period. The risk on the days with daily maximum temperature being higher than 95th percentile was 1.53 (95%CI: 1.16−2.02) times higher, and within the perinatal causes, the temperature−mortality relation was stronger for cardiovascular, respiratory, digestive system, and hemorrhagic and hematologic disorders. Based on these findings, the authors speculated that the mechanism of the heat effect for infants would be the same as that for adults. However, some of these disorders should be classified as congenital malformation, and the heat during pregnancy could also had some effects.

For infants, toddlers, and school children, heat stroke is one of the most apparent consequences of heat exposure. In Japan, the data on ambulance car dispatch due to heat stroke have been made available by Fire and Disaster Management Agency of the Ministry of Internal Affairs and Communications. As shown in Table 4.1, 0−6-year-old and 7−17-year-

Table 4.1 Comparison between ambulance visits and population by age (2015).

Age	Population (%)	# ambulance visits (%)
0–6	6	1
7–17	10	13
18–64	57	36
65+	27	50
Total	100	100

Note: Ambulance visit data were for May–September only.

old group occupied 1% and 13% of the number of ambulance visits, whereas the populations for these groups occupied 6% and 10% in 2015. This implies that younger children are under good supervision of their parents, but the school age children have more opportunities to attend physically demanding activities, e.g., physical education or baseball games. As for the severity, approximately 70% did not need hospitalization and only 0.2% (105 subjects) died. A simple semiparametric regression model using the aforementioned data for Tokyo and smoothing spline (degree of freedom = 4) model showed a hockey stick–type relation between daily maximum temperature and number of ambulance visits (Fig. 4.1). For children aged 7–17 years, apparently because of the opportunity of playing outside with harder physical exertion, the risk was higher than the lower

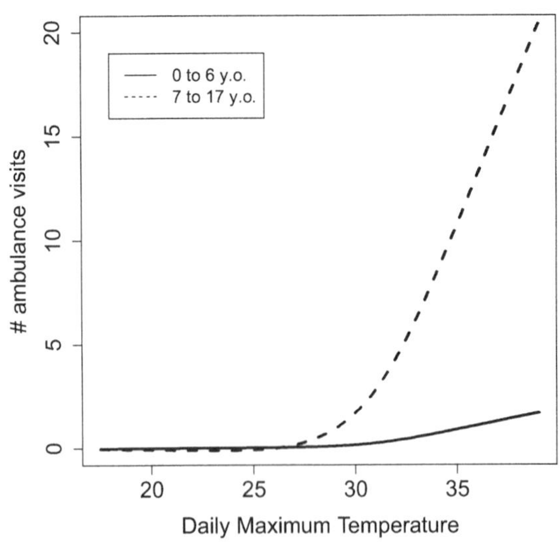

Figure 4.1 Relation between daily maximum temperature and number of ambulance visits by age group (Tokyo, June–September 2010–19).

age group; the number of ambulance visits for the same heat exposure is much higher for 7–17 age group even after considering that the population size is twice as large as that of the 0–6 age group.

Emergency department visits, not restricted to heat stroke, were investigated by Sheffield et al. (2018) in their time-stratified case–crossover analysis, they found that an interquartile range of daily maximum temperature was associated with 16.6% (95%CI: 3.0–31.9) increased risk for heat-specific diagnosis, 10.1% (8.2–11.9) for general symptom, 4.9% (3.9–5.9) for infectious diseases, and 5.1% (3.8–6.4) for injury. Another report by van Loenhout et al. (2018) used distributed lag nonlinear models assuming overdispersion; they reported hockey stick–type relation, i.e., elevated risk of emergency room visit beyond 21°C for heat-related and respiratory diseases among 0–14-year-old group in the Netherlands.

3.2 Heat-related mortality

The hardest challenge of the research on the heat-related mortality among children is its small number of daily deaths. Mainly due to this problem, a systematic review conducted in 2014 showed inconclusive results; some found elevated risks and others found no elevated risks (Xu et al., 2014). Even when the Tokyo data set with the population more than 10 million since the 1960s and long observation period from 1972 to 2015 was used, the conditional Poisson regression analyses incorporating 7 days of lag showed unclear relation, although some heat risk was suggested (Fig. 4.2).

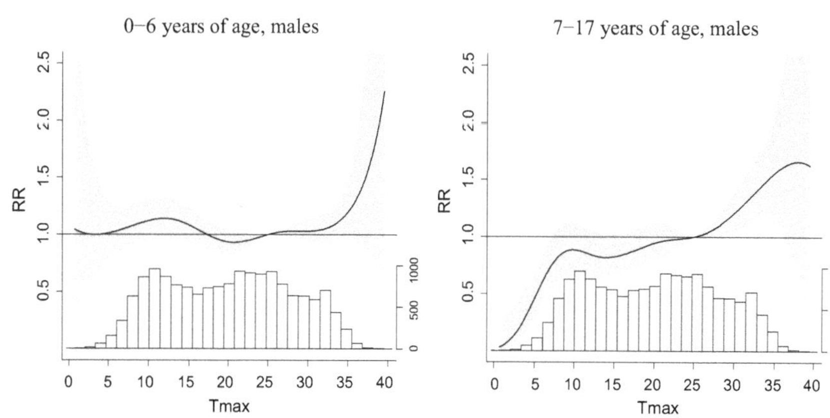

Figure 4.2 Relation between daily maximum temperature and mortality for selected age and sex groups (Tokyo, 1972–2015).

Table 4.2 Mortality rate ratio from external causes by age group and sex in Kyushu, Japan, 1972–90.

Daily maximum temperature	0–4 years old		5–9 years old		10–14 years old	
	Males	Females	Males	Females	Males	Females
23≤ <28	1.0	1.0	1.0	1.0	1.0	1.0
28≤ <33	1.1 (1.0 −1.2)	1.1 (0.97 −1.2)	1.3 (1.1 −1.4)	1.4 (1.2 −1.7)	1.8 (1.5 −2.1)	1.2 (0.90 −1.7)
33≤	1.2 (1.0 −1.4)	1.0 (0.87 −1.3)	1.8 (1.5 −2.2)	1.9 (1.4 −2.5)	3.1 (2.4 −3.9)	1.5 (0.94 −2.5)

Note: Year was controlled for by Mantel-Haenszel method.
Data from Honda Y, Ono M, Sasaki A, Uchiyama I. (1995). Relationship between daily high temperature and mortality in kyushu, Japan. *Nihon Koshu Eisei Zasshi, 42*, 260–268.

Another study, although old, with a relatively large study held in Kyushu, Japan, with the population size of more than 13 million in 1990 showed significant results (Honda et al., 1995). In this study, the daily maximum temperatures were classified into 5°C categories, and mortality rates for these categories were calculated. For age-specific analyses, they obtained year-controlled Mantel-Haenszel rate ratios shown in Table 4.2. Compared with the 23–28°C category, higher temperature categories showed elevated risks. They also showed that these elevated risks were attributed to external causes. This elevated risk would be considered as "risk shift"; during the summer vacation, children would have gone swimming or other activities with higher-than-usual risk. This interpretation is compatible with much higher risk among 10–14-year-old male group, with less parental guidance and more risky behaviors.

In conclusion, we have limited evidence that showed elevated risk of children's mortality due to high temperature. To overcome the aforementioned size problem, multicity pooled analyses are sorely needed for children's mortality impact assessment.

References

Avalos, L. A., Chen, H., Li, D. K., & Basu, R. (2017). The impact of high apparent temperature on spontaneous preterm delivery: A case-crossover study. *Environmental Health, 16*, 5.

Bar-Or, O. (1980). Climate and the exercising child — a review. *International Journal of Sports Medicine, 1*, 53–65.

Bar-Or, O. (1998). Effects of age and gender on sweating pattern during exercise. *International Journal of Sports Medicine, 19*(Suppl. 2), S106–S107.

Basagana, X., Sartini, C., Barrera-Gomez, J., Dadvand, P., Cunillera, J., Ostro, B., et al. (2011). Heat waves and cause-specific mortality at all ages. *Epidemiology, 22*, 765–772.

Basu, R., Sarovar, V., & Malig, B. J. (2016). Association between high ambient temperature and risk of stillbirth in California. *American Journal of Epidemiology, 183*, 894–901.

Charkoudian, N., & Johnson, J. M. (2000). Female reproductive hormones and thermoregulatory control of skin blood flow. *Exercise Sport Science Review, 28*, 108–112.

Council on Sports, M., Fitness, Council on School, H., Bergeron, M. F., Devore, C., Rice, S. G., & American Academy of, Pediatrics. (2011). Policy statement-climatic heat stress and exercising children and adolescents. *Pediatrics, 128*, e741–747.

Cureton, K. J., Sloniger, M. A., Black, D. M., McCormack, W. P., & Rowe, D. A. (1997). Metabolic determinants of the age-related improvement in one-mile run/walk performance in youth. *Medicine and Science inSports Exercise, 29*, 259–267.

Daanen, H. A. M., Racinais, S., & Periard, J. D. (2018). Heat acclimation decay and re-induction: A systematic review and meta-analysis. *Sports Medicine, 48*, 409–430.

Delamarche, P., Bittel, J., Lacour, J. R., & Flandrois, R. (1990). Thermoregulation at rest and during exercise in prepubertal boys. *European Journal of Applied Physiology and Occupational Physiology, 60*, 436–440.

Dreiling, C. E., Carman, F. S., 3rd, & Brown, D. E. (1991). Maternal endocrine and fetal metabolic responses to heat stress. *Journal of Dairy Science, 74*, 312–327.

Drinkwater, B. L., Kupprat, I. C., Denton, J. E., Crist, J. L., & Horvath, S. M. (1977). Response of prepubertal girls and college women to work in the heat. *Journal of Applied Physiology, Respiratory, Environmental and Exercise Physiology, 43*, 1046–1053.

Falk, B. (1998). Effects of thermal stress during rest and exercise in the paediatric population. *Sports Medicine, 25*, 221–240.

Fujii, N., Dervis, S., Sigal, R. J., & Kenny, G. P. (2016). Type 1 diabetes modulates cyclooxygenase- and nitric oxide-dependent mechanisms governing sweating but not cutaneous vasodilation during exercise in the heat. *American Journal of Physiology-Regulatory, Integrative and Compaative Physiology, 311*, R1076–R1084.

Fujii, N., Honda, Y., Ogawa, T., Tsuji, B., Kondo, N., Koga, S., & Nishiyasu, T (2012). Short-term exercise-heat acclimation enhances skin vasodilation but not hyperthermic hyperpnea in humans exercising in a hot environment. *European Journal of Applied Physiology, 112*, 295–307.

Fujii, N., McGinn, R., Stapleton, J. M., Paull, G., Meade, R. D., & Kenny, G. P. (2014). Evidence for cyclooxygenase-dependent sweating in young males during intermittent exercise in the heat. *Journal of Physiology, 592*, 5327–5339.

Fujii, N., Tsuji, B., Honda, Y., Kondo, N., & Nishiyasu, T. (2015). Effect of short-term exercise-heat acclimation on ventilatory and cerebral blood flow responses to passive heating at rest in humans. *Journal of Applied Physiology, 119*, 435–444 (1985).

Gagnon, D., & Kenny, G. P. (2012). Sex differences in thermoeffector responses during exercise at fixed requirements for heat loss. *Journal of Applied Physiology, 113*, 746–757 (1985).

Gibson, O. R., James, C. A., Mee, J. A., Willmott, A. G. B., Turner, G., Hayes, M., & Neil, S. M. (2020). Heat alleviation strategies for athletic performance: A review and practitioner guidelines. *Temperature (Austin), 7*, 3–36.

Ha, S., Liu, D., Zhu, Y., Sherman, S., & Mendola, P. (2018). Acute associations between outdoor temperature and premature rupture of membranes. *Epidemiology, 29*, 175–182.

Ha, S., Liu, D., Zhu, Y., Soo Kim, S., Sherman, S., Grantz, K. L., et al. (2017). Ambient temperature and stillbirth: A multi-center retrospective cohort study. *Environmental Health Perspectives, 125*, 067011.

Heathcote, S. L., Hassmen, P., Zhou, S., & Stevens, C. J. (2018). Passive heating: Reviewing practical heat acclimation strategies for endurance athletes. *Frontiers in Physiology, 9*, 1851.

He, S., Kosatsky, T., Smargiassi, A., Bilodeau-Bertrand, M., & Auger, N. (2018). Heat and pregnancy-related emergencies: Risk of placental abruption during hot weather. *Environment International, 111*, 295–300.

Ho, C. W., Beard, J. L., Farrell, P. A., Minson, C. T., & Kenney, W. L. (1997). Age, fitness, and regional blood flow during exercise in the heat. *Journal of Applied Physiology, 82*, 1126–1135.

Holowatz, L. A., & Kenney, W. L. (2007). Local ascorbate administration augments no- and non-no-dependent reflex cutaneous vasodilation in hypertensive humans. *American Journal of Physiology Heart and Circul;atory Physiology, 293*, H1090–H1096.

Holowatz, L. A., & Kenney, W. L. (2010). Peripheral mechanisms of thermoregulatory control of skin blood flow in aged humans. *Journal of Applied Physiology, 109*, 1538–1544.

Honda, Y., Kondo, M., McGregor, G., Kim, H., Guo, YL., Hales, S., Kovats, RS., et al. (2014). *Quantitative risk assessment of the effects of climate change on selected causes of death, 2030s and 2050s*. WHO.

Honda, Y., Ono, M., Sasaki, A., & Uchiyama, I. (1995). [relationship between daily high temperature and mortality in kyushu, Japan]. *Nihon Koshu Eisei Zasshi, 42*, 260–268.

Inbar, O., Morris, N., Epstein, Y., & Gass, G. (2004). Comparison of thermoregulatory responses to exercise in dry heat among prepubertal boys, young adults and older males. *Experimental Physiology, 89*, 691–700.

Inoue, Y., Havenith, G., Kenney, W. L., Loomis, J. L., & Buskirk, E. R. (1999). Exercise- and methylcholine-induced sweating responses in older and younger men: Effect of heat acclimation and aerobic fitness. *International Journal of Biometeorology, 42*, 210–216.

Inoue, Y., Ichinose-Kuwahara, T., Nakamura, S., Ueda, H., Yasumatsu, H., Kondo, N., & Araki, T (2009). Cutaneous vasodilation response to a linear increase in air temperature from 28 degrees c to 40 degrees c in prepubertal boys and young men. *Journal of Physiological Anthropology, 28*, 137–144.

Inoue, Y., Kuwahara, T., & Araki, T. (2004). Maturation- and aging-related changes in heat loss effector function. *Journal of Physiological Anthropology Applied Human Science, 23*, 289–294.

Inoue, Y., Shibasaki, M., Ueda, H., & Ishizashi, H. (1999). Mechanisms underlying the age-related decrement in the human sweating response. *European Journal of Applied Physiology and Occupational Physiology, 79*, 121–126.

Inoue, Y., Tanaka, Y., Omori, K., Kuwahara, T., Ogura, Y., & Ueda, H. (2005). Sex- and menstrual cycle-related differences in sweating and cutaneous blood flow in response to passive heat exposure. *European Journal of Applied Physiology, 94*, 323–332.

Kawabata, A. (1950). Studies on the function of human sweat organs. *Journal of Mie Medical College, 1*, 25–41.

Kellogg, D. L., Jr., Morris, S. R., Rodriguez, S. B., Liu, Y., Grossmann, M., Stagni, G., & Shepherd, A. M. (1998). Thermoregulatory reflexes and cutaneous active vasodilation during heat stress in hypertensive humans. *Journal of Applied Physiology, 85*, 175–180 (1985).

Kenney, W. L., Tankersley, C. G., Newswanger, D. L., Hyde, D. E., Puhl, S. M., & Turner, N. L. (1990). Age and hypohydration independently influence the peripheral vascular response to heat stress. *Journal of Applied Physiology, 68*, 1902–1908.

Kloog, I., Melly, S. J., Coull, B. A., Nordio, F., & Schwartz, J. D. (2015). Using satellite-based spatiotemporal resolved air temperature exposure to study the association between ambient air temperature and birth outcomes in Massachusetts. *Environmental Health Perspectives, 123*, 1053–1058.

Kuwahara, T., Inoue, Y., Abe, M., Sato, Y., & Kondo, N. (2005). Effects of menstrual cycle and physical training on heat loss responses during dynamic exercise at moderate intensity in a temperate environment. *American Journal of Physiology Regulatory-Integrative and Comparative Physiology, 288*, R1347–R1353.

Lamarche, D. T., Notley, S. R., Louie, J. C., Poirier, M. P., & Kenny, G. P. (2017). Fitness-related differences in the rate of whole-body evaporative heat loss in exercising men are heat-load dependent. *Experimental Physiology, 103*(1), 101–110.

Landing, B. H., Wells, T. R., & Williamson, M. L. (1970). Anatomy of eccrine sweat glands in children with chronic renal insufficiency and other fatal chronic diseases. *American Journal of Clinical Pathology, 54*, 15–21.

Larose, J., Boulay, P., Sigal, R. J., Wright, H. E., & Kenny, G. P. (2013). Age-related decrements in heat dissipation during physical activity occur as early as the age of 40. *PLoS One, 8*, e83148.

Leites, G. T., Cunha, G. S., Obeid, J., Wilk, B., Meyer, F., & Timmons, B. W. (2016). Thermoregulation in boys and men exercising at the same heat production per unit body mass. *European Journal of Applied Physiology, 116*, 1411–1419.

van Loenhout, J. A. F., Delbiso, T. D., Kiriliouk, A., Rodriguez-Llanes, J. M., Segers, J., & Guha-Sapir, D. (2018). Heat and emergency room admissions in The Netherlands. *BMC Public Health, 18*, 108.

Nielsen, M. (1938). Die regulation der körpertemperatur bei muskelarbeit. *Acta Physiology (Oxf), 79*(2), 193–230.

Notley, S. R., Akerman, A. P., Meade, R. D., McGarr, G. W., & Kenny, G. P. (2020). Exercise thermoregulation in prepubertal children: A brief methodological review. *Medical Science and Sports Exercise, 52*(11), 2412–2422.

Notley, S. R., Lamarche, D. T., Meade, R. D., Flouris, A. D., & Kenny, G. P. (2019). Revisiting the influence of individual factors on heat exchange during exercise in dry heat using direct calorimetry. *Experimental Physiology, 104*, 1038–1050.

Otani, H., Kaya, M., Tamaki, A., Watson, P., & Maughan, R. J. (2016). Effects of solar radiation on endurance exercise capacity in a hot environment. *European Journal of Applied Physiology, 116*, 769–779.

Periard, J. D., Racinais, S., & Sawka, M. N. (2015). Adaptations and mechanisms of human heat acclimation: Applications for competitive athletes and sports. *Scandian Journal of Medical Science and Sports, 25*(Suppl. 1), 20–38.

Rivera-Brown, A. M., Rowland, T. W., Ramirez-Marrero, F. A., Santacana, G., & Vann, A. (2006). Exercise tolerance in a hot and humid climate in heat-acclimatized girls and women. *International Journal of Sports and Medicine, 27*, 943–950.

Rowland, T. W., Auchinachie, J. A., Keenan, T. J., & Green, G. M. (1987). Physiologic responses to treadmill running in adult and prepubertal males. *International Journal of Sports and Medicine, 8*, 292–297.

Rowland, T., Hagenbuch, S., Pober, D., & Garrison, A. (2008). Exercise tolerance and thermoregulatory responses during cycling in boys and men. *Medical Science and Sports Exercise, 40*, 282–287.

Rowland, T. W., Staab, J. S., Unnithan, V. B., Rambusch, J. M., & Siconolfi, S. F. (1990). Mechanical efficiency during cycling in prepubertal and adult males. *International Journal of Sports and Medicine, 11*, 452–455.

Sato, K., & Sato, F. (1983). Individual variations in structure and function of human eccrine sweat gland. *American Journal of Physiology, 245*, R203–R208.

Sawka, M. N., Leon, L. R., Montain, S. J., & Sonna, L. A. (2011). Integrated physiological mechanisms of exercise performance, adaptation, and maladaptation to heat stress. *Comprative Physiology, 1*, 1883–1928.

Sheffield, P. E., Herrera, M. T., Kinnee, E. J., & Clougherty, J. E. (2018). Not so little differences: Variation in hot weather risk to young children in New York city. *Public Health, 161*, 119–126.
Shibasaki, M., Inoue, Y., & Kondo, N. (1997). Mechanisms of underdeveloped sweating responses in prepubertal boys. *European Journal of Applied Physiology Occup Physiol, 76*, 340–345.
Shibasaki, M., Inoue, Y., Kondo, N., Aoki, K., & Hirata, K. (1999). Relationship between skin blood flow and sweating rate in prepubertal boys and young men. *Acta Physiology Scandian, 167*, 105–110.
Smolander, J., Bar-Or, O., Korhonen, O., & Ilmarinen, J. (1992). Thermoregulation during rest and exercise in the cold in pre- and early pubescent boys and in young men. *Journal of Applied Physiology, 72*, 1589–1594 (1985).
Sokolnicki, L. A., Strom, N. A., Roberts, S. K., Kingsley-Berg, S. A., Basu, A., & Charkoudian, N. (2009). Skin blood flow and nitric oxide during body heating in type 2 diabetes mellitus. *Journal of Applied Physiology, 106*, 566–570 (1985).
Stephenson, L. A., & Kolka, M. A. (1993). Thermoregulation in women. *Exercise Sports and Science Review, 21*, 231–262.
Van Zutphen, A. R., Lin, S., Fletcher, B. A., & Hwang, S.-A. (2012). A population-based case–control study of extreme summer temperature and birth defects. *Environmental Health Perspective, 120*, 1443–1449.
Wagner, J. A., Robinson, S., & Marino, R. P. (1974). Age and temperature regulation of humans in neutral and cold environments. *Journal of Applied Physiology, 37*, 562–565.
Wagner, J. A., Robinson, S., Tzankoff, S. P., & Marino, R. P. (1972). Heat tolerance and acclimatization to work in the heat in relation to age. *Journal of Applied Physiology, 33*, 616–622.
Wick, D. E., Roberts, S. K., Basu, A., Sandroni, P., Fealey, R. D., Sletten, D., & Charkoudian, N. (2006). Delayed threshold for active cutaneous vasodilation in patients with type 2 diabetes mellitus. *Journal of Applied Physiology, 100*, 637–641.
Woloschuk, A., Hodges, G. J., Massarotto, R. J., Klentrou, P., & Falk, B. (2020). The skin blood flow response to exercise in boys and men and the role of nitric oxide. *European Journal of Applied Physiology, 120*, 753–762.
Xu, Z., Sheffield, P. E., Su, H., Wang, X., Bi, Y., & Tong, S. (2014). The impact of heat waves on children's health: A systematic review. *International Journal of Biometeorology, 58*, 239–247.

CHAPTER 5

Heat exposure and elderly health

Niilo R.I. Ryti[1,2,3] and Jouni J.K. Jaakkola[1,2,3,4]
[1]Center for Environmental and Respiratory Health Research (CERH), University of Oulu, Finland; [2]Biocenter Oulu, University of Oulu, Finland; [3]Medical Research Center Oulu, University of Oulu and Oulu University Hospital, Oulu, Finland; [4]Finnish Meteorological Institute, Helsinki, Finland

1. Thermoregulation in the elderly

Temperature is a critical determinant of cellular function and survival in humans. High temperature of the cellular microenvironment disturbs cellular membrane fluidity, ion fluxes, and enzyme activity. High temperature causes protein denaturation and cell death, leading to organ dysfunction, inflammatory response and disturbances in local and systemic homeostasis. The core body temperature of humans is optimally maintained within a narrow range of approximately 4°C (Tansey & Johnson, 2015) with complex regulative mechanisms by balancing heat losses and metabolic heat production.

The human body is in constant interaction with its surrounding microclimate. According to the second law of thermodynamics, the direction of heat transfer is from warm body to cold body. Exchange of thermal energy between the human body and environment occurs via conduction, convection, evaporation, and radiation. A mathematical description of the heat balance of the human body can be presented as follows:

$$S = [M - W] - [RES - C - R - K - E]. \tag{5.1}$$

S represents tissue heat content; metabolic energy production (*M*) minus mechanical energy (*W*) signifies internal heat production; heat exchange is represented by evaporation and convection in the respiratory tract (RES), and convection (*C*), radiation (*R*), conduction (*K*), and evaporation (*E*) in the skin.

A simplified presentation is as follows:

$$S = [\text{Heat Production}] - [\text{Heat Loss}]. \tag{5.2}$$

Yet another way of presenting heat balance takes into account heat gained from the environment:

$$\begin{aligned}\text{Heat Storage} = \text{Metabolism} - \text{Work} - \text{Evaporation} \pm \\ \text{Radiation} \pm \text{Conduction} \pm \text{Convection}\end{aligned} \tag{5.3}$$

Heat Exposure and Human Health in the Context of Climate Change
ISBN 978-0-12-819080-7
https://doi.org/10.1016/B978-0-12-819080-7.15001-3
© 2023 Elsevier Inc.
All rights reserved.

Thermoneutral zone is defined as the range of ambient temperature at which temperature regulation is achieved by control of sensible (dry) heat loss, i.e., without regulatory changes in metabolic heat production or evaporative heat loss (IUPS Thermal Commission, 2001). For a resting naked human, the thermoneutral zone is between air temperatures +23°C and +26°C (Castellani et al., 2016; Stocks et al., 2004). When the ambient temperature exceeds this, a set of specific physiological processes initiate to avoid hyperthermia and the related pathology. This chapter reviews how aging affects those physiological processes, and how in return the elderly may be at an increased risk of serious heat illness and other medical problems during heat exposure. The chapter focuses on the biological model for age-related impairments in thermoregulation and tolerance for thermal strain. The text is organised under two main sections: 1) thermoregulation, i.e., the maintenance of constant body temperature under variable internal and/or external heat loads; 2) health effects of heat exposure, i.e., direct health effects of heat when thermoregulation fails, and indirect health effects caused by the strain of thermoregulation itself. We hope that the relatively high level of detail in this presentation is useful to a wide range of professionals from scientists to clinicians. The biological model is complemented by a brief review of key epidemiological evidence and a synthesis of knowledge in the end of the chapter.

1.1 Neuroanatomical basis of thermoregulation in the elderly

To maintain body temperature at an optimal level, the human thermoregulatory system relies on autonomic and endocrine physiological responses and behavioral adjustments, which seem to occur in specific sequences depending on the type of exposure and circumstances (Schlader et al., 2018). Autonomic thermoregulation is driven by thermal sensation, while the main driver of thermoregulatory behavior is thermal comfort, which involves a different circuitry in the central nervous system (Flouris, 2011; Jung et al., 2022; Yahiro et al., 2017). Advances have been made in identifying the different parts of the peripheral and central nervous system involved in thermoregulation, and the neuroanatomy of thermoregulation is reasonably well understood (Flouris, 2011; Morrison, 2016; Morrison & Nakamura, 2011; Tan & Knight, 2018).

Thermal sensors are members of the transient receptor potential (TRP) family of cation channels and reside in the free nerve endings of sensory neurons. Each type of TRP sensor responds to a specific bandwidth of temperature, and the temperature-dependent discharge frequencies and partial overlaps in the bandwidths allow effective discrimination of thermal information (Tansey & Johnson, 2015). Specific sensors respond to cold temperatures or a change toward cooler, while others respond to hot temperatures or a change toward warmer. Both types are found in the periphery and the core, with distributions varying by anatomical area.

The signals from the skin enter the spinal dorsal horn to synapse with second-order ascending neurons and travel via lateral parabrachial nucleus (LPB) to the median preoptic nucleus (MnPO) of the preoptic area of the hypothalamus (POA). Other signals from the vagal and splanchnic nerves converge at the solitary nucleus in the brainstem, and this thermal information is integrated with thermal information from the skin in the LPB (Tansey & Johnson, 2015). From the MnPO in POA, neurons project to the medial preoptic area of the hypothalamus, where they activate inhibitory input traveling to rostral raphe. This inhibitory signal influences output to cutaneous blood vessels, skeletal muscle, and brown adipose tissue (Fig. 5.1). The LPB also mediates signals required for behavioral thermoregulation. Although the central circuitry of behavioral thermoregulation is less well understood, recent findings show that the signals from LPB may project to the lateral hypothalamus for coding of thermal punishment and reward needed for behavioral thermoregulation (Jung et al., 2022; Yahiro et al., 2017).

POA is still considered the central processing center of autonomic thermoregulation. For decades, the theory was that body temperature is compared to a set point optimal temperature, and a difference between the actual and optimal temperature level dictates thermoregulatory action. This set-point theory was later updated with a more nuanced balance point theory (Morrison & Nakamura, 2011; Romanovsky, 2007). Instead of a single controller, the several distinct thermoeffector loops are independently influenced by central and peripheral temperatures. They are able to activate in concert or in specific hierarchical sequence, which can be explained by their varying threshold temperatures (Tansey & Johnson, 2015).

Aging affects the neuroanatomy and neurological function related to thermoregulation. Altered nerve density and changes in the electrophysiological and neurochemical properties of the afferent pathway are just some of the suggested culprits. Aging is associated with progressive decrease in thermal sensitivity and perception (Blatteis, 2012). Several studies show that the thresholds for detecting thermal stimuli increase with age, i.e., a stimulus of greater intensity is needed for a sensation to occur. Reduced density of TRP receptors is perhaps driven by processes related to aging of the skin and long-term changes in skin vasculature. Peripheral fiber loss and lower conduction velocity may also play part in altered thermosensation and related responses. Because aging is known to alter the brain on molecular,

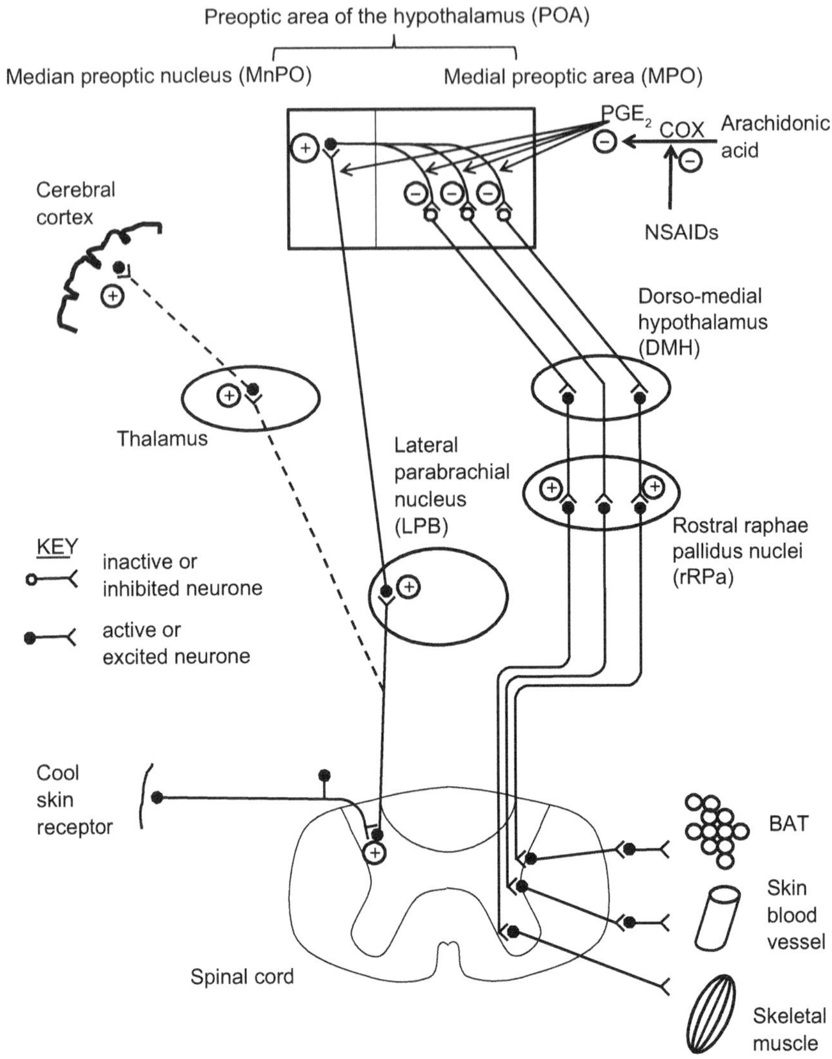

Figure 5.1 ***Central circuitry mediating the responses to heat exposure.*** *BAT*, brown adipose tissue; *DMH*, dorsomedial hypothalamus; *LPB*, lateral parabrachial nucleus; *MnPO*, median preoptic nucleus; *MPO*, medial preoptic area; *POA*, preoptic area of the hypothalamus; *rRPa*, rostral raphe pallidus nucleus. *(Reproduced with permission from Tansey, E., Johnson, C.D., (2015). Recent advances in thermoregulation.* Advances in Physiology Education, 39 *(3), 139–48. https://doi.org/10.1152/advan.00126.2014.)*

macroanatomical, and functional level (Peters, 2006), it seems justified to hypothesize that the thermoregulatory circuits in the central nervous system are also subject to changes related to biological aging. However, this has been less investigated.

Age-related reduction in thermal sensitivity follows a distal-to-proximal anatomical pattern, occurs earlier in the lower than in the upper extremities and core, and is more pronounced for hot than for cold temperatures (Blatteis, 2012; Guergova & Dufour, 2011; Stevens & Choo, 1998). Neuropathology of aging is important in the thermophysiological context, because the many reflex, autonomic, and behavioral thermoregulatory mechanisms need afferent input to activate.

1.2 Age-related changes in the skin blood flow response

Increased skin blood flow (SkBF) is the major thermoregulatory response to both external heat exposure and increased core temperature (Tansey & Johnson, 2015; Wong & Hollowed, 2016). It is part of the autonomic thermoregulatory responses in humans. Increased SkBF facilitates convective heat transfer from the core to periphery, while the increased skin temperature enhances conductive heat loss to the environment (Tan et al., 2020). The SkBF response is activated by minuscale increases in skin temperature, i.e., it is very sensitive (Schlader et al., 2018). The increased SkBF is initially achieved by withdrawal of sympathetic activity, which leads to passive vasodilation and the opening of arteriovenous anastomoses in glabrous nonhairy skin. When passive vasodilation is insufficient, it is complemented by active vasodilation in nonglabrous hairy skin. Sympathetic cholinergic nerves become activated and release acetylcholine (ACh), which stimulates the production of the local vasodilator nitric oxide (NO). Several transmitters such as substance P, histamine, neurokinin, vasoactive intestinal peptide, and calcitonin gene—related peptide are coreleased with ACh, and active vasodilation is the result of a complex interplay between these and other currently unidentified transmitters (Blatteis, 2012; Tan et al., 2020).

Aging is associated with a progressive attenuation of the cutaneous vasodilator response (Blatteis, 2012; Kenney, 2017; Kenney & Munce, 2003; Meade et al., 2019). For a given change in core temperature, older individuals exhibit smaller increases in SkBF than younger individuals, even when subjects are matched for aerobic fitness, acclimation, and hydration status. Studies have shown a delay in the initiation of the response and a reduced maximal vasodilator capacity. These phenomena can be explained by the structural and biomedical changes related to aging (Kenney & Munce, 2003). For example, flattening of the rete pegs in the epidermal-dermal junction is a common characteristic of the aged skin. This structural change is associated with disorganization and collapse of the vessels in the superficial vascular plexus and papillary dermis (Holowatz et al., 2011;

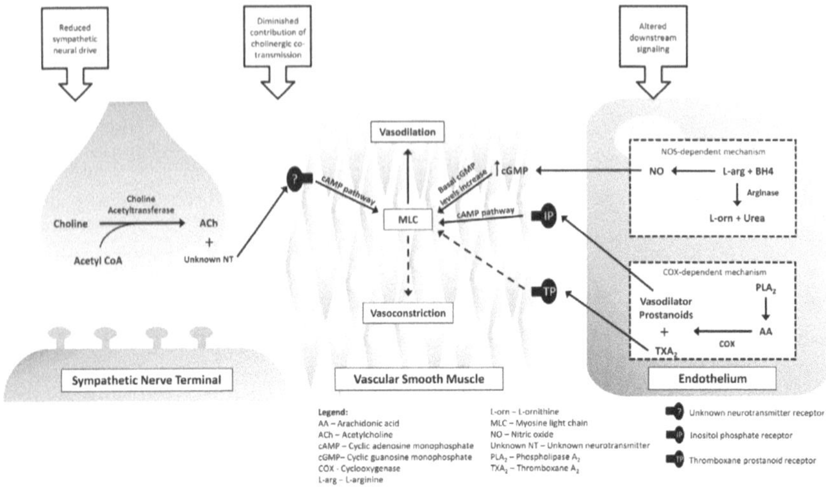

Figure 5.2 ***Age-related impairments in the cutaneous active vasodilation response in the context of the cholinergic cotransmitter hypothesis.*** Following an increase in skin sympathetic neural activity, ACh and unknown neurotransmitters are coreleased from the sympathetic cholinergic nerves. Subsequent activation of an also unidentified receptor modulates cutaneous vasodilation through cyclic adenosine monophosphate (cAMP) pathways along with NOS-mediated increase in cyclic guanosine monophosphate (cGMP) in the vascular smooth muscle. In response to heat exposure, cyclooxygenase synthesizes various vasodilator prostanoids (PGI2, PGE, PGF, PGD) and vasoconstrictor substances (thromboxane A2) which bind to IP receptors and TP receptors, respectively, to modulate cutaneous vasodilation. Multiple age-associated alterations along the efferent sympathetic reflex are indicated by the arrowed tab boxes at the top of the diagram. *(Reproduced with permission from: Tan, C.C.S., Chin, L.K.K., Low, I.C.C., (2020). Thermoregulation in the aging population and practical strategies to overcome a warmer tomorrow.* Proteomics, 20 *(5–6), e1800468. https://doi.org/10.1002/pmic.201800468.)*

Kenney & Munce, 2003; Montagna & Carlisle, 1979). Cutaneous microvascular smooth muscle hypertrophy and inward vessel remodeling may also limit the maximal vascular conductance in the elderly (Holowatz et al., 2011).

Experimental evidence indicates that a decreased bioavailability of NO plays a role in the attenuated SkBF response in the elderly. Age-related decrease in NO synthesis is probably due to reduced availability of L-arginine, which is a substrate for the NO synthase. It has also been suggested that accumulation of reactive oxygen species, which cause NO breakdown, contributes to the attenuated SkBF response in the elderly (Blatteis, 2012; Holowatz et al., 2011, Kelm, 1999). However, this is only a part of the story. Reduced sympathetic neural drive, diminished cholinergic cotransmission, altered second-messenger signaling, and K + -channel modulation all explain the attenuated reflex cutaneous vasodilation in the elderly, although many details of these processes remain to be elucidated (Tan et al., 2020). Fig. 5.2 shows a schematic presentation of age-related impairments in SkBF response in the context of the *cholinergic cotransmitter hypothesis* (from Tan et al., 2020).

The heat-driven changes in cutaneous blood flow do not happen in isolation. Rather, they are part of an integrated cardiovascular response (Crandall & Wilson, 2015) partly mediated by the excitation of the sympathetic nerves to the heart and many vascular beds (Blatteis, 2012). Cutaneous vasodilatation is complemented by redistribution of blood flow from splanchnic and renal vascular beds to the skin. The splanchnic and renal vascular beds receive approximately 50% of blood flow during preprandial normothermic conditions and thus constitute an important reserve for thermoregulatory purposes. This action is complemented by an increased cardiac output, which can double during heat stress. The current understanding is that the increase in cardiac output during heat exposure is achieved by increasing the heart rate while maintaining stroke volume. In turn, maintaining stroke volume in the face of reduced cardiac preload, cardiac contractility must be increased (Crandall & Wilson, 2015). The increased cardiac minute work means increased myocardial oxygen consumption, and the increased metabolic demands of the heart can predispose to indirect health effects described later in this chapter.

There is some experimental evidence that the integrated cardiovascular responses to heat stress are modified by age. For example, a study by Minson et al. (1998) showed that the elderly subjects responded to passive heating with a smaller increase in cardiac output than younger subjects. This was attributed to smaller stroke volume. The elderly were able to maintain similar heart rate than younger subjects, but this occurred by attaining a greater proportion of their chronotropic reserve. Redistribution of blood from splanchnic and renal vascular beds was also less effective in the older subjects. Together these findings amounted to reduced SkBF response in the face of heat stress.

1.3 Age-related changes in sudomotor activity and perspiration

Another major thermoregulatory response to heat stress is eccrine sweat gland activation via sympathetic cholinergic fibers, and the consequential sweating (Baker, 2019; Kenney & Munce, 2003). This too is part of the group of autonomic thermoregulatory responses in humans. Sweating is initiated almost simultaneously with active vasodilation, but although ACh plays an important part in both, these two responses seem to be mutually independent phenomena. Sweat generated by eccrine sweat glands collects to the surface of the skin and starts evaporating. The enthalpy of vaporization is the amount of energy that must be added to a liquid substance to transform a quantity of it to gas. Enthalpy of vaporization is particularly high for water. One gram of sweat liberates 2426J of thermal energy upon evaporation at +30°C (Wenger, 1972). It is important to understand that evaporative cooling is the only form of heat exchange which facilitates heat loss under conditions where heat fluxes by conduction, convection, and radiation are from a warmer environment to the cooler body (Formula 5.3). Thus, the ability to sweat in a hot environment is critical for effective thermoregulation.

Eccrine glands are present in millions with regionally varying distributions around the surface of the body. Effectiveness of sweating depends on the surface area of sweating, humidity of air, vapor pressure gradient, and convective air currents around the body. To achieve the same absolute rate of evaporation as an individual with a smaller surface area, an individual with large body surface area requires less sweat to be produced per unit body surface area (Balmain et al., 2018). In very humid environments, some sweat may drip from the skin, i.e., it does not evaporate from the body surface, leading to lower efficiency of the sweating response (Baker, 2019).

Several studies show that aging is associated with impaired sweating response to heat stress. The temperature threshold for activation of the sweating response seems to be higher in the elderly and the sweating response is delayed. Compared with young gender-matched subjects, the elderly exhibit lower overall and local sweating rates at multiple body sites during passive heating. Sweating rate is the product of the density of active sweat glands and the secretion rate per gland (Baker, 2019). The absolute number of sweat glands remains constant after early childhood, i.e., it is not known to decrease by aging. There is evidence showing that the reduced sweating rates in the elderly are not explained by a lower density or a smaller absolute number of activated sweat glands. Instead, it seems to be due to lower sweat output per gland. Diminished sweat gland cholinergic sensitivity and altered second-messenger signaling likely play a part (Tan et al., 2020). Eccrine sweat gland atrophy and other structural changes brought by photoaging and lifelong exposure to UV radiation may also contribute to this phenomenon to the extent of influencing individual sweat gland output (Blatteis, 2012; Kenney & Munce, 2003). When the skin ages, glandular function seems to decline in a peripheral-to-central direction.

1.4 Age-related changes in behavioral thermoregulation

Whereas human survival in extremely hot conditions is limited by the boundaries of human physiology and laws of physics, the potential of behavioral thermoregulation is virtually infinite. Indeed, it can be argued that the greatest thermoregulatory asset in humans is behavior. Examples of behavioral thermoregulatory responses to heat stress are as follows:

- Displacement from the hot environment;
- Changing orientation to the direction of wind and sun;
- Changing the activity level to reduce metabolic heat production;
- Creating convective air flows around the body;
- Spreading fluid on the skin surface to increase evaporative heat loss;
- Changing posture to optimize the surface-to-mass ratio and hence the rate of heat exchange between the body and the environment;
- Changing the clothing, e.g., removing clothing to enable convective air flows around the body, or adding clothing to shield the body from radiation;
- Changing food consumption behavior;
- Generation of more suitable microclimates by means of technology, e.g., air conditioners.

Behavioral thermoregulation is mainly driven by thermal comfort (Flouris, 2011). Thermal comfort is defined as the state of mind that expresses satisfaction with the surrounding environment. It should be distinguished from the driver of the autonomic thermoregulatory responses, thermal sensation, which is the perception of the stimulation of peripheral and central thermosensors (ASHRAE, 1997). The concept of thermal comfort is widely applied in engineering sciences and building physics using the so-called Fanger's Predicted Mean Vote model (van Hoof, 2008; van Hoof & Hensen, 2006). Neuroscientifically speaking, warm and cold stimuli are attributed affective components such as feeling pleasant or unpleasant. The processing of thermal stimuli occurs in different areas of the brain depending on whether it's related to the sensory properties (such as intensity) or to the affective properties (such as unpleasantness) and the related emotional experience (Flouris, 2011; Jung et al., 2022; Rolls et al., 2008; Yahiro et al., 2017). When an individual who feels hot and uncomfortable is cooled, skin temperature that is normally perceived as cold and uncomfortable is perceived as cold and comfortable. Generally speaking, a non—nociceptic thermal stimulus is felt agreeable when it serves to minimize body heat storage, and felt unagreeable when it acts to increase body heat storage (Formula 5.3) (Flouris, 2011; Gagge et al., 1967).

Thermal comfort zones differ between younger adults and the elderly. The extent and clinical relevance of this is unclear, however (van Hoof et al., 2017). Perhaps one of the most relevant findings is that during physical exertion in a hot environment, the elderly seem to experience less discomfort than younger adults (Millyard et al., 2020). In a study by Waldock et al. (2018), the elderly had limited recognition of the increased heat stress, even when it was objectively measured and factually present. Combined with delays in autonomic thermoregulation, this could put the elderly at an increased risk of accumulating heat storage during heat exposure.

It should be mentioned that not all behavioral thermoregulation is beneficial. In fact, some thermoregulatory behavior can be harmful (Gagnon et al., 2017; Ryti et al., 2020), which underlines the fact that guidelines should have solid footing in scientific understanding. Given the potential of behavioral thermoregulation in preventing temperature-related illness, the topic should be subject to further research.

1.5 Certain medical conditions affecting thermoregulation in the elderly

Several factors such as chronic diseases and medications can impair effective thermoregulation (Cabanac & Brinnel, 1987; Kenny et al., 2010). Although such factors may not be age-related per se, their high prevalence in the elderly populations warrants some consideration. Diabetes is a specific modifier of thermoregulatory responses to heat exposure in humans (Kenny et al., 2016). Diabetics have a reduced capability to maintain core temperature during thermal stress, and are at a high risk of heat exhaustion, heat stroke, and death during heat waves (Kenny et al., 2010). Many of the common complications of diabetes such as peripheral neuropathy and cardiovascular diseases limit effective thermoregulation. The SkBF response to heat exposure is both delayed and attenuated in diabetics, and it seems to be triggered at higher temperature threshold. A reduced sweating response consisting of a decreased sweating rate and changes in the threshold of onset of the response have also been documented. These changes may be related to diabetes-related autonomic dysfunction and peripheral neuropathy. In addition, glycemic control is influenced by heat exposure. According to a recent global estimation (Sinclair et al., 2020), nearly 20% of people aged 65–99 years have diabetes. Diabetes is therefore an important comorbidity in the context of heat exposure in the elderly.

Obesity is a common comorbidity with type II diabetes. Prevalence of obesity is high in the elderly populations, especially in industrialized countries, and body mass index—related deaths and loss of disability-adjusted life years increase by age (Dai et al., 2020; The GBD 2015 Obesity Collaborators, 2017). In overweight and obese adults, fatal heat-stroke is several times more frequent than in individuals with normal body weight (Kenny et al., 2010). Obesity-related small-fiber neuropathies may contribute to reduced thermal sensitivity to heat stimuli, and obese people have lower capacity for heat dissipation due to smaller surface-area-to-mass ratio. Adipose tissue has lower thermal capacity than fat-free mass, so a given heat input per unit body mass will cause a greater increase in average tissue temperature. Adipose tissue has lower thermal conductivity than other issues in the body, which limits conductive heat transfer from core to the skin. For the same absolute workload, obese people exhibit greater increase in core temperature (Kenny et al., 2010).

Many cardiovascular diseases debilitate thermoregulation during heat challenges. Examples include atrial fibrillation, coronary artery disease, and heart failure, all of which can limit the ability to increase cardiac output in the integrated cardiovascular response to core temperature elevation (Kenny et al., 2010). The prevalence of these diseases is highest in the elderly populations.

Several medications can have detrimental effect on thermoregulation during heat exposure (Kalisch Ellett et al., 2016; Kuzuya, 2013; Leon & Bouchama, 2015; Stöllberger et al., 2009). Angiotensin-converting enzyme inhibitors and angiotensin receptor blockers reduce the sensation of thirst. Pain medications and sedatives can attenuate the patient's ability to recognize heat exposure. Antihistamines and anticholinergic medications, such as those used to treat Parkinson's disease, reduce the ability to sweat. Low-dose acetylsalicylic acid therapy blunts the reflex cutaneous vasodilation through COX- and NO synthase-dependent mechanisms (Kenney, 2017). β-blockers too can reduce the SkBF response. Selective serotonin uptake inhibitors and some antipsychotics can impair thermoregulation by directly influencing the hypothalamus. Diuretics increase fluid loss and predispose to dehydration.

Dementia, lowered cognition, reduced awareness, handicaps, mental health problems, loneliness, substance abuse, homelessness, and living in an institution can limit behavioral thermoregulation. To some extent, behavioral thermoregulation requires freedom from major physical, social, and economical limitations.

1.6 Summary of the age-related changes in thermoregulation

Aging is associated with anatomical, physiological, and biochemical changes that influence thermoregulation. Thresholds for detecting thermal stimuli and heat load change with age. The SkBF response, the integrated cardiovascular response, and the sweating response become delayed and blunted. Altered thermal comfort level may render behavioral thermoregulation less effective. Age-related comorbidities, medications, and socioeconomic factors can further limit effective thermoregulation. Overall, the evidence is in favor of the hypothesis that aging impairs thermoregulation.

Finally, we present the one factor that impairs all major forms of thermoregulation including the SkBF response, sweating response, and behavioral thermoregulation: dehydration. The sensation of thirst reduces with aging, which increases the risk of inadequate fluid intake during heat exposure (Kenney & Chiu, 2001). Inadequate fluid intake leads to dehydration. When healthy young and old individuals are subjected to experimentally induced dehydration, the older individuals show lower subjective levels of thirst which leads to reduced water intake and a reduced plasma volume. Dehydration is a major contributing factor to the increased risk of hyperthermia in the elderly (Blatteis, 2012; Kenny et al., 2010).

2. Health effects of heat exposure in the elderly

We propose that the health effects of heat exposure are classified into two distinct classes:

1. Health effects resulting from increased tissue temperatures. These can be called *direct health effects* of heat exposure, and on a systemic level they are a direct result of thermoregulatory insufficiency and/or failure;
2. Health effects resulting from the additional physiological strain caused by the thermoregulatory responses to heat exposure. These can be called *indirect health effects* of heat exposure, and they can occur with or without a rise in tissue temperature.

The following part of this chapter describes both types of health effects, with a special focus on heat stress from external sources.

2.1 Direct health effects of heat exposure in the elderly

The inability to maintain suitable core temperature during heat exposure leads to hyperthermia and heat illness. Hyperthermia and heat illness can be attributed to different circumstances:

(1) All thermoregulatory mechanisms function normally, but the body is overwhelmed by extreme heat from external sources;

(2) All thermoregulatory mechanisms function normally, but the body is overwhelmed by internal heat production during extreme physical exertion in hot environment;
(3) Thermoregulatory mechanisms are impaired, which results in an increased body temperature in neutral or moderately hot ambient temperature;
(4) Internal heat production is increased by specific pathology such as major systemic infection or inflammation, hypothalamic tumor, et cetera; and
(5) A combination of these.

Direct consequences of thermoregulatory failure in hot conditions are heatstroke (ICD-11 NF01.0), heat syncope (ICD-11 NF01.1), heat exhaustion due to fluid depletion (ICD-11 NF01.2), transient heat fatigue (ICD-11 NF01.3), and other or unspecified effects of heat (ICD-11 NF01.Y-Z). Table 5.1 presents known risk factors of serious heat illness.

Of particular concern is heat stroke, which is associated with high mortality rate and permanent neurological damage in survivors (Argaud et al., 2007; Bouchama & Knochel, 2002; Leon & Bouchama, 2015; Varghese et al., 2005). The etiology and risk factors of heat stroke vary according to age. The classic form of heat stroke is related to a prolonged or extreme heat exposure, and it is more common among the elderly and those with chronic diseases. The exertional form of heat stroke is more common in the young and healthy adults and is a result of, for example, strenuous exercise in hot weather. Table 5.2 compares classic and exertional heat stroke by various characteristics, including age.

Table 5.1 Risk factors for serious heat illness.

Risk factor	Effect
Environmental factors	
Heat waves	Prolonged environmental heat exposure
High humidity	Impedes evaporative cooling
Lack of movement	Impedes convective cooling
Urban heat islands	Higher intensity, longer duration environmental heat exposure
Lack of air conditioning	Impedes evaporative, convective cooling

Continued

Table 5.1 Risk factors for serious heat illness.—cont'd

Risk factor	Effect
Medications (classic)	
Diuretics	Dehydration and salt, potassium, and calcium depletion
Anticholinergics	Impaired sweating
β-blockers	Reduced blood pressure; reduced skin blood flow
Antihistamines	Impaired sweating
Antidepressants	Increased heat production
Alcohol	Diuresis, impaired vasomotor reflexes
NSAIDs	Increased gut and liver toxicity; acute kidney injury
Drug use (exertional)	
Cocaine	Impairs sweating, cutaneous vasodilation, and heat perception
Ergogenic aids (ephedrine)	Increased activity and heat production, impaired sweating
Ecstasy (MDMA)	Increased heart rate and blood pressure; sweating or chills; increased muscle tension
Methylphenidate (ritalin)	Increased heart rate and blood pressure; excessive sweating
Compromised health status	
Viral or bacterial infection	Augmented hyperthermic response; immunocompromised
Fever	Augmented hyperthermic response
Skin disorders	Local inflammation; impaired sweating
Cardiovascular insufficiency	Orthostatic intolerance
Genetic conditions	
Malignant hyperthermia	Increased metabolic heat production
TLR4 polymorphisms	Endotoxin hyporesponsiveness

MDMA, 3,4-methylenedioxymethamphetamine; *NSAIDs*, nonsteroidal anti-inflammatory drugs; *TLR*, toll-like receptor.
Modified with permission from Leon, L.R., Bouchama, A., (2015). Heat stroke. Comprehensive Physiology, 5 (2), 611-647. doi: 10.1002/cphy.c140017.

Much work has been done to better understand the mechanisms through which heat stroke causes multiorgan failure and death. The main pathways include thermal tissue injury and apoptotic and necrotic cell death, coagulopathies, and immunological responses to increased gut

Table 5.2 Comparison of classic and exertional heat stroke.

Characteristic	Classic heat stroke	Exertional heat stroke
Age	Elderly, very young	Young (15–50 years)
Health	Often chronically ill	Typically healthy
Febrile illness	Unusual	Common
Weather condition	Heat wave	Temperate or hot
Activity	Sedentary	Sustained or heavy exertion
Medications or drug use	Diuretics, β-blockers, antihistamines, antidepressants	Ergogenic aids, ecstasy, cocaine
Sweating	Often absent	Often present
Acid–base disturbance	Mixed respiratory alkalosis and metabolic acidosis	Severe metabolic acidosis
Calcium	Normal	Hypocalcemia
Potassium	Normal	Hyperkalemia, hypokalemia (c.a. 30%)
Phosphate	Hypophosphatemia	Hyperphosphatemia
Blood glucose	Hyperglycemia	Hypoglycemia
Rhabdomyolysis	Rarely severe	Often severe
Acute renal failure	Uncommon (c.a. 5%)	Common (c.a. 25%)
DIC	Mild	Severe
CK	Mild elevation	Marked elevation
ASAT, ALAT	Mild elevation	Marked elevation

ALAT, alanine aminotransferase; *ASAT*, aspartate aminotransferase; *CK*, creatine kinase; *DIC*, disseminated intravascular coagulation.
Modified with permission from Leon, L.R., Bouchama, A., (2015). Heat stroke. *Comprehensive Physiology*, *5* (2), 611-647. doi: 10.1002/cphy.c140017.

permeability and neurotoxins (Leon & Bouchama, 2015). These factors trigger systemic inflammatory response syndrome leading to multiorgan failure and death. Heat stroke is commonly treated in the ICU. Because intensive care is resource-heavy and can also prolong human suffering in futile cases, ICU admissions are influenced by probability of survival and acceptable outcome. In some health care settings, age negatively influences the ICU admission criteria. This makes it ever more important to focus on heat stroke prevention in the elderly populations.

The age distribution of heat stroke patients seems to vary from country to country, but reliable comparisons are difficult to make because of varying definitions and recording practices (Hifumi et al., 2018). According to some estimates, 30%–50% of all heatstroke patients hospitalized in Japan are aged 65 or more (Kuzuya, 2013). During the 2003 heat waves in Europe, 84% of all patients treated for heat stroke at the University Hospital of Lyon, France, were over 70 years old. 33% were 85 years old or older, and 96% of the patients had at least one coexisting medical condition (Argaud et al., 2007).

2.2 Indirect health effects of heat exposure in the elderly

In addition to the direct health effects of increased tissue temperatures, thermoregulation causes physiological strain (Ebi et al., 2021), and probably leads to indirect health effects via multiple pathways. Heat exposure and thermoregulation increase the cardiorespiratory workload. The increased rate pressure product signifies increased myocardial oxygen consumption, which predisposes to myocardial ischemia in patients with coronary artery disease. This could perhaps partially explain observed associations between heat waves and acute myocardial infarction in epidemiological studies (Sun et al., 2018). Persistent tachycardia and the increased myocardial workload can also trigger or worsen heart failure, which is a condition with a particularly poor prognosis and dramatic impact on the quality of life. In return, heart failure inhibits effective thermoregulation. Heat exposure and thermoregulation impose increased demands on the respiratory system, and the elderly seem particularly vulnerable. Aging is associated with changes in pulmonary mechanics, clearance from large and small airways, gas exchange, ventilatory control, respiratory muscle strength, and chest wall rigidity and compliance (Aalami et al., 2003; Janssens, 2005; Pride, 2005; Svartengren et al., 2005). The fraction of the lung consisting of alveolar ducts and airspace increases by aging, with the effective surface area for gas exchange decreasing 15% by the age of 70 years (Aalami et al., 2003). Together with age-induced diffusion block, anatomical shunt, and ventilation—perfusion mismatch, these changes decrease the respiratory reserve in the elderly (Aalami et al., 2003). This could perhaps partially explain observations of associations between heat waves and a variety of respiratory outcomes, such as hospitalization for chronic obstructive pulmonary disease and respiratory infections in the elderly (Anderson et al., 2013).

The kidneys undergo structural and functional changes when aging. Glomerulosclerosis, cortical tissue loss, and tubular senescence blunt the secretion and reabsorption of solutes. The capacity to reabsorb sodium and to secrete potassium and hydrogen ions decreases. Renin production at the juxtaglomerular apparatus is decreased and the aldosterone response is blunted (Aalami et al., 2003). The elderly present with lower levels of antidiuretic hormone (ADH), and the ADH response is attenuated. These and other changes predispose the elderly to electrolyte imbalances, acid–base imbalances, and disturbances of water homeostasis (Aalami et al., 2003; Cowen et al., 2013). Thermoregulation causes a particular problem when sweating is combined with inadequate water intake. Sweating consumes large amounts of water, and although sweating also consumes plasma $Na+$ and $Cl-$ ions, sweat is hypotonic and profuse sweating can lead to dehydration, i.e., hypernatremia (Tansey & Johnson, 2015). The elderly are at a high risk of developing dehydration, tolerate dehydration poorly, and restore fluid balance in a slowed manner after dehydration (Kenney & Chiu, 2001). Growing dehydration causes a progression of health effects including fatigue, headache, muscle cramps, oliguria, acute kidney injury (AKI), delirium, and coma. In addition, dehydration impairs all forms of thermoregulation. Recurrent dehydration increases the risk of chronic kidney disease which is problematic as the renal function and glomerular filtration rate deteriorate by age anyway (Aalami et al., 2003).

Another challenge introduced by heat stress and thermoregulation is central hypovolemia, which can be compounded by dehydration. Central hypovolemia can precipitate orthostatic hypotension and increase the risk of syncope and falling (Crandall & Wilson, 2015). Traumatic injuries are among the leading causes of death in the elderly, and prevention of falls is especially important in this group. Combination of relative and absolute hypovolemia—perhaps most when combined with dehydration—can cause renal hypoperfusion and AKI during heat exposure. Perhaps this could partially explain observations of associations between heat waves and AKI (de Lorenzo & Liaño, 2017). Comorbidities and medications that impair renal blood flow or autoregulation can further promote AKI during the heat challenge. AKI and kidney failure can also lead to increased plasma concentrations of many drugs or their metabolites, predisposing patients to overdose-related adverse health effects. The risk of AKI is greatest in the elderly.

In the above mentioned paragraphs, we have presented just a few simplified examples of the interplay between heat exposure, thermoregulation, the elderly physiology, and comorbidities. Reality is infinitely more complex, with dynamic interactions within and between all organ systems. Most importantly, it seems that while heat exposure can disturb the balance of many chronic diseases, chronic diseases can also impair effective thermoregulation leading to a vicious circle of health effects.

Box 5.1 Heat exposure and kidney in a nutshell

- Part of the normal thermoregulatory response to heat exposure is redistribution of blood flow from splanchnic and renal vascular beds to the skin. Normally, renal autoregulation is able to maintain adequate tissue perfusion and glomerular filtration rate. During hypovolemia and/or severe dehydration, renal blood flow deteriorates until it is inadequate for supporting the vitality of the organ.
- Kidney failure can also lead to increased plasma concentrations of several drugs or their metabolites, predisposing patients to overdose-related adverse health effects.
- In more severe heat illness, disseminated intravascular coagulation can cause microembolisms in vital organs including the kidneys.
- Heat exposure has been associated with nephrolithiasis, i.e., kidney stones, which may be due to low urine volumes and high urine concentration. Urinary tract infections (UTIs) can occur due to a decreased efferent flow of fluid in the tract, and several studies show associations between hot weather and hospitalizations due to UTI in the elderly. Although UTIs are usually relatively easy to treat, in the elderly they are heavily associated with delirium. UTI can progress into urosepsis.

Fig. 5.3 presents a simplified version of the biological model for the role of aging in the heat-related health effects.

2.3 Summary of the adverse health effects of heat exposure in the elderly

Thermoregulatory insufficiency during heat exposure leads to hyperthermia and thermal illness. The elderly are at an increased risk of these direct health effects including serious thermal illness, heat stroke, and death during heat exposure.

Thermophysiological responses to heat exposure predispose the elderly to disturbances in homeostasis, particularly when certain comorbidities exist. These indirect health effects are not caused by tissue warming, but they can occur concurrently with the direct health effects.

Individuals with diminished physiological reserve and reduced tolerance for homeostatic disturbances are likely at an increased risk of both kinds of health effects.

The risk of dehydration is particularly high in the elderly, and its comprehensive health effects likely contribute to the increased mortality during heat waves even in the absence of chronic diseases.

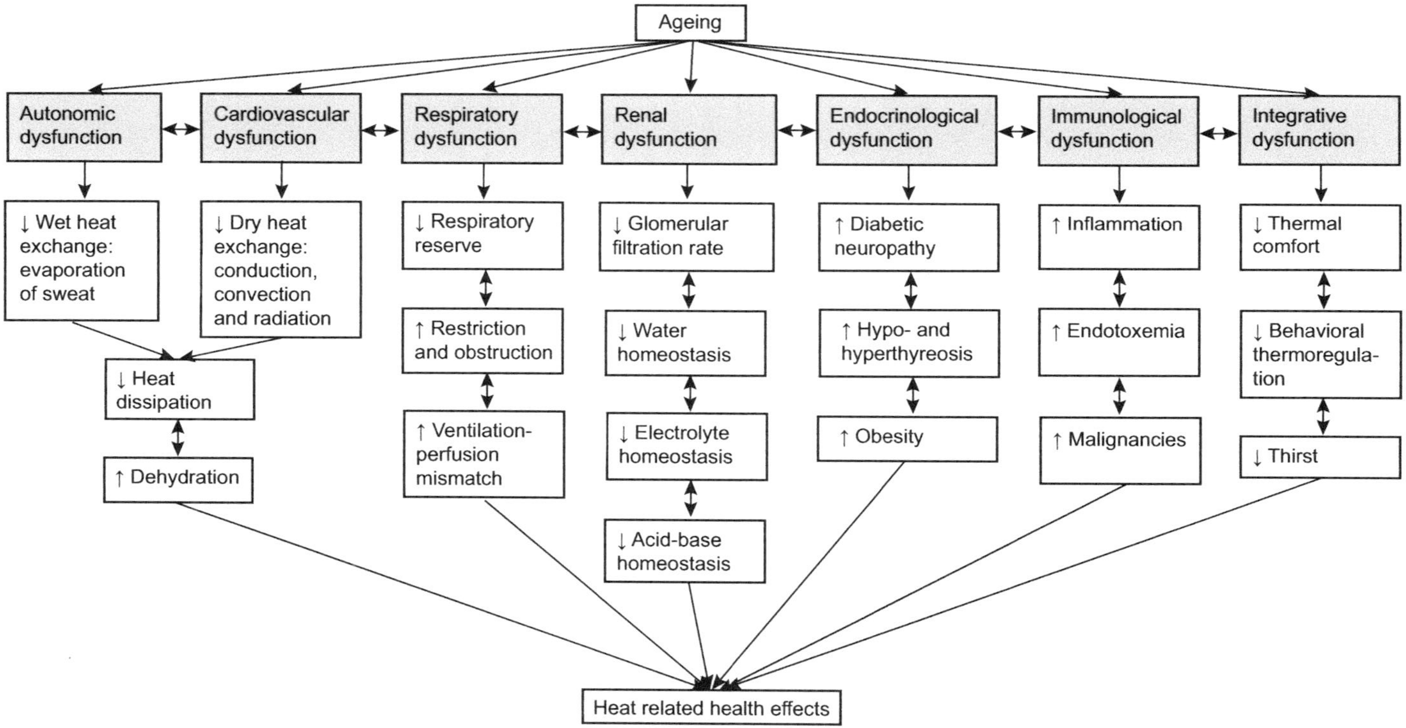

Figure 5.3 ***A biological model for the role of aging in the heat-related health effects.*** All of the organ systems are in interaction with each other (*top row*), and pathology in one system is usually accompanied by compensatory action and strain in the others. The arrows do not represent chronological sequences. For example, dehydration affects all organ systems.

3. Limitations of the biological model

In this chapter, we have reviewed mechanisms by which aging influences on one hand thermoregulation during heat exposure, and how it on another hand influences tolerance for the physiological stress that it causes. While the biological model seems reasonable in the light of empirical evidence and clinical practice, some limitations need to be recognized. Primarily, inference is based on small experimental studies with limited external validity. It can also be difficult to differentiate between biological aging and age-related medical conditions in some research settings. It is not always possible to ascribe vulnerability to the effects of heat exposure to aging or comorbidities alone. Comorbidities are not always accounted for in the studies. The age groups vary between studies, sometimes being overtly narrow or overtly wide. Some experimental studies include only elderly subjects, whereas fewer studies compare the physiological responses to heat exposure in different age groups. In many studies, only one sex is included. Sample sizes are generally small, and the varying schemes of selecting the study subjects and control subjects make comparisons between studies difficult. Generalizing the results is hard. Most importantly, the clinical relevance of many of the presented details remains to be proven. This review and presentation of evidence also has several limitations. We focused primarily on review articles and secondarily on key experimental studies. Many details of the biological model also had to be constructed applying indirect inference, where gaps in the knowledge needed to be filled with evidence from thermophysiological responses in the younger adult populations.

Evidence-based synthesis of this topic is not without challenges. Therefore, it is particularly useful to complement the analysis with a review of epidemiological evidence.

4. The epidemiological perspective

There is consistent epidemiological evidence that increase in daily average temperature above and below the range of local optimal temperature increases the total mortality (Gasparrini et al., 2015). A study of 384 locations in 13 countries estimated that 0.42% of total mortality is attributable to temperatures above the optimal, whereas 7.72% is attributable to temperatures below optimal. Epidemiological studies provide also consistent evidence of the effects of heat waves on mortality. The previous sections provide basis for the hypothesis that the elderly are more sensitive to the effects of both high ambient temperatures and heat waves, i.e., several consecutive days of high ambient temperature.

In epidemiologic studies of heat-related mortality, different age groups, both young and elderly, have been identified as being more severely influenced and often defined as being more "vulnerable." Empirical evidence of vulnerability is present as differences in the magnitude of effect estimates between the compared age groups. In epidemiologic terminology, vulnerability manifests as heterogeneity in the effect estimates.

Majority of studies on the effects of high temperature have presented effect estimates in different age groups, but few studies have formally assessed to what extent observed differences could be explained by chance.

Metaanalyses of the results of several epidemiologic studies or all the studies from a systematic review provided additional methodological possibilities to elaborate on the epidemiologic evidence of "vulnerable" groups. There are two approaches for evaluating the vulnerability using results from several or all available studies: metaanalysis of the ratios of relative risk (RRR), and metaregression with the effect estimate as the dependent variable and age (categorical or continuous) and other indicators of vulnerability as independent variables.

Benmarhnia et al. (2015) addressed vulnerability to heat-related mortality among elderly and other potential groups of vulnerability in a systematic review of studies up to January 2015. They identified 61 studies fulfilling the inclusion criteria, most of these studies provided information on differences of effects among elderly compared to nonelderly. In the metaanalysis of the RRRs from 29 studies, the summary–effect estimate for individuals aged >65 years, compared to individuals aged between 15 and 64 years, was 1.02 (95% CI = 1.01–1.03). The summary–effect estimate for the contrast of >75 years versus 15–74 years was 1.04 (95% CI = 1.02–1.07). There was large heterogeneity in the summary RRRs (all $I^2 > 50\%$). In metaregression of the lnRRR, only the age contrast, but not local temperature, latitude, or study period, was found statistically significant. In summary, the effect of high temperature was approximately 2% and 4% stronger among individuals 65+ and 75+ years compared to nonelderly (15–64 years and 15–74 years, respectively).

Luo et al. (2019) summarized the evidence of effects of heat and cold exposures on mortality in China, including 45 studies, 33 in English and 11 in Chinese. The heat effects rate ratio was 1.014 (95% CI = 1.010–1.019) for individuals of age <65 and 1.029 (95% CI = 1.017–1.041). The point estimate of RRR was 1.02, i.e., identical with that of Benmarhania and colleagues (2015).

There is also epidemiological evidence of associations between hot weather, heat waves and high indoor temperature, and cause-specific mortality and morbidity in the elderly for a variety of outcomes related to different organ systems (Anderson et al., 2013; Bunker et al., 2016; de Lorenzo & Liaño, 2017; Lavados et al., 2018; Sun et al., 2018; Tham et al., 2020; Åström et al., 2011), to give a few examples.

5. Synthesis of knowledge

The global life expectancy continues to rise, and the fertility level continues to fall. Consequently, the fraction of elderly people of the world population is expected to rise (United Nations 2019). At the same time, the number and severity of heat waves is predicted to increase under the many climate change scenarios (https://www.epa.gov/climate-indicators/climate-change-indicators-heat-waves). In the context of elderly health, these trends are particularly troubling (Vu et al., 2019).

This chapter reviewed experimental evidence on how aging affects thermoregulation, and how age-related pathologies and comorbidities could compromise tolerance to ambient heat exposure, eventually leading to increased morbidity and mortality during hot weather and heat waves. While the biological model that was developed throughout the text is fundamentally limited by its constituents, it seems coherent as a whole and is in line with the current opinion. Overall, the body of evidence provides a reasonable framework for the synthesis.

We summarize that the evidence is in favour of the hypothesis that the elderly are at a particularly high risk of adverse health effects during heat exposure. The elderly seem vulnerable during hot weather and heat waves, which is observed as increased mortality and morbidity in different populations around the world. These findings suggest that definitive action is needed to safeguard the elderly against heat exposure in the current era and decades to come.

References

Aalami, O. O., Fang, T. D., Song, H. M., & Nacamuli, R. (2003). Physiological features of aging persons. *Archives of Surgery, 138*(10), 1068–1076. https://doi.org/10.1001/archsurg.138.10.1068

Anderson, G. B., Dominici, F., Wang, Y., McCormack, M. C., Bell, M. L., & Peng, R. D. (2013). Heat-related emergency hospitalizations for respiratory diseases in the medicare population. *American Journal of Respiratory and Critical Care Medicine, 187*(10), 1098–1103. https://doi.org/10.1164/rccm.201211-1969OC

Argaud, L., Ferry, T., Le, Q. H., Marfisi, A., Ciorba, D., Achache, P., Ducluzeau, R., & Robert, D. (2007). Short- and long-term outcomes of heatstroke following the 2003 heat wave in Lyon, France. *Archives of Internal Medicine, 167*(20), 2177–2183. https://doi.org/10.1001/archinte.167.20.ioi70147

ASHRAE. (1997). Thermal comfort. In *ASHRAE handbook of fundamentals, chapter 8*. Atlanta: American Society of Heating, Refrigerating and Air-Conditioning Engineers.

Åström, D. O., Forsberg, B., & Rocklöv, J. (2011). Heat wave impact on morbidity and mortality in the elderly population: A review of recent studies. *Maturitas, 69*(2), 99–105. https://doi.org/10.1016/j.maturitas.2011.03.008

Baker, L. B. (2019). Physiology of sweat gland function: The roles of sweating and sweat composition in human health. *Temperature (Austin), 6*(3), 211–259. https://doi.org/10.1080/23328940.2019.1632145

Balmain, B. N., Sabapathy, S., Louis, M., & Morris, N. R. (2018). Aging and thermoregulatory control: The clinical implications of exercising under heat stress in older individuals. *BioMed Research International, 2018*, 8306154. https://doi.org/10.1155/2018/8306154

Benmarhnia, T., Deguen, S., Kaufman, J., & Smargiassi, A. (2015). Vulnerability to heat-related mortality: A systematic review, meta-analysis, and meta-regression analysis. *Epidemiology, 26*(6), 781–793. https://doi.org/10.1097/EDE.0000000000000375

Blatteis, C. M. (2012). Age-dependent changes in temperature regulation—A mini review. *Gerontology, 58*(4), 289–295. https://doi.org/10.1159/000333148

Bouchama, A., & Knochel, J. P. (2002). Heat stroke. *The New England Journal of Medicine, 346*(25), 1978–1988. https://doi.org/10.1056/NEJMra011089

Bunker, A., Wildenhain, J., Vandenbergh, A., Henschke, N., Rocklöv, J., Hajat, S., & Sauerborn, R. (2016). Effects of air temperature on climate-sensitive mortality and morbidity outcomes in the elderly; a systematic review and meta-analysis of epidemiological evidence. *EBioMedicine, 6*, 258–268. https://doi.org/10.1016/j.ebiom.2016.02.034

Cabanac, M., & Brinnel, H. (1987). The pathology of human temperature regulation: Thermiatrics. *Experientia, 43*(1), 19–27. https://doi.org/10.1007/BF01940348

Castellani, J. W., & Young, A. J. (2016). Human physiological responses to cold exposure: Acute responses and acclimatization to prolonged exposure. *Autonomic Neuroscience, 196*, 63–74.

Cowen, L. E., Hodak, S. P., & Verbalis, J. G. (2013). Age-associated abnormalities of water homeostasis. *Endocrinology and Metabolism Clinics of North America, 42*(2), 349–370. https://doi.org/10.1016/j.ecl.2013.02.005

Crandall, C. G., & Wilson, T. E. (2015). Human cardiovascular responses to passive heat stress. *Comprehensive Physiology, 5*(1), 17–43. https://doi.org/10.1002/cphy.c140015

Dai, H., Alsalhe, T. A., Chalghaf, N., Riccò, M., Bragazzi, L., & Wu, J. (2020). The global burden of disease attributable to high body mass index in 195 countries and territories, 1990–2017: An analysis of the Global Burden of Disease Study. *PLoS Medicine, 17*(7), e1003198. https://doi.org/10.1371/journal.pmed.1003198

Ebi, K. L., Capon, A., Berry, P., Broderick, C., de Dear, R., Havenith, G., Honda, Y., Kovats, R. S., Ma, W., Malik, A., & Morris, N. B. (2021). Hot weather and heat extremes: health risks. *The Lancet, 398*(10301), 698–708.

Flouris, A. D. (2011). Functional architecture of behavioural thermoregulation. *European Journal of Applied Physiology, 111*(1), 1–8. https://doi.org/10.1007/s00421-010-1602-8

Gagge, A. P., Stolwijk, J. A., & Hardy, J. D. (1967). Comfort and thermal sensations and associated physiological responses at various ambient temperatures. *Environmental Research, 1*(1), 1–20. https://doi.org/10.1016/0013-9351(67)90002-3

Gagnon, D., Romero, S. A., Cramer, M. N., Kouda, K., Poh, P. Y. S., Ngo, H., Jay, O., & Crandall, C. G. (2017). Age modulates physiological responses during fan use under

extreme heat and humidity. *Medicine & Science in Sports & Exercise, 49*(11), 2333–2342. https://doi.org/10.1249/MSS.0000000000001348
Gasparrini, A., Guo, Y., Hashizume, M., Lavigne, E., Zanobetti, A., Schwartz, J., Tobias, A., Tong, S., Rocklöv, J., Forsberg, B., Leone, M., De Sario, M., Bell, M. L., Guo, Y. L., Chang-fu, Wu, Kan, H., Yi, S.-M., de Sousa Zanotti Stagliorio Coelho, M., Saldiva, P. H. N., … Armstrong, B. (2015). Mortality risk attributable to high and low ambient temperature: A multicountry observational study. *Lancet, 386*(9991), 369–375. https://doi.org/10.1016/S0140-6736(14)62114-0, 12.
Guergova, S., & Dufour, A. (2011). Thermal sensitivity in the elderly: A review. *Ageing Research Reviews, 10*(1), 80–92. https://doi.org/10.1016/j.arr.2010.04.009
Hifumi, T., Kondo, Y., Shimizu, K., & Miyake, Y. (2018). Heat stroke. *Journal of Intensive Care Medicine, 6*, 30. https://doi.org/10.1186/s40560-018-0298-4
Holowatz, L. A., Thompson-Torgerson, C., & Kenney, L. (2011). Aging and the control of human skin blood flow. *Frontiers in Bioscience, 15*(2), 718–739. https://doi.org/10.2741/3642
van Hoof, J. (2008). Forty years of Fanger's model of thermal comfort: Comfort for all? *Indoor Air, 18*(3), 182–201. https://doi.org/10.1111/j.1600-0668.2007.00516.x
van Hoof, J., & Hensen, J. L. M. (2006). Thermal comfort and older adults. *Gerontechnology, 4*(4), 223–228. https://doi.org/10.4017/gt.2006.04.04.006.00
van Hoof, J., Schellen, L., Soebarto, V., Wong, J. K. W., & Kazak, J. K. (2017). Ten questions concerning thermal comfort and ageing. *Building and Environment, 120*, 123–133. https://doi.org/10.1016/j.buildenv.2017.05.008
IUPS Thermal Commission. (2001). Glossary of terms for thermal physiology. Third edition. Revised by the commission for thermal physiology of the international union of physiological sciences (IUPS thermal commission). *The Japanese Journal of Physiology, 51*, 245–280.
Janssens, J.-P. (2005). Aging of the respiratory system: Impact on pulmonary function tests and adaptation to exertion. *Clinics in. Chest Medicine, 26*(3), 469–484.
Jung, S., Lee, M., Kim, D.-Y., Son, C., Ahn, B. H., Heo, G., Park, J., Kim, M., Park, H.,-E., Koo, D.-J., Park, J. H., Lee, J. W., Choe, H. K., & Kim, S.-Y. (2022). A forebrain neural substrate for behavioral thermoregulation. *Neuron, 110*(2), 266–279. https://doi.org/10.1016/j.neuron.2021.09.039
Kalisch Ellett, L. M., Pratt, N. L., Le Blanc, V. T., Westaway, K., & Roughead, E. E. (2016). Increased risk of hospital admission for dehydration or heat-related illness after initiation of medicines: A sequence symmetry analysis. *Journal of Clinical Pharmacy and Therapeutics, 41*(5), 503–507. https://doi.org/10.1111/jcpt.12418
Kenney, W. L. (2017). Edward F. Adolph distinguished lecture: Skin-deep insights into vascular aging. *Journal of Applied Physiology, 123*(5), 1024–1038. https://doi.org/10.1152/japplphysiol.00589.2017
Kenney, W. L., & Chiu, P. (2001). Influence of age on thirst and fluid intake. *Medicine & Science in Sports & Exercise, 33*(9), 1524–1532. https://doi.org/10.1097/00005768-200109000-00016
Kenney, W. L., & Munce, T. A. (2003). Invited review: Aging and human temperature regulation. *Journal of Applied Physiology, 95*(6), 2598–2603. https://doi.org/10.1152/japplphysiol.00202.2003
Kenny, G. P., Sigal, R. J., & McGinn, R. (2016). Body temperature regulation in diabetes. *Temperature, 3*(1), 119–145. https://doi.org/10.1080/23328940.2015.1131506
Kenny, G. P., Yardley, J., Brown, C., Sigal, R. J., & Jay, O. (2010). Heat stress in older individuals and patients with common chronic diseases. *Canadian Medical Association Journal, 182*(10), 1053–1060. https://doi.org/10.1503/cmaj.081050
Kelm, M. (1999). Nitric oxide metabolism and breakdown. *Biochimica et Biophysica Acta, 1411*(2-3), 273–289.

Kuzuya, M. (2013). Heatstroke in older adults. *Japan Medical Association Journal, 56*(3), 193–198.
Lavados, P. M., Olavarria, V. V., & Hoffmeister, L. (2018). Ambient temperature and stroke risk: Evidence supporting a short-term effect at a population level from acute environmental exposures. *Stroke, 49*(1), 255–261. https://doi.org/10.1161/STROKEAHA.117.017838
Leon, L. R., & Bouchama, A. (2015). Heat stroke. *Comprehensive Physiology, 5*(2), 611–647. https://doi.org/10.1002/cphy.c140017
de Lorenzo, A., & Liaño, F. (2017). High temperatures and nephrology: The climate change problem. *Nefrologia, 37*(5), 492–500. https://doi.org/10.1016/j.nefro.2016.12.008
Luo, Q., Li, S., Guo, Y., Han, X., & Jaakkola, J. J. K. (2019). A systematic review and meta-analysis of the association between daily mean temperature and mortality in China. *Environmental Research, 173*, 281–299. https://doi.org/10.1016/j.envres.2019.03.044
Meade, R. D., Notley, S. R., & Kenny, G. P. (2019). Aging and human heat dissipation during exercise-heat stress: An update and future directions. *Current Opinion in Physiology, 10*, 219–225. https://doi.org/10.1016/j.cophys.2019.07.003
Millyard, A., Layden, J. D., Pyne, D. B., Edwards, A. M., & Bloxham, S. R. (2020). Impairments to thermoregulation in the elderly during heat exposure events. *Gerontology and Geriatric Medicine, 6*. https://doi.org/10.1177/2333721420932432, 2333721420932432.
Minson, C. T., Wladkowski, S. L., Cardell, A. F., Pawelczyk, J. A., & Kenney, W. L. (1998). Age alters the cardiovascular response to direct passive heating. *Journal of Applied Physiology, 84*(4), 1323–1332. https://doi.org/10.1152/jappl.1998.84.4.1323
Montagna, W., & Carlisle, K. (1979). Structural changes in aging human skin. *Journal of Investigative Dermatology, 73*(1), 47–53. https://doi.org/10.1111/1523-1747.ep12532761
Morrison, S. F. (2016). *Central control of body temperature*. https://doi.org/10.12688/f1000research.7958.1. F1000Res. 5, F1000 Faculty Rev-880.
Morrison, S. F., & Nakamura, K. (2011). Central neural pathways for thermoregulation. *Frontiers in Bioscience, 16*(1), 74–104. https://doi.org/10.2741/3677
Peters, R. (2006). Ageing and the brain. *Postgraduate Medical Journal, 82*(964), 84–88. https://doi.org/10.1136/pgmj.2005.036665
Pride, N. B. (2005). Ageing and changes in lung mechanics [editorial]. *European Respiratory Journal, 26*(4), 563–565. https://doi.org/10.1183/09031936.05.00079805, 2005.
Rolls, E. T., Grabenhorst, F., & Parris, B. A. (2008). Warm pleasant feelings in the brain. *Neuroimage, 41*(4), 1504–1513. https://doi.org/10.1016/j.neuroimage.2008.03.005
Romanovsky, A. A. (2007). Thermoregulation: Some concepts have changed. Functional architecture of the thermoregulatory system. *American Journal of Physiology. Regulatory, Integrative and Comparative Physiology, 292*(1), R37–R46. https://doi.org/10.1152/ajpregu.00668.2006
Ryti, N. R. I., Korpelainen, A., Seppänen, O., & Jaakkola, J. J. K. (2020). Paradoxical home temperatures during cold weather: A proof-of-concept study. *International Journal of Biometeorology, 64*, 2065–2076. https://doi.org/10.1007/s00484-020-01998-7
Schlader, Z. J., Sackett, J. R., Sarker, S., & Johnson, B. D. (2018). Orderly recruitment of thermoeffectors in resting humans. *American Journal of Physiology. Regulatory, Integrative and Comparative Physiology, 314*(2), R171–R180. https://doi.org/10.1152/ajpregu.00324.2017
Sinclair, A., Saeedi, P., Kaundai, A., Karuranga, S., Malanda, B., & Williams, R. (2020). Diabetes and global ageing among 65–99-year-old adults: Findings from the international diabetes federation diabetes atlas, 9th edition. *Diabetes Research and Clinical Practice, 162*, 108078. https://doi.org/10.1016/j.diabres.2020.108078

Stevens, J. C., & Choo, K. K. (1998). Temperature sensitivity of the body surface over the life span. *Somatosensory & Motor Research, 15*(1), 13–28. https://doi.org/10.1080/08990229870925

Stocks, J. M., Taylor, N. A., Tipton, M. J., & Greenleaf, J. E. (2004). Human physiological responses to cold exposure. *Aviation, Space, and Environmental Medicine, 75*(5), 444–457.

Stöllberger, C., Lutz, W., & Finsterer, J. (2009). Heat-related side-effects of neurological and non-neurological medication may increase heatwave fatalities. *European Journal of Neurology, 16*(7), 879–882. https://doi.org/10.1111/j.1468-1331.2009.02581.x

Sun, Z., Chen, C., Xu, D., & Li, T. (2018). Effects of ambient temperature on myocardial infarction: A systematic review and meta-analysis. *Environmental Pollution, 241*, 1106–1114. https://doi.org/10.1016/j.envpol.2018.06.045

Svartengren, M., Falk, R., & Philipson, K. (2005). Long-term clearance from small airways decreases with age. *European Respiratory Journal, 26*(4), 609–615. https://doi.org/10.1183/09031936.05.00002105

Tan, C. L., & Knight, Z. A. (2018). Regulation of body temperature by the nervous system. *Neuron, 98*(1), P31–48. https://doi.org/10.1016/j.neuron.2018.02.022

Tan, C. C. S., Chin, L. K. K., & Low, I. C. C. (2020). Thermoregulation in the aging population and practical strategies to overcome a warmer tomorrow. *Proteomics, 20*(5–6), e1800468. https://doi.org/10.1002/pmic.201800468

Tansey, E. A., & Johnson, C. D. (2015). Recent advances in thermoregulation. *Advances in Physiology Education, 39*(3), 139–148. https://doi.org/10.1152/advan.00126.2014

Tham, S., Thompson, R., Landeg, O., Murray, K. A., & Waite, T. (2020). Indoor temperature and health: A global systematic review. *Public Health, 179*, 9–17.

The GBD 2015 Obesity Collaborators. (2017). Health effects of overweight and obesity in 195 countries over 25 years. *The New England Journal of Medicine, 377*(1), 13–27. https://doi.org/10.1056/NEJMoa1614362

United Nations, & Department of Economic and Social Affairs, Population Division. (2019). *World population prospects 2019: Highlights (ST/ESA/SER.A/423). UN, New York*. Available from: https://population.un.org/wpp/Publications/Files/WPP2019_Highlights.pdf. (Accessed 12 February 2021).

Varghese, G. M., John, G., Thomas, K., Abraham, O. C., & Mathai, D. (2005). Predictors of multi-organ dysfunction in heatstroke. *Emergency Medicine Journal, 22*(3), 185–187. https://doi.org/10.1136/emj.2003.009365

Vu, A., Rutherford, S., & Phung, D. (2019). Heat health prevention measures and adaptation in older populations—A systematic review. *International Journal of Environmental Research and Public Health, 16*(22), 4370. https://doi.org/10.3390/ijerph16224370

Waldock, K. A. M., Hayes, M., Watt, P. W., & Maxwell, N. S. (2018). Physiological and perceptual responses in the elderly to simulated daily living activities in UK summer climatic conditions. *Public Health, 161*, 163–170. https://doi.org/10.1016/j.puhe.2018.04.012

Wenger, C. B. (1972). Heat of evaporation of sweat: Thermodynamic considerations. *Journal of Applied Physiology, 32*(4), 456–459. https://doi.org/10.1152/jappl.1972.32.4.456

Wong, B. J., & Hollowed, C. G. (2016). Current concepts of active vasodilation in human skin. *Temperature, 4*(1), 41–59. https://doi.org/10.1080/23328940.2016.1200203

Yahiro, T., Kataoka, N., Nakamura, Y., & Nakamura, K. (2017). The lateral parabrachial nucleus, but not the thalamus, mediates thermosensory pathways for behavioural thermoregulation. *Scientific Reports,* 7(1), 5031. https://doi.org/10.1038/s41598-017-05327-8

CHAPTER 6

Heat exposure and occupational health

Dino Pisaniello[1] and Rosario Di Corleto[2]
[1]School of Public Health, University of Adelaide, Adelaide, Australia; [2]Principal Consultant, Ashgrove, Brisbane, QLD, Australia

1. Introduction

Occupational exposure to heat as a hazard has been recorded in historical texts, dating back to ancient civilizations. This hazard has been linked to adverse outcomes ranging from the simplest of health conditions such as dehydration through to severe heat strokes and fatalities. This is even though key aspects of homeothermic balance were not well understood.

In more recent times, there has been an increasing awareness of the indirect impacts of heat exposure in working populations. These include impacts such as

- increased fatigue,
- impairment of cognitive function and fine motor skills,
- increased incidence of physical injury,
- reduction in productivity of the workforce, and
- increased costs.

In conjunction with these potential impacts, there are three key areas that need to be addressed for risk assessment and control:

1. the work or task being undertaken,
2. the worker (as an individual), and
3. ultimately, the work environment.

While the first two risk factors play key roles, it is important to note that climate change and the projected increased incidence of thermal extremes have become major influences.

As the influence increases, so too will the negative consequences that have been noted before. In this chapter, evidence from relevant literature is presented from both qualitative and quantitative perspectives, so as to inform occupational health risk management in a warming climate.

Heat Exposure and Human Health in the Context of Climate Change
ISBN 978-0-12-819080-7
https://doi.org/10.1016/B978-0-12-819080-7.00004-5
© 2023 Elsevier Inc.
All rights reserved.

2. Occupational health hazards and impacts

2.1 Some historical notes

The impact of heat on exertional tasks and the working environment has often been mentioned in history. One of the earliest reference is in the Bible, in the book of Judith, Chapter VIII, verses 2 and 3, to deaths from sunstroke where Judith's husband, "Manasseh who died in the time of the barley harvest; for he was standing over them that bound sheaves in the field; and the heat came upon his head, and he died in Bethulia…." (Wakefield et al., 1927).

Dating back to 3000 BCE the Romans, Egyptians and Greeks had linked the appearance of Sirius, the dog star, with "ushering in the 'dog days' of summer … [and] bringing on fever in men and madness in dogs" (Goldman, 2001, p. 4).

There is mention of exposure to heat and its control in the military for example:

"The Roman legionary used an early form of auxiliary cooling, inserting rushes into his head gear and keeping them wet with water...[and used] auxiliaries to carry as much of the legionnaires' load, and do as much of the engineering digging as possible, thus sparing the fighting edge of the legion" (Goldman, 2001, p. 5).

In 400 BCE, Herodotus details Spartans and Athenians affected by the sun, and armies of Alexander the Great taking precautions (Casa et al., 2010).

In a British naval report in 1841, there was a report of deaths on the Frigate, The Liverpool, " …. *a species of coup de soleil*" in which there was a loss of three officers and 30 men (Wakefield et al., 1927).

Over time, heat stroke has been referred to in a number of ways such as insolatio, ictus solaris, and the fore mentioned coup de soleil.

Studies on the impact of heat (in exertional and nonexertional situations) have been undertaken by various researchers over the years. In 1774, Sir Charles Blagden in his writings stated "Further experiments and observations in a heated room" describe where personnel were exposed to temperatures from 240°F (114°C) to 261°F (127°C) without ill effect, and in 1842, Claude Bernard began experiments to assess the mechanism of death due to the action of heat (Sayers et al., 2011).

The impact of heat and humidity on the worker featured in a number of papers and studies in the earlier part of the 20th century (Moss et al., 1923; Vernon, 1932; Young et al., 1920). This was also the period in which the effective temperature index (ET) was published (Houghten & Yaglou, 1924). Though developed more for thermal comfort, it became one of the key tools used for the assessment of thermal stress at the workplace. ET is still used as a limit in some mining regulations to this day. For example,

> "A person must not work in a place at the mine where the effective temperature exceeds 29.4°C .... ." (Queensland Coal Mining Safety and Health Regulations, 2017).

A key milestone in the assessment of heat stress was the 1957 investigation into casualties during training in the US Navy. This ultimately led to the development of the now well-known wet bulb globe temperature index (WBGT) (Yaglou & Minard, 1957).

Numerous other indices have been developed since then ranging from the more simplistic empirically based through to the more complex rational indices based on the thermal balance equation such as the predicted heat strain (PHS) (ISO 7933 2018), the thermal work limit (TWL) (Brake & Bates, 2002), and the environmental stress index (ESI) (Moran et al., 2005).

Apart from frank heat-induced illness associated with elevated core body temperature, other adverse impacts were examined. Early investigations carried out on accident rates were conducted by Vernon during World War I in the munitions industries, and further research by Osborne and Vernon indicated a relationship between temperature and accident rates (Osborne et al., 1922). This is illustrated in Fig. 6.1, which also shows effects in cold

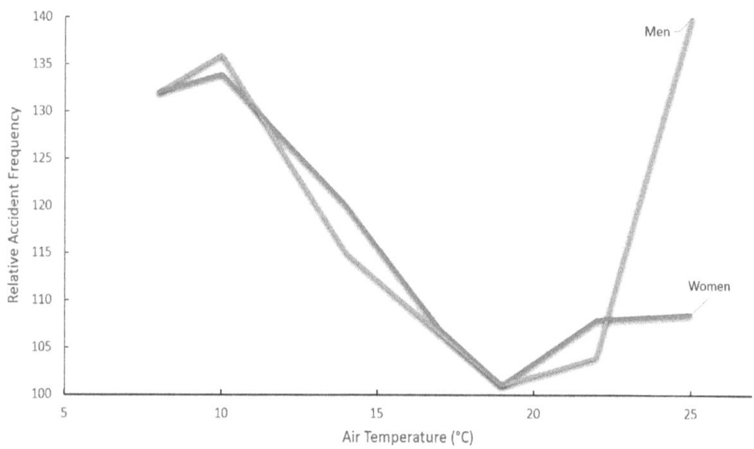

Figure 6.1 Accident frequency related to workplace temperature for munitions workers. *(Adapted from Osborne, E., Vernon, H. M., & Muscio, B. 1922. The influence of temperature and other conditions on the frequency of industrial accidents. In Two contributions to the study of accident causation. London: H.M. Stationery Office.)*

conditions. Later studies have shown a deterioration of short-term memory, cognitive skills (Ganio et al., 2011; Wilson & Morley, 2003), and psychomotor skills (Cian et al., 2000) as a result of exposure to heat and dehydration. This has the potential to affect the work quality and productivity (Flouris et al., 2018; Kjellstrom et al., 2014), for example, via altered behavior and physiological changes. It can also influence the rate of accidents via interactions with existing hazards (Xiang et al., 2014b).

While there have been significant advances made in the control and management of both hot and cold environments since these early days, there continues to be an issue with the occurrence of thermally related injuries and illnesses in current times. Systematic understanding of occupational heat-induced mortality, morbidity, costs, and productivity losses is required, and recent evidence in support of this understanding is outlined later in this chapter. This new knowledge can underpin educational, policy, regulatory, and other interventions. Acknowledging that management of heat-related effects entails risk factors associated with "work," "worker," and "workplace," it is important to firstly consider contributors to heat balance, and then how these might be disrupted by the aforementioned risk factor categories.

2.2 Thermal balance

The thermal balance equation is key to the understanding of how an individual copes with work in the thermal environment. It not only provides clues as to the cause of the harm but can also be used in the identification of controls.

The equation in its simplest form is represented by

$$M \pm C \pm R - E = \pm S$$

S = the amount of heat gained or lost by the body and will determine the overall heat load on the body. So, where S is positive, there is a heat gain, and if this continues to increase, it can result in heat stress (hyperthermia). In reverse, if S is negative, then there is a loss of heat, and this can end up as cold stress (hypothermia). The ideal is that S is zero, no net heat gain or loss.

E = the heat exchange via evaporation, for example, when we sweat, the sweat is evaporated off the skin, which has a cooling effect, hence the negative indicating a heat loss.

R = the heat exchange via radiation, this can be from hot sources such as the sun or surfaces such as walls, the ground and hot objects such as ovens and molten metals in some industries.

C = heat exchange via convection. An example of convection is where the air temperature is higher than the skin temperature leading to a heat gain, i.e., desert environments have both radiation and convection heat.

M = heat gained via a metabolic load, i.e., when an individual performs physical work, it results in a gain of heat for the body. If we look at muscular activity, muscle contraction is about 20% (Kenny & Jay, 2013; Sawka et al., 2011). Where an individual uses energy at a rate of 100 W, 20 W is used to perform the external work, but the remaining 80 W of energy is given off as heat (Casa, 2018). The body must then lose this heat or run the risk of an overall increase in internal temperature.

The innate biological/avoidance mechanisms and organ/system effects that come into play as a result of heat stress and thermal dysregulation have been described in other chapters. Critical to this is the exposed skin and hydration. Suffice to say that the human body has evolved means of adapting and protecting itself, including acclimatization, but work and workplaces are diverse, sometimes entailing extreme heat (including radiant heat), hazardous microclimates, and emergency situations including fire-fighting and military activity.

The following section profiles the evidence for the adverse impacts on working populations and relationships with metrics of heat. Although thermal comfort is a key consideration for many workplaces such as offices, the focus is on industrial environments.

2.3 Scale of the occupational heat—health issue—population studies of the impact of heat exposure

2.3.1 Morbidity (injury and illness)

2.3.1.1 Deaths

Periods of extreme ambient heat have been associated with increased fatalities among workers. Most studies have reported on the number of fatalities from heat illness (heat stroke) rather than fatal traumatic injury in hot conditions (e.g., falls due to loss of grip). Limited studies have estimated the relationship between summer temperatures and number of fatalities in the workplace (Xiang et al., 2014a).

In a study of deaths from work-related heat stress, Gun (2019) noted the infrequency of such events in the United States and Australia. However, 7 out of 13 Australian deaths occurred in workers within 1 week of hiring, emphasizing the importance of acclimatization and self-pacing.

2.3.1.2 Illness and injury

Several reviews summarize the impacts of heat on workers' health (Xiang et al., 2014a; Bonafede et al., 2016; Varghese et al., 2018; Levi et al., 2018; Binazzi et al., 2019; Spector et al., 2019; Fatima et al., 2021). Epidemiological studies relating high temperature to worker health outcomes characterize the relationship by evaluating heat in one of two ways: (1) ambient temperature or (2) heat wave days or events. In general, all the studies show a positive association between heat exposure and work–injury outcomes, despite a likelihood of underreporting (Xiang et al., 2014a).

Risk estimates associated with hot conditions, derived from population studies, are described in Section 6.3.

2.3.2 Productivity

While the consequences relating to the impact of heat strain on the worker have been predominately focused on the physiological impacts such as morbidity and mortality, the economic impact also needs to be considered. Results from a metaanalysis undertaken to identify the potential impacts of heat strain on worker productivity as part of the European Union Horizon 2020 Research and Innovation Programme (Flouris et al., 2018) indicated that occupational heat strain has a significant impact on productivity outcomes. Productivity losses were estimated to increase at an approximate rate of 2.6% per degree WBGT beyond 24°C WBGT.

Comparable results in productivity losses of up to 5% C WBGT (Sahu et al., 2013) have been identified with similar production losses shown in other regions (Dunne et al., 2013; Kjellstrom et al., 2009; Zander et al., 2015), which experience increasing climatic temperatures. In a paper prepared by the secretariat of the Climate Vulnerable Forum published by the United Nations Development Programme, it indicated:

"Business-as-usual warming (4°C) could yield output reductions for some sectors in excess of 20% during the second half of the twentieth century" (UNDP, 2016, p. 6).

Although previously not a focus of the health professional in the control of heat exposures in the occupational environment, it is obvious that successful control programs in this area will also provide additional benefits to the business case for increased resources to address the risk from heat exposures in the workplace.

2.3.3 Costs

There are limited studies addressing the costs resulting from workplace heat exposures. These have often relied on workers compensation claims data sets. Injury claims and days lost have been found to increase with rising heat stress (Martinez-Solanas et al., 2018). Ma et al. (2019) quantified the attributable fractions of injury claims and subsequent insurance payouts associated with calculated daily maximum WBGT (above the heat stress limit) in Guangzhou, China. The risks of injury claims and insurance payouts increased 4.8% and 4.1%, respectively. Male workers, those with low educational attainment, and small enterprises were at elevated risk. Interestingly, changes in labor protection policy reported in one study were associated with reduced injuries (13%) and insurance payouts (24%) (Su et al., 2020). Xiang et al. (2018) investigated the correlates of occupational heat-induced illness (OHI) costs in South Australia. A 1°C increase in maximum temperature above about 33°C was associated with a 42% increase in medical costs and a 75% increase in days lost due to OHI, respectively. Relatively higher OHI burdens were found in men, those aged 25–44 years, new workers, medium-size businesses, and those employed in the mining industry. In a review of global evidence, Borg et al. (2021) found that low- and middle-income countries and countries with warmer climates had greater losses as a proportion of GDP. Greater costs per worker were observed in outdoor industries, medium-sized businesses, among males and workers aged 25–44 years.

3. Heat stress risk assessment

3.1 Risk factors

Assessment of the risk associated with heat stress in the workplace requires a multilayered approach. The factors that impact on the eventual consequence are far more varied than the assessment of just temperature.

There are three key areas that require consideration in the thermal risk assessment.

1. The task being undertaken, which will look at metabolic load, complexity of task, posture, and personal protective equipment (PPE) (*work*).
2. The individual, their health, and general well-being including medications, etc. (*worker*).
3. The environmental parameters, which include the ambient temperature, humidity, air flow, and radiant heat (*workplace*).

Each of these components plays a significant role on the final heat load and the body's thermal balance equation.

3.1.1 Task aspects (work)

As mentioned earlier, metabolic load can increase the internal temperature and be a key contributor to subsequent heat strain. The intensity of the task may only require the undertaking of very light activities as such checks of equipment using a clipboard, which would involve in the vicinity of 90 W/m^2. Alternatively, it could involve a laborer undertaking heavy manual work such as intense shoveling or axe work, which is rated at greater than 230 W/m^2. This variation of internal heat load cannot be ignored when assessing the work environment for heat stress.

The task may require the use of heavy clothing and personal protective equipment that may disrupt normal heat exchange. A fully encapsulating plastic suit is an extreme example. Working with hot tools can contribute to heat stress in certain circumstances (Lao et al., 2016).

Consideration should also be given to the complexity of the task. Is it simple and straight forward or repetitive with minimal actions, or is it complex and intricate requiring high levels of concentration and precision?

3.1.2 Individual characteristics (worker)

Human physiology presents with a very large range of variation; this can include but is not limited to the following:

- Height, weight, and BMI
- Age
- Gender
- Physical fitness
- General well-being
- Acclimatization
- Hydration status

These, combined with additional variables that can impact on the individual such as the use of prescribed medications, alcohol, and illicit drugs can result in a vast number of differing situations. All can and will impact on how an individual will cope with the heat in a work environment.

Medical conditions such as renal, neuromuscular, respiratory disorder, diabetes, hypertension, fainting spells, or previous susceptibility to heat illness may also increase risk (Brake et al., 1998; Hanson & Graveling, 1997). Those more at risk might be excluded from certain work conditions or be medically assessed more frequently.

Short-term disorders and minor illnesses such as colds or flu, fevers, diarrhea, and vomiting should also be considered. These illnesses will impact on the individual's ability to cope with heat stress and potentially make them more susceptible to an onset of heat illness (Di Corleto et al., 2013).

It should be noted that many heat stress limits are set without taking into account these variables. Looking at the profile of most subjects used in testing heat stress indices and across general research in this area in the vast majority of cases they are fit young individuals. When undertaking a thermal risk assessment, it is important to consider these parameters to ensure that no one is put at risk. This will require an understanding of the exposure group and their general demographics.

An additional consideration in this area is the utilization of preemployment medicals with appropriate questions relating to conditions that may predispose a worker to the risk of heat stress in high heat roles. There is guidance available from a number of sources and standards such as ISO 12894 (2001).

Apart from the physiological aspects, worker vulnerability can be considered in terms of new worker status, migrant status, and risk taking behavior (Xiang et al., 2014B; Varghese et al., 2019a; Ricco et al., 2019; Gun, 2019).

3.1.3 Environmental parameters (workplace)

While these are the most obvious, they are often not fully addressed. The dry bulb temperature also known as the air temperature is sometimes used as the sole point of reference when undertaking assessments, and this can lead to quite erroneous assessments.

Humidity must be considered as this will determine the efficacy of the sweat evaporation process, which is a key aspect of the body's cooling system. High humidities limit the amount of sweat that can be evaporated and hence the amount of effective cooling. Conversely, low humidities allow for increased evaporation of sweat and hence more effective cooling.

Air flow also plays a key role in this equation as the increased airflow can enhance the evaporation process at temperatures below approximately 42°C. Under some conditions, if the temperatures become higher than 42°C (this can vary), the body's sweating capacity and the evaporation rate cannot fully compensate for the added heat load from the increased air flow, and hence the physiological heat load increases rather than decreases.

Solar radiation, radiant heat from hot processes such as smelting or from uninsulated process vessels, and pipework can add an additional load to the work environment. There needs to be consideration given to the radiant heat load from these sources. It is not unusual to see globe temperatures greater than 10°C above the ambient dry bulb temperature when working in the open without shade cover.

3.2 Assessment of heat stress exposure and risk

A three-tiered approach for the risk assessment of exposure to heat has been designed such that it may be applied to a number of varying scenarios where there is a potential risk to heat stress. The approach as suggested in the Australian Institute of Occupational Hygienists heat stress guideline (Di Corleto et al., 2013) involves a three-stage process dependent on the severity and complexity of the situation. It allows for the application of the right tool for a specific task utilizing a variation of risk assessment approaches.

The first level or the basic thermal risk assessment was primarily designed as a qualitative assessment, which does not require specific technical skills in its application or interpretation. It was originally developed for use by frontline employees to assist them to better understand the mechanisms and the many different impacts associated with thermal assessments. By asking the assessor to rate specific parameters that impact on the thermal stress, its intent was to show that there are many aspects of the work environment that need to be considered, not just a single dry bulb temperature. In the original version of the basic thermal assessment, there was in fact no actual measurement required and the use of a basic heat stress index in the equation was added to help balance off the subjective and qualitative nature of the assessment with some quantification.

The second step of the process begins to look more toward a quantitative risk approach and requires the measurement of a number of environmental and personal parameters such as dry bulb and globe temperatures, relative humidity, air velocity, metabolic workload, and clothing insulation. These are dependent on the heat stress index used for the next level of the analysis. Rational indices such as PHS, TWL, and ESI are usually employed at this stage as they employ a combination of the parameters from the three key sections described earlier. They tend to be more complex in nature than the empirical indices such as WBGT and effective temperature (ET) for this reason, but with the advent of modern computers, tablets, and phone applications, many of these issues can now be overcome. Whether using an empirical or rational index, the process involves a number of measurements and calculations to determine an outcome and apply allowable exposure times for groups of people exposed. When using these indices, it is important to note that they are usually applied with certain qualifications: e.g., health, hydration, fitness, and in some cases age. It could therefore be viewed more as a semiquantitative risk assessment.

The third step approaches from a more individually targeted quantitative risk perspective in that it utilizes measurements based on an individual's strain and reactions to the thermal stress to which they are being exposed. This direct measurement takes into account the variables (age, fitness, etc.) for which assumptions are made in the second step of the assessment which utilizes indices. This approach measures the actual strain on the individual rather than basing it around a group. The assessment provides a more accurate result and a greater level of confidence in relation to the impact on the individual regardless of the conditions and, importantly, includes the impact of PPE. There are a number of physiological criteria which can be used to assess the physiological strain being experienced by the individual, which can include

- core temperature,
- heart rate,
- hydration levels, and
- recovery rate.

There are international guides such as ISO 9886 (2004) and the ACGIH Threshold Limit Values (ACGIH, 2021) that can be utilized to determine criteria at which point a heat exposure must be discontinued. For example, when:

- Core temperature exceeds 38°C or 38.5°C for medically supervised acclimatised personnel.
- "Heart Rate Limit (HRL) = 185−0.65A", where A = Age in years; Noting that, individual variability can range up to 20 bpm from this average so this level could present a risk for some individuals.
- HRL, sustained = 180 − age. (For individuals with assessed normal cardiac performance).
- Greater than 1.5% weight loss over a shift: OR.
- 24-hour urinary sodium excretion is less than 50 mmoles.

No matter which limiting values are used, interpretation requires discussion with the workers affected and may require the services of a specialist such as an occupational hygienist or occupational physician.

It should be noted that the differing levels of risk assessment do require increasing levels of technical expertise, and while a level 1 assessment could be undertaken by a variety of individuals with limited technical skills, the use of a level 3 assessment should be restricted to someone with specialist knowledge and skills. It is important that the appropriate tool be selected and applied depending on the scenario and skills of the assessor.

3.3 Occupational population studies of heat–health risks

3.3.1 Illness risks

In an epidemiological review of 55 studies reporting the health impacts of workplace heat exposure, Xiang et al. (2014A) concluded that manual workers in low–middle-income countries in tropical regions are at elevated risk. In particular, this includes farmers, construction workers, firefighters, miners, soldiers, and manufacturing workers working around process-generated heat. They noted significant underreporting and a low rate of workers compensation claims. One explanation provided is that most heat illnesses are mild and individuals recover rapidly through simple treatment or rest on site without hospitalization. Overall, there appears to be a shortage of good epidemiological evidence on the relationship of mild heat illness and temperature metrics. However, a systematic review and meta-analysis by Flouris et al. (2018) estimated that 15% of individuals who typically or frequently worked under heat stress (minimum of 6 h per day, 5 days per week, for 2 months of the year) would experience kidney disease or acute kidney injury.

3.3.2 Injury risks

One of the earliest papers in this area was by Fogleman et al. (2005). Compared with 21°C, injury risks were higher when temperatures were greater than 32°C. There was an elevated risk for younger workers. Morabito et al. (2006) found elevated injury risks at higher ambient temperatures. Later studies of high temperature also found higher injury risks (Tawatsupa et al., 2013; Xiang et al., 2014B; Adam-Poupart et al., 2015; McInnes et al., 2017; Sheng et al., 2018; Ricco et al., 2018; Martinez-Solanas et al., 2018; Marinaccio et al., 2019). In the Marianaccio et al. (2019) study, there were significant effects for high temperatures with a relative risk 1.17 (95% CI: 1.14–1.21).

Studies using heatwave metrics showed similar findings (Xiang et al., 2014C; Rameezdeen et al., 2017; Varghese et al., 2018, 2019b, 2019c).

This was also evident in studies using ambient WBGT (Garzon-Villalba et al., 2016) and Humidex (Calkins et al., 2019; Spector et al., 2016).

In a recent systematic review and metaanalysis, the pooled results suggested that the overall risk of occupational injury increased by 1% (RR 1.010, 95% CI: 1.009–1.011) for 1°C increase in temperature above reference values and 17.4% (RR 1.174, 95% CI: 1.057–1.291) during heatwaves (Fatima et al., 2021). Young workers (age $<$ 35 years) and male workers had elevated injury risks during hot temperatures, as well as heatwaves.

4. Prevention, control, and adaptation

As with most hazards encountered in the workplace, the management of the risk associated with heat exposure needs to be addressed via a hierarchy of controls. Ideally the approach should be to implement controls from the higher levels, i.e., elimination, substitution, and engineering. However, there are situations where the lower-level approaches such as administrative or PPE may be required to be adopted. Indeed, often a layered approach is necessary depending on risk factors such as the nature of the tasks being undertaken, vulnerability of workers, and the climatic region. In addition, the priority of each will depend on the risk assessment, clarifying the relative importance of heat exchange mechanisms.

4.1 The hierarchy of heat hazard control

Examples of the hierarchy in the context of heat–health are given in the following:

Elimination/substitution:

- Avoidance of high air temperature and solar radiation, by rescheduling work to the early morning or if practical night shift to avoid the hottest times of the day.
- Avoidance of solar radiation and potential control of an indoor environment, by moving equipment into a workshop for maintenance rather than in the field.
- Reduction of internal temperatures by using reflective or light-colored external cladding and roofing.

- Designing buildings to help dissipate the heat from the structure. Good air flow can be utilized through positioning of windows, shutters, and roof design, i.e., "chimney effects."
- Reducing radiant heat transmission by insulating hot processed vessels and pipework.
- Reducing humidity by minimizing steam leaks in process plants, by design.

Engineering:

- Reflective shields or insulating barriers can be utilized where a process produces radiated heat such as furnaces, ovens, and in smelting. These barriers may incorporate highly reflective surfaces.
- Shade cover to prevent the exposure to solar radiation can be simple and portable.
- Increasing air circulation can improve the evaporation of sweat (the body's principal cooling mechanism) particularly in high humidities where more air needs to be moved. This can be simply accomplished by utilizing air movers such as fans. As mentioned earlier, caution should be exercised when increasing air movement in temperatures above approximately higher than 42°C as the increased heat load may exceed the benefit obtained from sweat evaporation.
- Cooled air from "chiller" units, air conditioners, or large vortex tubes can be utilized in situations where cooling is required.
- Reducing the metabolic workload by mechanizing tasks.

The benefits of engineering controls can be demonstrated via the use of the PHS heat index to model the impact of introducing the controls on predicted body core temperature.

4.1.1 Scenario

Employee is undertaking a task in a temporary workshop which is an unventilated shipping container. It is estimated that the task will take approximately 2 h to complete but may take longer. Some monitoring has been undertaken and the results are as follows:

- Dry bulb = 40.9°C
- Globe = 45.3°C
- Rel humidity = 22.6%
- Air velocity = 0.2 m/s
- Metabolic workload = 175 W/m^2
- Posture = standing

- Clothing is a single-layer disposable cotton overall, an insulation factor of 0.8 clo
- Acclimatized worker

By utilizing the aforementioned data in PHS, the resultant predicted core temperature limit of 38°C will be reached at approximately 79 min (Fig. 6.2).

By increasing the airflow in the temporary workshop from 0.2 m/s to 2.0 m/s, the resultant graph shows a flattening of the curve below 38°C, resulting in a more favorable outcome (Fig. 6.3).

4.1.2 Administrative controls

In an Australian survey of 180 occupational hygienists and relevant specialists, 19% were dissatisfied with current heat stress prevention measures. Barriers to effective heat stress risk management were reported to be lack of awareness (68%), insufficient training (56%), unsatisfactory management commitment (52%), and low compliance with prevention policies (40%) (Xiang et al., 2015). Qualitative studies also highlight organizational issues and the importance of protective behaviors (Hansen et al., 2018, 2020). This suggests that there needs to be a greater emphasis on administrative controls.

Administrative controls represent an essential layer for vulnerable workers and high heat hazard workplaces and tasks. Such controls encompass the following:

- Ready access to cool palatable drinking water to maintain hydration.

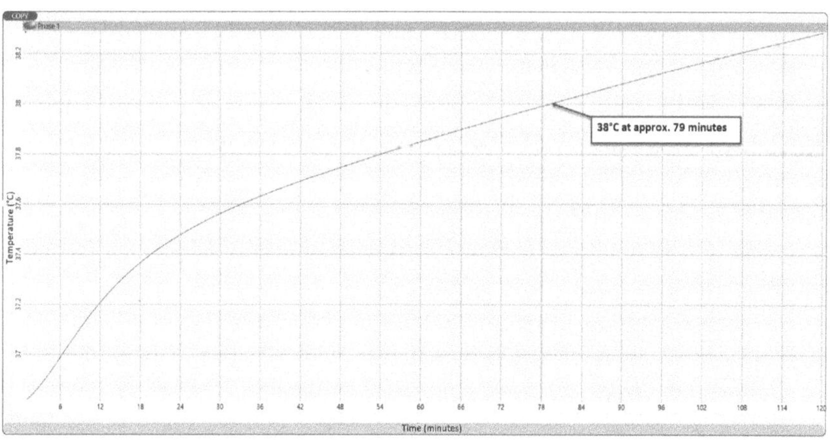

Figure 6.2 Predicted heat strain graph showing predicted core temperature for the scenario.

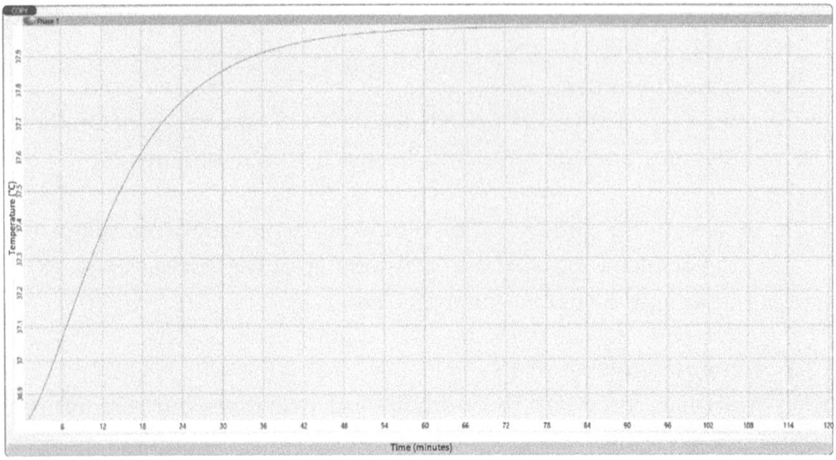

Figure 6.3 Predicted heat strain graph after increase in airflow in scenario.

- Employees need to rest, rehydrate, and recuperate. In some cases, resting in the work environment is the only option, and it can provide some relief for the worker. In a hot environment, the level of recovery is much slower and less efficient than in an air-conditioned environment. The latter could be a basic inexpensive portable structure with an air conditioner, water supply, and seating. There are also scenarios where work groups undertake tasks that are predominantly carried out in the outdoor work environment and are constantly moving from one location to another. In these situations, access to airconditioned vehicles or at a minimum a simple shade structure readily available from hardware stores or large umbrellas can provide relief from solar radiation.
- In some situations where engineering controls are insufficient to protect the individual or control options are limited, work—rest regimes are sometimes necessary. Heat stress indices, such as WBGT, PHS, or thermal work limit (TWL), can assist in determining duration of work and rest periods.
- Following protocols that allow unacclimatized personnel time to acclimatize with light tasks allocated for the first few days in the field.
- Training: Workers need to understand heat stress, to identify symptoms and the potential onset of heat-related illness and the controls available.

- Employ a "buddy system" where coworkers maintain an awareness of their work colleagues and watch for potential onset of symptoms.
- Adopting a self-determination or pacing approach of the work to meet the conditions.

  Personal protective equipment (PPE):
- Cooling vests with either "phase change" cooling inserts (not ice) have been successfully utilized under some conditions. Ice or chilled water can result in contraction of the blood vessels, reducing the cooling effect of the garment, and so is not the preferred approach.
- Adopting the appropriate type of clothing suited to the environment and task, i.e.:
  - Lightweight material and vented to allow air flow.
  - For tasks where radiant heat is an issue, the use of reflective material and/or slightly heavier weight weave to provide some insulation.

4.2 Occupational population studies of heat—health/injury management

Protocols for heat—health management are common and are often industry-specific, such as in construction, professional sports, and the military. However, high-quality peer-reviewed literature on the effectiveness of interventions in real-world settings is relatively sparse.

In an empirical study, self-pacing as a protective mechanism against heat strain was demonstrated to be effective (Miller et al., 2011). There was evidence that this applied even for comparatively uneducated workers.

Rowlinson and Jia (2014) used a PHS model to develop localized, threshold-based heat stress management guidelines for the construction industry. This was aimed at both health protection and productivity. Yang and Chan (2017) reviewed heat stress intervention research in construction and identified a range of methodological limitations, including arbitrary sampling methods. A research framework was proposed.

Chicas et al. (2020) reviewed cooling intervention studies among outdoor workers in various industries. There was evidence that adverse health effects may be mitigated by multilayered approaches, namely targeted cooling interventions and administrative controls such as rest cycles. Ioannou et al. (2021) monitored 238 experienced and acclimatized individuals in various industrial sectors. Heat strain mitigating strategies focusing on hydration, work—rest cycles, and ventilated clothing were found to be effective.

The effectiveness of heat-related injury prevention approaches has not yet been established (Spector et al., 2019) although a conceptual framework for injury occurrence in hot conditions has been described (Varghese et al., 2018).

5. Concluding remarks

The impact of heat exposure in the occupational environment has been recognized for millennia. It is undeniable that this risk has and will continue to have a deleterious impact on the health of the working population and productivity across all types of industries. Studies outlined in this chapter have detailed the increasing level of illness, injury, and fatalities that continue to occur.

While climate change and the ever increasing occurrence of episodes of extreme temperatures will further exacerbate the situation, this impact is not just a result of increasing temperatures. It is a combination of several risk factors contained with the areas of the type of work, the worker, and the workplace.

Many of these are preventable or readily mitigated if identified and addressed early in the process. However, the lack of awareness of the tools to assess and the general heat stress prevention measures available continue to be a barrier in the effective control of the thermal environment.

The raising of this awareness in all levels of the workforce would be a major step toward the reduction of heat-related illnesses and injuries in the future.

References

ACGIH. (2021). Heat stress and heat strain. In *Threshold limit values for chemical substances and physical agents*. Cincinnati, OH: American Conference of Governmental Industrial Hygienists.

Adam-Poupart, A., Smargiassi, A., Busque, M. A., Duguay, P., Fournier, M., Zayed, J., & Labrèche, F. (2015). Effect of summer outdoor temperatures on work-related injuries in Quebec (Canada). *Occupational and Environmental Medicine, 72*(5), 338–345.

Binazzi, A., Levi, M., Bonafede, M., Bugani, M., Messeri, A., Morabito, M., Marinaccio, A., & Baldasseroni, A. (2019). Evaluation of the impact of heat stress on the occurrence of occupational injuries: Meta-analysis of observational studies. *American Journal of Industrial Medicine, 62*(3), 233–243.

Bonafede, M., Marinaccio, A., Asta, F., Schifano, P., Michelozzi, P., & Vecchi, S. (2016). The association between extreme weather conditions and work-related injuries and diseases. A systematic review of epidemiological studies. *Annali dell'Istituto superiore di sanita, 52*(3), 357–367.

Borg, M. A., Xiang, J., Anikeeva, O., Pisaniello, D., Hansen, A., Zander, K., Dear, K., Sim, M. R., & Bi, P. (2021). Occupational heat stress and economic burden: A review of global evidence. *Environmental Research, 195*, 110781.

Brake, D. J., & Bates, G. P. (2002). Limiting metabolic rate (thermal work limit) as an index of thermal stress. *Applied Occupational and Environmental Hygiene, 17*(3), 176–186.

Brake, D. J., Donoghue, A. M., & Bates, G. P. (1998). A new generation of health and safety protocols for working in heat. In *Proceedings of Queensland mining industry health and safety conference: New opportunities* (pp. 91–100). Queensland: Yeppoon, 30 August-2 September 1998.

Calkins, M. M., Bonauto, D., Hajat, A., Lieblich, M., Seixas, N., Sheppard, L., & Spector, J. T. (2019). A case-crossover study of heat exposure and injury risk among outdoor construction workers in Washington State. *Scandinavian Journal of Work, Environment & Health, 45*(6), 588–599.

Casa, D. J. (Ed.). (2018). *Sport and physical activity in the heat. Maximising performance and Safety*. Springer International Publishing.

Casa, D. J., Armstrong, L. E., Carter, R., III., Lopez, R., McDermott, B., & Scriber, K. (2010). Historical perspectives on medical care for heat stroke, Part 1: Ancient times through the nineteenth century. A review of the literature. *Athletic Training and Sports Health Care, 2*(3), 132–138.

Chicas, R., Xiuhtecutli, N., Dickman, N. E., Scammell, M. L., Steenland, K., Hertzberg, V. S., & McCauley, L. (2020). Cooling intervention studies among outdoor occupational groups: A review of the literature. *American Journal of Industrial Medicine, 63*(11), 988–1007.

Cian, C., Koulmann, N., Barraud, P. A., Raphel, C., Jimenez, C., & Melin, B. (2000). influence of variations in body hydration on cognitive function: Effect of hyper-hydration, heat stress, and exercise-induced dehydration. *Journal of Psychophysiology, 14*, 29–36.

Di Corleto, R., Firth, I., & Matè, J. (2013). *A guide to managing heat stress: Developed for use in the Australian environment*. Tullamarine, Victoria: Australian Institute of Occupational Hygienists (AIOH).

Dunne, J. P., Stouffer, R. J., & John, J. G. (2013). Reductions in labour capacity from heat stress under climate warming. *Nature Climate Change, 3*, 563–566.

Fatima, S. H., Rothmore, P., Giles, L. C., Varghese, B. M., & Bi, P. (2021). Extreme heat and occupational injuries in different climate zones: A systematic review and meta-analysis of epidemiological evidence. *Environment International, 148*, 106384.

Flouris, A. D., Dinas, P. C., Gloannou, L., Nybo, L., Havenith, G., Kenny, G. P., & Kjellstrom, T. (2018). Workers' health and productivity under occupational heat strain: A systematic review and meta-analysis. *Lancet Planet Health, 2*(12), e521–e531.

Fogleman, M., Fakhrzadeh, L., & Bernard, T. E. (2005). The relationship between outdoor thermal conditions and acute injury in an aluminum smelter. *International Journal of Industrial Ergonomics, 35*(1), 47–55.

Ganio, M. S., Armstrong, L. E., Casa, D. J., McDermott, B. P., Lee, E. C., Yamamoto, L. M., Marzano, S., Lopez, R. M., Jimenez, L., Le Bellego, L., Chevillotte, E., & Lieberman, H. R. (2011). Mild dehydration impairs cognitive performance and mood of men. *British Journal of Nutrition, 106*(10), 1535–1543.

Garzon-Villalba, X. P., Mbah, A., Wu, Y., Hiles, M., Moore, H., Schwartz, S. W., & Bernard, T. E. (2016). Exertional heat illness and acute injury related to ambient wet bulb globe temperature. *American Journal of Industrial Medicine, 59*(12), 1169–1176.

Goldman R. F. (2001). Introduction to Heat-Related Problems in Military Operations. In: Pandolf KB, Burr RE, eds. *Medical Aspects of Harsh Environments*. Washington: Borden Institute: 3-49.

Gun, R. (2019). Deaths in Australia from work-related heat stress, 2000–2015. *International Journal of Environmental Research and Public Health, 16*(19), 3601.

Hansen, A., Pisaniello, D., Varghese, B., Rowett, S., Hanson-Easey, S., Bi, P., & Nitschke, M. (2018). What can we **learn** about workplace heat stress management from a safety regulator complaints database? *International Journal of Environmental Research and Public Health, 15*(3), 459.

Hansen, A. L., Williams, S., Hanson-Easey, S., Varghese, B. M., Bi, P., Heyworth, J., Nitschke, M., Rowett, S., Sim, M. R., & Pisaniello, D. L. (2020). Using a qualitative phenomenological approach to inform the etiology and prevention of occupational heat-related injuries in Australia. *International Journal of Environmental Research and Public Health, 17*(3), 846.

Hanson, M. A., & Graveling, R. A. (1997). *Development of a code of practice for work in hot and humid conditions in coal mines (Research Report TM 97/06).* Edinburgh: Institute of Occupational Medicine.

Houghten, F. C., & Yaglou, C. P. (1924). Cooling effect in human beings produced by various air velocities. *Transactions of the American Society of Heating and Ventilating Engineers, 30,* 193.

Ioannou, L. G., Mantzios, K., Tsoutsoubi, L., Nintou, E., Vliora, M., Gkiata, P., Dallas, C. N., Gkikas, G., Agaliotis, G., Sfakianakis, K., Kapnia, A. K., Testa, D. J., Amorim, T., Dinas, P. C., Mayor, T. S., Gao, C., Nybo, L., & Flouris, A. D. (2021). Occupational heat stress: Multi-country observations and interventions. *International Journal of Environmental Research and Public Health, 18*(12), 6303.

ISO 9886. (2004). *Ergonomics — evaluation of thermal strain by physiological measurements.* Geneva: International Organization for Standardization.

Kenny, G. P., & Jay, O. (2013). Thermometry, calorimetry, and mean body temperature during heat stress. *Comprehensive Physiology, 3*(4), 1689–1719.

Kjellstrom, T., Kovats, R. S., Lloyd, S. J., Holt, T., & Tol, R. S. (2009). The direct impact of climate change on regional labor productivity. *Archives of Environmental & Occupational Health, 64*(4), 217–227.

Kjellstrom, T., Lemke, B., Otto, M., Hyatt, O., Briggs, D., & Freyberg, C. (2014). *Threats to occupational health, labor productivity and the economy from increasing heat during climate change: An emerging global health risk and a challenge to sustainable development and social equity.* Mapua: Health and Environment International Trust.

Lao, J., Hansen, A., Nitschke, M., Hanson-Easey, S., & Pisaniello, D. (2016). Working smart: An exploration of council workers' experiences and perceptions of heat in Adelaide, South Australia. *Safety Science, 82,* 228–235.

Levi, M., Kjellstrom, T., & Baldasseroni, A. (2018). Impact of climate change on occupational health and productivity: A systematic literature review focusing on workplace heat. *La Medicina del lavoro, 109*(3), 163–179.

Marinaccio, A., Scortichini, M., Gariazzo, C., Leva, A., Bonafede, M., De' Donato, F. K., Stafoggia, M., Viegi, G., Michelozzi, P., & Collaborative Group, B. E. E. P. (2019). Nationwide epidemiological study for estimating the effect of extreme outdoor temperature on occupational injuries in Italy. *Environment International, 133*(Pt A), 105176.

Martínez-Solanas, È., López-Ruiz, M., Wellenius, G. A., Gasparrini, A., Sunyer, J., Benavides, F. G., & Basagaña, X. (2018). Evaluation of the impact of ambient temperatures on occupational injuries in Spain. *Environmental Health Perspectives, 126*(6), 067002.

Ma, R., Zhong, S., Morabito, M., Hajat, S., Xu, Z., He, Y., Bao, J., Sheng, R., Li, C., Fu, C., & Huang, C. (2019). Estimation of work-related injury and economic burden attributable to heat stress in Guangzhou, China. *The Science of the Total Environment, 666,* 147–154.

McInnes, J. A., Akram, M., MacFarlane, E. M., Keegel, T., Sim, M. R., & Smith, P. (2017). Association between high ambient temperature and acute work-related injury: A case-crossover analysis using workers' compensation claims data. *Scandinavian Journal of Work, Environment & Health, 43*(1), 86–94.

Miller, V., Bates, G., Schneider, J. D., & Thomsen, J. (2011). Self-pacing as a protective mechanism against the effects of heat stress. *Annals of Occupational Hygiene, 55*(5), 548–555.

Morabito, M., Cecchi, L., Crisci, A., Modesti, P. A., & Orlandini, S. (2006). Relationship between work-related accidents and hot weather conditions in Tuscany (central Italy). *Industrial Health, 44*(3), 458–464.

Moran, D. S., Pandolf, K. B., Epstein, Y., Heled, Y., Shapiro, Y., & Gonzalez, R. R. (2005). Validation of the environmental stress index (ESI) for physiological variables. In Y. Tochihara, & T. Ohnaka (Eds.), *Environmental ergonomics* (pp. 487–494). Amsterdam: Elsevier.

Moss, K. N. (1923). Some effects of high air temperatures and muscular exertion upon colliers. In *Proceedings of the Royal Society of London. Series B, Containing papers of a biological character* (pp. 181–200). Vol. 95, No. 666 (Aug. 1, 1923).

Osborne, E., Vernon, H. M., & Muscio, B. (1922). The influence of temperature and other conditions on the frequency of industrial accidents. In *Two contributions to the study of accident causation*. London: H.M. Stationery Office.

Queensland Coal Mining Safety and Health Regulation. (2017). Chapter 4 Underground mines. Part 12 Working environment. Division 2, Heat stress management. Section 369(3).

Rameezdeen, R., & Elmualim, A. (2017). The impact of heat waves on occurrence and severity of construction accidents. *International Journal of Environmental Research and Public Health, 14*(1), 70.

Riccò, M., Garbarino, S., & Bragazzi, N. L. (2019). Migrant workers from the Eastern-Mediterranean region and occupational injuries: A retrospective database-based analysis from North-Eastern Italy. *International Journal of Environmental Research and Public Health, 16*(4), 673.

Rowlinson, S., & Jia, Y. A. (2014). Application of the predicted heat strain model in development of localized, threshold-based heat stress management guidelines for the construction industry. *Annals of Occupational Hygiene, 58*(3), 326–339.

Sahu, S., Sett, M., & Kjellstrom, T. (2013). Heat exposure, cardiovascular stress and work productivity in rice harvesters in India: Implications for a climate change future. *Industrial Health, 51*(4), 424–431.

Sawka, M. N., Leon, L. R., Montain, S. J., & Sonna, L. A. (2011). Integrated physiological mechanisms of exercise performance, adaptation, and maladaptation to heat stress. *Comprehensive Physiology, 1*(4), 1883–1928.

Sayers, R. R., & Davenport, S. J. (2011). Review of the literature on the physiological effects of abnormal temperatures and humidities, 1896–1970 *Public Health Reports, 42*(No. 14), 933–996. Apr. 8, 1927).

Sheng, R., Li, C., Wang, Q., Yang, L., Bao, J., Wang, K., Ma, R., Gao, C., Lin, S., Zhang, Y., Bi, P., Fu, C., & Huang, C. (2018). Does hot weather affect work-related injury? A case-crossover study in Guangzhou, China. *International Journal of Hygiene and Environmental Health, 221*(3), 423–428.

Spector, J. T., Bonauto, D. K., Sheppard, L., Busch-Isaksen, T., Calkins, M., Adams, D., Lieblich, M., & Fenske, R. A. (2016). A case-crossover study of heat exposure and injury risk in outdoor agricultural workers. *PloS One, 11*(10). e0164498.

Spector, J. T., Masuda, Y. J., Wolff, N. H., Calkins, M., & Seixas, N. (2019). Heat exposure and occupational injuries: Review of the literature and implications. *Current Environmental Health Reports, 6*(4), 286–296.

Su, Y., Cheng, L., Cai, W., Lee, J., Zhong, S., Chen, S., Li, T., Huang, X., & Huang, C. (2020). Evaluating the effectiveness of labor protection policy on occupational injuries caused by extreme heat in a large subtropical city of China. *Environmental Research, 186*, 109532.

Tawatsupa, B., Yiengprugsawan, V., Kjellstrom, T., Berecki-Gisolf, J., Seubsman, S. A., & Sleigh, A. (2013). Association between heat stress and occupational injury among Thai workers: Findings of the Thai Cohort Study. *Industrial Health, 51*(1), 34–46.

UNDP. (2016). *Climate change and labour: Impacts of heat in the workplace. Climate change, workplace environmental conditions, occupational Health risks, and productivity – an emerging global challenge to decent work, sustainable development and social equity*. Paper prepared by Climate Vulnerable Forum, United Nations Development Programme.

Varghese, B. M., Barnett, A. G., Hansen, A. L., Bi, P., Hanson-Easey, S., Heyworth, J. S., Sim, M. R., & Pisaniello, D. L. (2019a). The effects of ambient temperatures on the risk of work-related injuries and illnesses: Evidence from Adelaide, Australia 2003–2013. *Environmental Research, 170*, 101–109.

Varghese, B. M., Barnett, A. G., Hansen, A. L., Bi, P., Nairn, J., Rowett, S., Nitschke, M., Hanson-Easey, S., Heyworth, J. S., Sim, M. R., & Pisaniello, D. L. (2019c). Characterising the impact of heatwaves on work-related injuries and illnesses in three Australian cities using a standard heatwave definition-Excess Heat Factor (EHF). *Journal of Exposure Science & Environmental Epidemiology, 29*(6), 821–830.

Varghese, B. M., Hansen, A., Bi, P., & Pisaniello, D. (2018). Are workers at risk of occupational injuries due to heat exposure? A comprehensive literature review. *Safety Science, 110*, 380–392. Part A.

Varghese, B. M., Hansen, A., Nitschke, M., Nairn, J., Hanson-Easey, S., Bi, P., & Pisaniello, D. (2019b). Heatwave and work-related injuries and illnesses in Adelaide, Australia: A case-crossover analysis using the excess heat factor (EHF) as a universal heatwave index. *International Archives of Occupational and Environmental Health, 92*(2), 263–272.

Vernon, H. M. (1932). The measurement of radiant heat in relation to human comfort. *Journal of Industrial Hygiene, 14*, 95–111.

Wakefield, E. G., & Hall, W. W. (1927). Heat injuries: A preparatory study for experimental heat-stroke. *Journal of the American Medical Association, 89*, 92–95.

Wilson, M. M., & Morley, J. E. (2003). Impaired cognitive function and mental performance in mild dehydration. *European Journal of Clinical Nutrition, 57*(Suppl. 2), S24–S29.

Xiang, J., Bi, P., Pisaniello, D., & Hansen, A. (2014a). Health impacts of workplace heat exposure: An epidemiological review. *Industrial Health, 52*(2), 91–101.

Xiang, J., Bi, P., Pisaniello, D., & Hansen, A. (2014c). The impact of heatwaves on workers' health and safety in Adelaide, South Australia. *Environmental Research, 133*, 90–95.

Xiang, J., Bi, P., Pisaniello, D., Hansen, A., & Sullivan, T. (2014b). Association between high temperature and work-related injuries in Adelaide, South Australia, 2001–2010. *Occupational and Environmental Medicine, 71*(4), 246–252.

Xiang, J., Hansen, A., Pisaniello, D., & Bi, P. (2015). Perceptions of workplace heat exposure and controls among occupational hygienists and relevant specialists in Australia. *PloS One, 10*(8). e0135040.

Xiang, J., Hansen, A., Pisaniello, D., Dear, K., & Bi, P. (2018). Correlates of occupational heat-induced illness costs: Case study of South Australia 2000 to 2014. *Journal of Occupational and Environmental Medicine, 60*(9), e463–e469.

Yaglou, C. P., & Minard, D. (1957). Control of heat casualties at military training centres. *American Medical Association Archives of Industrial Health, 16*, 302–306. and 405 (corrections).

Yang, Y., & Chan, A. P. (2017). Heat stress intervention research in construction: Gaps and recommendations. *Industrial Health, 55*(3), 201–209.

Young, W. J., Breinl, A., Harris, J. J., & Osborne, W. A., (1920). Effect of exercise and humid heat upon pulse rate, blood pressure, body temperature, and blood concentration. Proceedings of the Royal Society of London. Series B, containing papers of a biological character, Vol. 91, No. 636 (Jan. 1), pp. 111–126.

Zander, K. K., Botzen, W. J. W., Oppermann, E., Kjellstrom, T., & Garnett, S. T. (2015). Heat stress causes substantial labour productivity loss in Australia. *Nature Climate Change, 5*, 647–651.

Further reading

Chrenko, F.A. (1974) (Ed.) Bedford's basic principles of ventilation and heating, H.K. Lewis.

ISO. (2001). Ergonomics of the thermal environment—medical supervision of individuals exposed to extreme hot or cold environments. Geneva: International Organization for Standardization.

Riccò, M. (2018). Air temperature exposure and agricultural occupational injuries in the Autonomous Province of Trento (2000–2013, North-Eastern Italy). *International Journal of Occupational Medicine and Environmental Health, 31*(3), 317–331.

Varghese, B. M., Barnett, A. G., Hansen, A. L., Bi, P., Heyworth, J. S., Sim, M. R., Hanson-Easey, S., Nitschke, M., Rowett, S., & Pisaniello, D. L. (2019d). Geographical variation in risk of work-related injuries and illnesses associated with ambient temperatures: A multi-city case-crossover study in Australia. *The Science of the Total Environment, 687*, 898–906, 2005–2016.

CHAPTER 7

Heat exposure and cardiorespiratory health

Shanshan Li, Pei Yu, Suying Guo and Yuming Guo
Climate, Air Quality Research Unit, School of Public Health and Preventive Medicine, Monash University, Melbourne, Australia

1. Introduction

The globe is experiencing a period of warming. The Special Report on climate change and land of the Intergovernmental Panel on Climate Change (IPCC) stated that, from 1850–1900 to 2006–15, mean land surface air temperature has increased by 1.53°C and the global mean surface (land and ocean) temperature has increased by 0.87°C (Intergovernmental Panel on Climate Change (IPCC)). The frequency of hot days, hot nights, and heatwaves have also increased over the previous 50 years. Heatwave is a type of sustained extreme heat, which could lead to more adverse health outcomes than high but not extreme temperature conditions. To date, there is no single global heatwave definition. In the research, the heatwave is mainly defined by the temperature metric (i.e., maximum temperature, mean temperature, apparent temperature, and heat index), the intensity (i.e., relative and absolute temperature threshold), and the duration (Xu et al., 2016). In general, most studies used maximum temperature as the temperature indicator adopted an absolute temperature value as the threshold, or used mean temperature as the temperature indicator adopted a relative threshold (percentiles) to define a heat wave. A heat episode with the duration over several days, ranging from 1 to 7 days in most studies, is defined as a heatwave. A study reviewed the different definitions of heat wave in the research, and the results were as given in Table 7.1 (Xu et al., 2016).

As the trend of global climate change, there are subsequent health impacts caused by the increasing extreme heat events (Mitchell et al., 2016), such as the 2003 European heat wave, which caused approximately 70,000 excess deaths (Robine et al., 2008). Many studies indicated that high temperatures were significantly associated with increased hospital admissions (Zhao et al., 2019; Phung et al., 2016; Arbuthnott & Hajat, 2017) and

Heat Exposure and Human Health in the Context of Climate Change
ISBN 978-0-12-819080-7
https://doi.org/10.1016/B978-0-12-819080-7.00005-7
© 2023 Elsevier Inc.
All rights reserved.

Table 7.1 Heatwave definitions used in the existing literature.

Temperature metric	Intensity	Duration
Maximum temperature	Absolute temperature: 25°C, 30°C, 32°C, 35°C, 37°C Relative threshold: 95th, 97.5th, 98th Exceeding the average maximum temperature by 5°C	2–6 days
Mean temperature	Relative threshold: 90th, 95th, 97th, 98th, 99th, 99.5th (some limited the temperature in the warm season)	2–5 days
Specific events	1995 Chicago heatwave 2003 European heatwave 2004 Brisbane heatwave 2006 California heatwave 2006 French heatwave 2010 Ahmedabad heatwave 2011 Sydney heatwave 2011 UK heatwave	Most ≥7 days

deaths (Kim et al., 2014; Luo et al., 2019; Tham et al., 2020). Globally, the heat was estimated to account for 0.42% of deaths in 384 locations of 13 countries during the period from 1985 to 2012 (Gasparrini et al., 2015). In the United Kingdom, it is estimated that around 2000 heat-related deaths occur annually, and the magnitude is predicted to rise to over 7000 by 2050 as a result of more frequent and severe heatwaves and the aging population (Hajat et al., 2014).

Although there is adaptation to the high temperature among the population, which shows a decreasing trend of disease risks and deaths over the recent several years (Ng et al., 2016), numerous studies have documented that the heat exposure has a significant impact on the cardiorespiratory health outcomes. There are many systematic review and metaanalysis studies (Xu et al., 2016; Phung et al., 2016; Arbuthnott & Hajat, 2017; Tham et al., 2020; Luo et al., 2019; Kim et al., 2014; Turner et al., 2012; Odame et al., 2018; Gronlund et al., 2018; Bunker et al., 2016; Astrom et al., 2011; Basu, 2009; Xu et al., 2012; Xu et al., 2014; Zanobetti & O'Neill, 2018), and most of them found results that the impact was adverse and that heatwaves had greater influence.

For the mortality, for example, a metaanalysis study indicated that for every 1°C increase beyond the reference temperature point (including temperature stratified by the predetermined percentiles, specific thresholds, and minimum mortality temperature), the rate of cardiovascular mortality increased by 3% (95% CI: 3%–4%) and the rate of respiratory mortality increased by 2% (95% CI: 1%–3%) in China (Luo et al., 2019).

For the morbidity, a study investigated the association between the heat and the morbidity for cardiorespiratory health in New York City, United States, during 1991–2004 (Lin et al., 2009). The results showed that the risk of hospital admissions increased by 2.7% (95% CI: 1.3%–4.2%) for respiratory diseases on the same day with every 1°C increase in daily mean temperature above 28.9°C and increased by 3.6% (95% CI: 0.3%–6.9%) for cardiovascular diseases (CVDs) with a 3-day delayed effect, with every 1°C increase in daily mean temperature above 29.4°C. Besides, the risk increased by 2.1% (95% CI: 1.1%–3.1%) and 1.4% (95% CI: 0.4%–2.4%) for respiratory admissions on the same day and at 1 day later, with each 1°C increase in daily mean apparent temperature above 31.7°C. The risk for cardiovascular admissions was also found by 2.5% (95% CI: 0.6%–4.4%), 2.1% (95% CI: 0.5%–3.9%), and 3.6% (95% CI: 1.9%–5.3%), at 1, 2, and 3 days later, respectively, with each 1°C increase in daily mean apparent temperature above 35.6°C.

There are also studies that mainly focus on the symptoms rather than the diseases. For example, in Finland, approximately 12% of the people aged 25–74 years reported the heat-related cardiorespiratory symptoms, including shortness of breath, prolonged cough or coughing bouts, wheezing of breath, increased excretion of mucus from the lungs, chest pain, and cardiac arrhythmias (Nayha et al., 2017).

Among the total population, some groups are more vulnerable to the high temperature, including the elderly, children, pregnant women, those with a lower socioeconomic status population, those with chronic health conditions, and outdoor workers (Gronlund et al., 2018; Zanobetti & O'Neill, 2018). Because of the aging society, the population of the elderly people become larger and their health attributes more to the global health burden, so that there are abundant studies that have chosen the elderly as the study group. For the elderly people, many studies showed that the heat and temperature variability could increase their mortality risk, especially from cardiovascular and respiratory diseases (Leyva et al., 2017). For example, a metaanalysis study reported that 1°C temperature rise could increase the cardiovascular mortality by 3.44% (95% CI: 3.10%–3.78%) and the respiratory mortality by 3.60% (95% CI: 3.18%–4.02%) (Bunker et al., 2016).

Despite ambient heat, indoor temperatures can also affect the health outcomes and should be considered when selecting the data of temperature condition to evaluate the heat impact on the population. Some studies suggested people in high-income countries spend approximately 90% of their time in indoor spaces (Klepeis et al., 2001; Schweizer et al., 2007), and over 25% of homes in the United Kingdom overheated during 2007–08 with more than 5% of occupied indoor hours exceeding the relevant threshold temperature, even in normal summer conditions (Beizaee et al., 2013). Tham et al. reviewed the past research and found that all the selected studies reported worsening respiratory symptoms with increasing higher indoor temperatures and that the effect was stronger in the presence of higher concentration of indoor air pollutants (Tham et al., 2020).

Cardiovascular and respiratory diseases are two of the key negative health outcomes associated with heat exposure. This chapter aims to summarize current epidemiological evidence regarding the effects of heat exposure on cardiorespiratory health outcomes, such as coronary heart disease, myocardial infarction (MI), heart failure, chronic obstructive pulmonary disease (COPD), and asthma.

2. Heat exposure and cardiovascular diseases

CVDs are actually among the main causes of death in the world (Go et al., 2013) and the main cause of death during the heatwave. An evaluation of deaths during two heatwaves in Milwaukee likewise revealed CVDs as the primary cause in 51% and 64% of total deaths in the two heatwaves, respectively (Welsskopf et al., 2002). The death can be attributed to the physiological reactions to the high temperature exposure, including increased core body temperature and heart rate, a shift in blood flow from central organs to skin, and greater sweating and associated dehydration, with insufficient external liquid supply (Coris et al., 2004).

As many studies have shown, the cardiovascular morbidity can also be influenced by the heat. A study in Brisbane, Australia, found that the ambulance attendances increased by 29.5% (95% CI: 0.4%–67.0%) for CVDs after the heat exposure under a daily maximum temperature at or above 37°C for two or more consecutive days during 2000–07, and the risk was greatest among people who were 65–74 years old (163.7%, 95% CI: 56.0%–345.8%) (Turner et al., 2013). Another study in Shanghai, China, reported similar result with risk increased by 8% (95% CI: 5%–11%) for CVDs during the period of at least seven consecutive days with daily maximum temperature above 35.0°C and daily mean temperatures above 97th percentile (Ma et al., 2011).

2.1 Heat exposure and cardiovascular mortality

2.1.1 Heat exposure and total cardiovascular mortality

Many studies have investigated associations between heat exposure and cardiovascular mortality. Although some reported contrasting or conflicting results (Basu et al., 2005; Bai et al., 2014; Michelozzi et al., 2009; Ho & Wong, 2019), most studies showed that the risk of CVD mortality increased with higher temperature. A meta-analysis study suggested that the combined RR for total CVD mortality was 1.11 (95% CI: 1.05–1.18) per 1°C increase in mean daily temperature in rural locations (Odame et al., 2018). A study in New York City, United States, reported the risk of CVD mortality increased by 6% (95% CI: 3%–8%) and 23.4% (95% CI: 19%–28%) on hot days (mean daily temperatures above 34°C in the dry areas and 32°C in the moist areas) and heatwave days (hot days for more than three consecutive days) in summer, respectively, during 1991–2004 (Sheridan & Lin, 2014).

Vulnerable groups are the popular cohort in the research, such as the elderly and children. For example, the elderly people, even in the absence of obvious heart diseases, may be more susceptible to heat, because they have reductions in the amount of blood pumped per heartbeat (stroke volume) and a decreased capacity for both vasodilation and vasoconstriction when they are exposed to the heat (Charkoudian, 2010; Kenney et al., 2014). A study in Beijing, China, reported mortality due to CVD increased by 24% among the elderly ($\geq$65 years) and increased by 22% among females during the heat periods when the daily mean temperature was above 27.7°C (93rd percentile) lasting for more than 5 days, compared with the condition under the minimum mortality temperature (Dong et al., 2016). In the same region, there was a 94% increase among the elderly and a 104% increase among females on the 10th day after the daily maximum temperatures began to be consecutively higher than 33°C (Yin & Wang, 2017).

The mortality can also be affected if the temperature suddenly increased in a short term. For example, a study in Brisbane, Australia, explored the association between daily temperature changes and the risk of CVD mortality and found that the risk increased 35% when the temperature increased more than 3°C in 24 h (Guo et al., 2011). A 1.9°C increase in maximum temperature change was found to be associated with 8% (95% CI: 2%–15%, lag 0) increase in total CVD mortality in eastern China (Cheng et al., 2014). Also, the 1.9°C increase in maximum temperature change corresponded to a 3% increase at lag 0 (95% CI: 0%–6%) among males and lag 3 (95%CI: 1%–5%) among elderly people (aged 65 years or older) (Cheng et al., 2014). Another study indicated that the number of deaths from CVDs would increase by 0.23% with an increment of monthly temperature by 1% (Liao et al., 2010).

2.1.2 Heat exposure and subtypes of cardiovascular mortality

In addition to the total CVD mortality, some studies focused on the deaths caused by specific diseases. Coronary heart disease (CHD) is the second leading cause of cardiovascular death in China and accounts for 22% of cardiovascular deaths in urban areas and 13% in rural areas (Zhang et al., 2008). There are a few studies with the health outcome of CHD in China. A study incorporated 272 Chinese cities from 2013 to 2015 and found the mortality risk increased 22% for total CVDs (95% CI: 16%–28%) and 19 for CHD (95% CI: 11%–28%) when the daily mean temperature was above 29°C (97.5th percentile) (Chen et al., 2018). The study in Beijing during 2000–01 also reported an 38% increase in CHD mortality (95% CI: 20%–60%) when the daily mean temperature was above 30.5°C (99th percentile), compared with 27.0°C (90th percentile) (Tian et al., 2012). Additionally, when the extreme high temperature lasted for more than 2 days, the risk increased by about 31% (Tian et al., 2013).

A study in Prague and southern Bohemia, Czech Republic, explored the deaths during 1994–2009. The results showed that the risk of the mortality due to CVD, ischemic heart disease (IHD), chronic ischemic heart disease (CIHD), and atherosclerosis–atherosclerotic vascular disease (ASVD) increased by 10.8% (95% CI: 7.5%–14.1%), 7.0% (95% CI: 2.1%–12.2%), 10.9% (95% CI: 4.7%–17.4%), and 19.4% (95% CI: 11.6%–27.7%), respectively, when the daily mean temperature was above 22.5°C (90th percentile) (Urban & Kyselý, 2014). A study in four cities of South Korea also found the similar association between the increased risk of cardiac failure and cardiac disease and the heat (Lim et al., 2012). In addition, a Chinese study reported that the risk of IHD mortality during 2004–08 significantly 17% increased at lag 0–2 (95% CI: 10%–25%), 24% at lag 0–13 (95% CI: 9%–39%), and 18% at lag 0–20 (95% CI: 0%–36%) days when the study group was exposed to the 99th percentile of the daily mean temperature, compared with the 90th percentile temperature (Guo et al., 2013). The mortality for cardiac disease and stroke was also found to increase by 3% (95% CI: 1%–5%) at temperatures above 42°C, among people aged 65 years and older (Harlan et al., 2014).

2.2 Heat exposure and cardiovascular morbidity

Sometimes the pattern of hospital admissions can be completely different from that of mortality, which means that the association between the heat exposure and CVD morbidity may be negative or not statistically significant (Linares & Diaz, 2008). For example, a study showed that increased indoor heat was not significantly associated with the cardiovascular distress calls in New York City, United States, from July to August 2013 (Uejio et al., 2016).

Most studies, however, indicated that the exposure to heat and heatwaves could significantly increase the risk for CVD morbidity, using the data of the hospital admissions, emergency visits, and distress calls. For example, a study in northern Italy from 2002 to 2006 found that the risk of daily emergency dispatches for CVDs increased by 7.93% (95% CI: 3.76%–12.10%) for every 1°C increase when the daily mean apparent temperature exceeded 30°C, and the risk was greater among people who were 75 years and older (11.82%, 95% CI: 6.18%–17.46%) (Alessandrini et al., 2011). During the 2003 heatwave in Paris, France, the daily mean temperature was 38.1°C from 4th to 12th August, with a 134% (95% CI: 60%–241%) increased risk of out-of-hospital cardiac arrest attended by mobile intensive care units among the total study participants (RR = 2.34, 95% CI: 1.60–3.41) among the older people over 60 years old , compared with the 2000–02 and 2004–05 (Empana et al., 2009). Another heatwave in California, United States, from 15th July to 1st August, 2006, slightly increased the risk of the emergency department visits by 2% (95% CI: 1%–3%) and 1% for hospitalizations (95% CI: 0%–2%) for CVD (Knowlton et al., 2009). As for the heatwave in Chicago, United States, in July 1995, the inpatient admission for CVD increased by 22.8% (95% CI: 7.2%–38.3%) (Semenza et al., 1999).

Despite the total CVD morbidity, some studies focus on the specific diseases. Most of them showed significant results of increased risk influenced by the heat, whereas some may demonstrate insignificant, conflicting or contrary results. For example, a study in Sydney, Australia, investigated the CVD risk during 1999–2009, when the daily mean temperature was above 25.3°C (90th percentile). Compared with days with daily mean temperature of 20.4°C, the risk of total CVDs increased 1% (95% CI: 0%–2%), whereas the ORs for specific diagnoses of CVDs were not statistically significant, including IHD, acute MI, acute rheumatic fever, chronic rheumatic heart disease, hypertensive disease, diseases of pulmonary circulation, diseases of arteries, arterioles, and capillaries (Vaneckova & Bambrick, 2013). There is another study focusing on hospitalizations in different cities of California during 1983–98. The results indicated different results in each city: the risk of acute MI, congestive heart failure, angina pectoris, and stroke increased by more than 3% with 3°C increase in minimum temperature in San Francisco and Sacramento, whereas the risk of three decreased in Los Angeles (Ebi et al., 2004).

2.2.1 Myocardial infarction

Although some studies did not find a significant increase in the risk for MI admissions with the increasing temperatures and the heatwave (Knowlton et al., 2009; Bhaskaran et al., 2010), there are research studies reporting statistically significant increase or decrease in the risk with the heat.

Of the present studies, some reported the association was positive between the temperature and the risk for MI. For example, an increased risk for hospital admission due to MI by 1.9% (95% CI: 0.5%–3.3%) for every 1°C increase above the estimated threshold of 20°C was found 1–6 h after the heat exposure in 11 conurbations of England and Wales (Bhaskaran et al., 2012). A study in South Korea demonstrated that the risk of acute MI increased by 7% (95% CI: 5%–10%) for every 1°C increase in maximum temperature above 31.5°C, and increased by 26% (95% CI: 8%–46%) for every 1°C increase in mean temperature above 28.5°C, after adjusting for other climate factors including humidity and air pollutants (Lee et al., 2014). Another study among elderly people in Denver, United States, during 1993–97 also showed that the risk of acute MI significantly increased by 17.5% (95% CI: 2.9%–34.4%) for every 1°C increase in maximum temperature ranging from 16.7 to 37.2°C (Koken et al., 2003). Besides, the change of maximum temperature from 28.3 to 34.2°C could increase the risk of hospitalization for acute MI by 18%. For the symptoms of MI, an increase 71% of ST-segment elevation MI among the older people over 60 years old (95% CI: 3%–186%) was observed during the 2003 heatwave in Paris, France, when the daily mean temperature was 38.1°C from 4th to 12th August (Empana et al., 2009).

In the contrast, there are some studies with the results of significantly decreased risk for MI. For example, a study in Moscow, Russia, found that the risk of hospital admissions for acute MI decreased by 32% (RR = 68%, 95% CI: 48%–95%) if the daily mean temperature was above 22.7°C (97th percentile) for at least five consecutive days. Additionally, the decrease in risk for every 1°C was even greater at higher temperatures (Shaposhnikov et al., 2014). Another study in rural Greek regions reported that during 2003–04, the risk of acute MI decreased 1% with every 1°C increase in daily mean (95% CI: 0%–2%), 1% at maximum (95% CI: 0%–2%), and 1% at minimum (95% CI: 0%–2%) temperature (Misailidou et al., 2006). In addition, the decreased risks were also found between the risk of unstable angina and the daily mean by 2% (95% CI: 1%–3%), maximum by 2% (95% CI: 1%–3%), and minimum by 3% (95% CI: 1%–4%) temperature (Misailidou et al., 2006).

2.2.2 Ischemic heart disease

To investigate the effects of heatwaves on specific CVDs, a study in Australia from 1992 to 2011 suggested that compared with the nonextreme temperatures (the middle 90th percentile of daily maximum temperature), the extreme heat which was over 95th percentile significantly increased the risk of IHD by 17% among people under 65 years old. A tendency for higher rates of IHD admissions on hotter days was also observed, although results were not significant (Webb et al., 2014). In another study in Adelaide, Australia, Nitschke et al. quantified the impact of heatwaves from 1993 to 2006 on morbidity and mortality using ambulance and hospital admission data. They found that the hospital admissions for IHD increased by 8% (95% CI: 1%–15%) among people between 65 and 74 years old, whereas the risk decreased 9% (95% CI: 3%–15%) among people aged 75 years and older during the heatwave period when the daily maximum temperature was above 35°C for three or more consecutive days (Nitschke et al., 2007). In California, United States, a study reported that during 2005–08, with every 5.6°C increase in daily mean apparent temperature from 14.1 to 30.1°C, the risk of IHD increased by 1.7% (95% CI: 0.2%–3.3%) (Basu et al., 2012). The increased risk for inpatient admissions (18.5%, 95% CI: 5.7%–31.4%) was also observed during the period of the heatwave in July 1995 in Chicago, United States, compared with the same period in 1994 (Semenza et al., 1999).

2.2.3 Heart failure

The effect of heat on heart failure is also unsure. An international and multicenter study with the data from Acute Study of Clinical Effectiveness of Nesiritide in Decompensated Heart Failure (ASCEND-HF) demonstrated that among the 7,141 participants, the risk of heart failure significantly increased with the 18% increased at maximum (95% CI: 6%–30%) and 21% at average (95% CI: 10%–32%) temperature on the 2nd–3rd day after the exposure (Das et al., 2014). Also, a change of maximum temperature from 28.3 to 34.2°C could increase the risk of congestive heart failure by 13% (Koken et al., 2003). However, a study found the risk reduced by 63% (95% CI: 14%–89%, $P = .047$) among females over 65 years old in hot weather (over 95th percentile of the daily maximum temperature) in Australia (Webb et al., 2014). Another study in the United States demonstrated that the risk for congestive heart failure among people aged 65 years or older significantly decreased when the daily mean temperature was above the 99th percentile for at least two consecutive days (Bobb et al., 2014).

2.2.4 Other cardiovascular diseases

Dawson et al. investigated the data of hospitalization in Scotland during 1990–2005 and found that the risk of ischemic stroke increased by 2.1% (95% CI: 0.7%–3.5%) per 1°C increase in daily mean temperature after 24 h of the exposure (Dawson et al., 2008).

Another study found that during 2005–08, with every 5.6°C increase in daily mean apparent temperature from 14.1 to 30.1°C in California, United States, the risk of ischemic stroke, cardiac dysrhythmias, and hypotension increased by 2.8% (95% CI: 0.9%–4.7%), 2.8% (95% CI: 0.9%–4.9%), and 12.7% (95% CI: 8.3%–17.4%), respectively (Basu et al., 2012).

A study in Taiwan, China, indicated the risk for acute coronary syndrome, one of the most common causes of death for patients with CVDs, slightly increased when the average daily temperature exceeded 29°C, and the risk significantly increased by 28.4% when the temperature increased from 17 to 27°C (Liang et al., 2008).

The 1995 Chicago heatwave increased the risk of inpatient admissions by 24.0% for hypertensive disease (95% CI: 1.1%–46.9%), by 33.8% for cardiac dysrhythmia (95% CI: 14.1%–53.4%), and by 22.1% for diseases of arteries, arterioles, and capillaries (95% CI: 5.3%–38.9%) (Semenza et al., 1999).

The temperature change was also reported to have an impact on specific CVDs. The change of maximum temperature from 28.3 to 34.2°C could increase the risk of hospitalization for coronary atherosclerosis by 13% and for pulmonary heart disease by 28% (Koken et al., 2003).

2.3 Physiological factors

Biological pathways may help explain mechanisms of the effect of heat on health, which gives tips to the future interventions.

An increase of air temperature could lead to an immediate and delayed increase in heart rate of 3–4 beats per minute among females and could result in ventricular ectopic beats (Schneider et al., 2008). Also, the platelet and red cell counts, blood viscosity, and plasma cholesterol levels were reported to increase during heat stress, which may help explain increased relative mortality under high temperature (Keatinge et al., 1986).

Further, biomarkers such as B-type natriuretic peptide (BNP) and C-reactive protein (CRP) can affect the severity and prognosis of heart failure. A study in the United States focused on the association between ambient temperature and the biomarkers of heart failure among patients with heart failure and systolic dysfunction and found that a 5°C increase in 4-day moving averages of apparent temperature was significantly associated with an 11.4% increase in BNP (95% CI: 1.2%–22.5%) and a 21.6% increase in CRP (95% CI: 2.5%–44.2%) (Wilker et al., 2012).

3. Heat exposure and respiratory diseases

3.1 Heat exposure and respiratory mortality

Compared with the CVDs, there are fewer studies focusing on the possible heat-related respiratory diseases. Unlike the CVDs, the respiratory mortality is not always found significantly increased with the high temperature (Guo et al., 2011; Ho & Wong, 2019; Lam, 2007). For instance, a study among rural population in Bangladesh reported that the percentage of cardiovascular mortality significantly increased by 62.9% (95% CI: 23.2%–115.2%) for every 1°C increase in temperature which was above 30°C, whereas the respiratory mortality did not show a significant association with the heat (Hashizume et al., 2009). Even so, most studies showed significant increase in the risk for respiratory mortality. For example, a metaanalysis study reported that the rate of respiratory mortality increased by 2% (95% CI: 1%–3%) with each 1°C increase beyond the reference temperature point in China (Luo et al., 2019). Among the reviewed studies, a study in 272 main Chinese cities from 2013 to 2015 reported a 36% increase in the mortality risk for total respiratory diseases (95% CI: 24%–48%) and 26% for COPD (95% CI: 14%–39%) when the daily mean temperature was above 29°C (97.5th percentile) (Chen et al., 2018). In addition, the risk of respiratory mortality was reported to increase by 8.6% (95% CI: 3.4%–14.1%) when the mean daily temperatures above 34°C in the dry areas and 32°C in the moist areas in New York City, United States, during 1991–2004 (Sheridan & Lin, 2014). Also, there are research studies about the specific-cause respiratory deaths. For example, in the Sonoran Desert of south-central Arizona, the mortality for COPD and asthma increases 5% at temperatures above 43°C (95% CI: 0%–10%) among people aged 65 years and older (Harlan et al., 2014).

3.2 Heat exposure and respiratory morbidity

Many studies showed that high temperatures had a significant effect on the increasing morbidity for respiratory diseases, with the data of hospital admissions, emergency room visits, and emergency dispatches.

A study in Washington, DC, United States, reported that risks of the emergency department visits and admissions among children aged 5–12 years increased by 0.2% (95% CI: 0.1%–0.4%) and 0.9% (95% CI: 0.4%–1.5%), respectively, with every 1°F (0.56°C) increase in daily maximum temperature, after adjusting the air pollutants and other environmental factors (Babin et al., 2007). Another American study reported that the respiratory hospitalizations in 221 counties increased by 4.3% (95% CI: 3.8%–4.8%) for each 5.6°C increase in daily mean summer temperature during 1999–2008 (Anderson et al., 2013). In the western Pacific region, a Chinese study in Shanghai also found the significant association between increased risk of respiratory diseases and the heat for at least seven consecutive days with daily maximum temperature above 35.0°C and daily mean temperatures above 97th percentile (Ma et al., 2011). Some specific diseases also have increased risk during the period of heat. For example, a study in South Korea indicated that the increasing temperature could accelerate the risk of hospital admissions due to COPD, asthma, and pneumonia (Lim et al., 2012).

Alessandrini et al. found that the risk of emergency dispatches for respiratory diseases was significantly associated with 1°C increase in the mean apparent temperature between 25 and 30°C (increased by 2.74%, 95% CI: 1.34%–4.14%), and the risk was greater when the temperature exceeded 30°C (increased by 4.97%, 95% CI: 0.40%–9.54%), especially among people who were 75 years and older (increased by 7.59%, 95% CI: 2.03%–13.14%) (Alessandrini et al., 2011). Another study suggested that the odds of making a respiratory distress call to paramedics increased (although not significantly) by 63% (95% CI: -2%—168%) when the indoor heat was above a 26°C threshold in New York City, United States (Uejio et al., 2016).

A study in Toronto, Canada, reported that the risk of daily emergency room visits for respiratory diseases increased by 22% (95% CI: 1%—46%) under the daily mean temperature of 28.4°C (99th percentile), compared with 18.2°C (75th percentile) during the period from 2002 to 2010 (Lavigne et al., 2014).

As previoursly discussed, the elderly people are thought to be particularly sensitive to the heat on the respiratory diseases. For example, in London, United Kingdom, found the risk for respiratory admissions among those who were over 75 years old increased by 10.86% (95% CI: 4.44%—17.67%) for each degree increase in daily mean temperature above 23°C (Kovats et al., 2004). Another study toward 12 European cities demonstrated similar associations in most of the cities, between an increase in hospital admissions due to respiratory causes and increasing temperatures, particularly in the older age groups (Michelozzi et al., 2009).

Sometimes even in the same region, different studies may report conflicting results. For example, a study in Brisbane, Australia, found that during 2000—07, a daily maximum temperature at or above 37°C for two or more consecutive days, increased the ambulance attendances by 48.7% (95% CI: 6.4%—107.7%) for respiratory diseases, especially among people between 65 and 74 years old (127.3%, 95% CI: 14.7%—350.3%) (Turner et al., 2013). In comparison, in Adelaide, Australia, the hospital admission for respiratory diseases was reported to be significantlya 16% decrease during heatwaves in children aged 0—4 years (95% CI: 3%—24%) and 10% in elder people aged 75 years and older (95% CI: 2%—18%) during the heatwaves when the daily maximum temperature was above 35°C for three or more consecutive days for three or more consecutive days, during 1993 and 2006 (Nitschke et al., 2007).

3.2.1 Chronic obstructive pulmonary disease

COPD is one of the common health outcomes in the studies with associations between the heat and the respiratory morbidity. A national study in Brazil reported a 5% hospitalization increased due to COPD (95% CI: 4%—6%) for every 5°C increase in ambient daily mean temperature during the 2000—15 hot seasons (Zhao et al., 2019). A study in 221 U.S counties also reported increased risk for COPD by 4.7% (95% CI: 3.9%—5.5%) with each 5.6°C increase in daily mean summer temperature during 1999–2008 (Anderson et al., May 15, 2013).

People aged 65 years and older, were the vulnerable group under the risk of COPD, particularly those who were over 75 years old (Zanobetti et al., 2012; Zhao et al., 2019), so that many studies chose the elderly to be the target population. By studying elderly people cohorts aged 65 years and older in the United States, Zanobetti et al. reported that decreased air quality in combination with heat stress increased the risk of COPD mortality by 5% (95% CI: 3%–7%) for every 1°C increase in summer temperature (Zanobetti et al., 2012).

3.2.2 Asthma

Asthma can be defined as a diffuse, obstructive lung disease with hyperreactivity of the airways to a variety of stimuli and a high degree of reversibility of the obstructive process (Hashimoto et al., 2004). Some studies have explored the association between the heat and the asthma. Unlike air pollutants, the increasing temperature was often reported to be associated with a decrease in the morbidity for asthma. A study in Malta found that the mean monthly ambient temperatures were negatively associated with the monthly admissions for asthma among children under 14 years old ($r = -0.71$, $P < .001$) and people who 15 years and older ($r = -0.36$, $P < .001$) during 1994–98 (Grech et al., 2002). A study in Tokyo, Japan, investigated the childhood asthma between 1998 and 2002 and found that the higher daily mean temperature and lower daily maximum temperature were significantly associated with the asthma among children aged 2–15 years (Hashimoto et al., 2004). Another study in Athens, Greece, showed the similar results that the risk of childhood asthma admissions among children aged 0–4 years was negatively related to the mean monthly air temperature and a decrease of 10°C in ambient temperature corresponded to an increase 31% of having asthma (Nastos et al., 2008). Loh et al. also reported the negative association using data from 1982 to 2008 in Singapore that an increase of 2°C in weekly maximum temperature from 27.6 to 34.9°C was associated with a decrease of 2.7 admissions for pediatric asthma per week (Loh et al., 2011). A study in Sydney, Australia, indicated that during 1999–2009, compared with days with daily mean temperature of 20.4°C, the risk of asthma decreased 12% (95% CI: 6%–17%) with the daily mean temperature above 27.5°C (99th percentile) (Vaneckova & Bambrick, 2013).

3.2.3 Respiratory allergy

A study on children under 18 years in the United States found that changes in average annual temperature from 1998 to 2006 with no evidence to influence the prevalence of respiratory allergy, and concluded childhood otolaryngologic disease would not be influenced by the global warming (Miller et al., 2012).

Another study utilized data from the 2007 National Survey of Children's Health in the United States and found that the prevalence of hay fever was higher among children exposed to a higher quartile mean annual temperature (14.2–22.7°C, 11.9–14.1°C, 8.5–11.8°C), compared with those under the lowest quartile mean annual temperature (5.9–8.4°C) after adjusting the demographic characteristics (Silverberg et al., 2015).

3.2.4 Respiratory infections and lung function

The acute respiratory infections increased with the heat. A study in Sydney, Australia, indicated that during 1999–2009, compared with days with daily mean temperature of 20.4°C, the risk of acute respiratory infections increased 4% (95% CI: 1%–8%) with the daily mean temperature above 25.3°C (90th percentile) with the daily mean temperature above 27.5°C (99th percentile) (Vaneckova & Bambrick, 2013). Another study on children under 18 years in Cameroon reported that during 2006–07, the risk of acute respiratory infections, including upper and lower respiratory tract infections, was significantly associated with the monthly maximum (RR = 1.94, 95% CI: 1.34–2.81) and minimum temperatures (RR = 0.72, 95% CI: 0.59–0.87) (Tchidjou et al., 2010). A study in 221 U.S counties also reported the similar increased risk for COPD by 4.7% (95% CI: 3.9%–5.5%) and the risk for respiratory tract infections increased by 4.1% (95% CI: 3.4%–4.7%) for each 5.6°C increase in daily mean summer temperature during 1999–2008 (Anderson et al., 2013).

For the pneumonia and other lung function diseases, increased risks were also reported to be associated with the heat. A study in California, United States, demonstrated that during 2005–08, with every 5.6°C increase in daily mean apparent temperature from 14.1 to 30.1°C, the risk of pneumonia increased by 1.9% (95% CI: 0.6%–3.3%) (Basu et al., 2012). Green et al. found a similar result with the data during 1999–2005 in California, with the risk of all types of pneumonia increased by 3.7% (95% CI: 1.7%–5.7%) when the mean apparent temperature on the same day increased 5.6°C between 18.1 and 25.3°C (Green et al., 2010). During the heatwave in Chicago in July 1995, the risk of inpatient admissions increased by 19.6% for pneumonia and influenza (95% CI: 4.7%–34.7%) and by 50.0% for emphysema (95% CI: 26.3%–73.7%), compared with the same period in July 1994 (Semenza et al., 1999).

3.2.5 Other respiratory diseases

Some studies used a specific score with several questions of relative symptoms to evaluate the respiratory health. A study on older individuals with COPD and impaired pulmonary function reported that a 5.5°C increase in indoor temperature was associated with a 38% (95% CI: 1%–67%) significant increase in the breathlessness, cough, and sputum scale (BCSS) score, after adjusting the indoor air pollution and relative humidity (McCormack et al., 2016).

A study on people with heart failure demonstrated that the indoor temperature was negatively associated with the oxygen saturation ($Rs = -0.05$) and shortness of breath at night ($Rs = -0.10$), whereas the outdoor temperature showed the contrary result (Rs for oxygen saturation: 0.01, Rs for shortness of breath at night: 0.15) (Goldberg et al., 2015).

A study among infants showed the significant association between the increasing indoor temperature and the frequency of central apneas during rapid eye movement sleep (Franco et al., 2000). A similar study also found a significant difference in infant respiratory rates at 20, 25, 30°C (Fox et al., 1991). Both studies, however, did not provide a measure of correlation or risk estimates.

4. Summary

In this chapter we reviewed the current epidemiological evidence regarding the impact of heat exposure on cardiorespiratory health. In general, heat and heatwaves were found to significantly increase the risk of total cardiovascular and total respiratory mortality and morbidity. Trends in risks for specific diseases were inconsistent. Positive associations between heat exposure and increased health risk were observed for some diseases, including IHD, COPD, respiratory allergy, acute respiratory infections, pneumonia, and other cardiorespiratory diseases. There was a suggestion that heat exposure may have a protective effect for some diseases, such as asthma. For some diseases trends were uncertain, with different studies finding both positive and negative associations, such as MI and heart failure. The results summarized in this chapter are expected to provide epidemiological evidence for future public health research and policy.

References

Alessandrini, E., Zauli Sajani, S., Scotto, F., Miglio, R., Marchesi, S., & Lauriola, P. (2011). Emergency ambulance dispatches and apparent temperature: A time series analysis in Emilia-Romagna, Italy. *Environmental Research, 111*(8), 1192–1200.

Anderson, G. B., Dominici, F., Wang, Y., McCormack, M. C., Bell, M. L., & Peng, R. D. (May 15, 2013). Heat-related emergency hospitalizations for respiratory diseases in the Medicare population. *American Journal of Respiratory and Critical Care Medicine, 187*(10), 1098–1103.

Arbuthnott, K. G., & Hajat, S. (December 5, 2017). The health effects of hotter summers and heat waves in the population of the United Kingdom: A review of the evidence. *Environmental Health, 16*(Suppl. 1), 119.

Astrom, D. O., Forsberg, B., & Rocklov, J. (2011). Heat wave impact on morbidity and mortality in the elderly population: A review of recent studies. *Maturitas, 69*(2), 99–105.

Babin, S. M., Burkom, H. S., Holtry, R. S., et al. (March 21, 2007). Pediatric patient asthma-related emergency department visits and admissions in Washington, DC, from 2001–2004, and associations with air quality, socio-economic status and age group. *Environmental Health, 6*, 9.

Bai, L., Woodward, A., & Liu, Q. (July 1, 2014). Temperature and mortality on the roof of the world: A time-series analysis in three Tibetan counties, China. *Science of The Total Environment, 485–486*, 41–48.

Basu, R. (September 16, 2009). High ambient temperature and mortality: A review of epidemiologic studies from 2001 to 2008. *Environmental Health, 8*, 40.

Basu, R., Dominici, F., & Samet, J. M. (2005). Temperature and mortality among the elderly in the United States: A comparison of epidemiologic methods. *Epidemiology, 16*(1), 58–66.

Basu, R., Pearson, D., Malig, B., Broadwin, R., & Green, R. (2012). The effect of high ambient temperature on emergency room visits. *Epidemiology, 23*(6), 813–820.

Beizaee, A., Lomas, K. J., & Firth, S. K. (2013). National survey of summertime temperatures and overheating risk in English homes. *Building and Environment, 65*, 1–17.

Bhaskaran, K., Armstrong, B., Hajat, S., Haines, A., Wilkinson, P., & Smeeth, L. (December 13, 2012). Heat and risk of myocardial infarction: Hourly level case-crossover analysis of MINAP database. *BMJ, 345*, e8050.

Bhaskaran, K., Hajat, S., Haines, A., Herrett, E., Wilkinson, P., & Smeeth, L. (August 10, 2010). Short term effects of temperature on risk of myocardial infarction in England and Wales: Time series regression analysis of the Myocardial Ischaemia National Audit Project (MINAP) registry. *BMJ, 341*, c3823.

Bobb, J. F., Obermeyer, Z., Wang, Y., & Dominici, F. (December 24–31, 2014). Cause-specific risk of hospital admission related to extreme heat in older adults. *JAMA, 312*(24), 2659–2667.

Bunker, A., Wildenhain, J., Vandenbergh, A., et al. (2016). Effects of air temperature on climate-sensitive mortality and morbidity outcomes in the elderly; a systematic review and meta-analysis of epidemiological evidence. *EBioMedicine, 6*, 258–268.

Charkoudian, N. (2010). Mechanisms and modifiers of reflex induced cutaneous vasodilation and vasoconstriction in humans. *Journal of Applied Physiology, 109*(4), 1221–1228, 1985.

Cheng, J., Zhu, R., Xu, Z., et al. (2014). Temperature variation between neighboring days and mortality: A distributed lag non-linear analysis. *International Journal of Public Health, 59*(6), 923–931.

Chen, R., Yin, P., Wang, L., et al. (October 8, 2018). Association between ambient temperature and mortality risk and burden: Time series study in 272 main Chinese cities. *BMJ, 363*, k3965.

Coris, E. E., Ramirez, A. M., & Durme, D. J. V. (2004). Heat illness in athletes: The dangerous combination of heat, humidity and exercise. *Sports Medicine, 34*(1), 9–16.

Das, D., Bakal, J. A., Westerhout, C. M., et al. (December 20, 2014). The association between meteorological events and acute heart failure: New insights from ASCEND-HF. *International Journal of Cardiology, 177*(3), 819–824.

Dawson, J., Weir, C., Wright, F., et al. (2008). Associations between meteorological variables and acute stroke hospital admissions in the west of Scotland. *Acta Neurologica Scandinavica, 117*(2), 85–89.

Dong, W., Zeng, Q., Ma, Y., Li, G., & Pan, X. (2016). Impact of heat wave definitions on the added effect of heat waves on cardiovascular mortality in Beijing, China. *International Journal of Environmental Research and Public Health, 13*(9).

Ebi, K. L., Exuzides, K. A., Lau, E., Kelsh, M., & Barnston, A. (2004). Weather changes associated with hospitalizations for cardiovascular diseases and stroke in California, 1983–1998. *International Journal of Biometeorology, 49*(1), 48–58.

Empana, J. P., Sauval, P., Ducimetiere, P., Tafflet, M., Carli, P., & Jouven, X. (2009). Increase in out-of-hospital cardiac arrest attended by the medical mobile intensive care units, but not myocardial infarction, during the 2003 heat wave in Paris, France. *Critical Care Medicine, 37*(12), 3079–3084.

Fox, G. P. P., O'Regan, M., & Matthews, T. G. (1991). The influence of ambient temperature on some commonly measured physiological variables in infants. *Irish Journal of Medical Science, 160*(8), 249–250.

Franco, P., Szliwowski, H., Dramaix, M., & Kahn, A. (2000). Influence of ambient temperature on sleep characteristics and autonomic nervous control in healthy infants. *Sleep, 23*(3), 1–7.

Gasparrini, A., Guo, Y., Hashizume, M., et al. (2015). Mortality risk attributable to high and low ambient temperature: A multicountry observational study. *The Lancet, 386*(9991), 369–375.

Goldberg, M. S., Wheeler, A. J., Burnett, R. T., et al. (2015). Physiological and perceived health effects from daily changes in air pollution and weather among persons with heart failure: A panel study. *Journal of Exposure Science & Environmental Epidemiology, 25*(2), 187–199.

Go, A. S., Mozaffarian, D., Roger, V. L., et al. (2013). Executive summary: Heart disease and stroke statistics—2013 update: A report from the American heart association. *Circulation, 127*(1), 143–152.

Grech, V., Balzan, M., Asciak, R. P., & Buhagiar, A. (2002). Seasonal variations in hospital admissions for asthma in Malta. *Journal of Asthma, 39*(3), 263–268.

Green, R. S., Basu, R., Malig, B., Broadwin, R., Kim, J. J., & Ostro, B. (April 2010). The effect of temperature on hospital admissions in nine California counties. *International Journal of Public Health, 55*(2), 113–121.

Gronlund, C. J., Sullivan, K. P., Kefelegn, Y., Cameron, L., & O'Neill, M. S. (2018). Climate change and temperature extremes: A review of heat- and cold-related morbidity and mortality concerns of municipalities. *Maturitas, 114*, 54–59.

Guo, Y., Barnett, A. G., Yu, W., et al. (2011). A large change in temperature between neighbouring days increases the risk of mortality. *PLoS One, 6*(2), e16511.

Guo, Y., Li, S., Zhang, Y., et al. (2013). Extremely cold and hot temperatures increase the risk of ischaemic heart disease mortality: Epidemiological evidence from China. *Heart, 99*(3), 195–203.

Hajat, S., Vardoulakis, S., Heaviside, C., & Eggen, B. (2014). Climate change effects on human health: Projections of temperature-related mortality for the UK during the 2020s, 2050s and 2080s. *Journal of Epidemiology & Community Health, 68*, 641–648.

Harlan, S., Chowell, G., Yang, S., et al. (2014). Heat-related deaths in hot cities: Estimates of human tolerance to high temperature thresholds. *International Journal of Environmental Research and Public Health, 11*(3), 3304–3326.

Hashimoto, M., Fukuda, T., Shimizu, T., et al. (2004). Influence of climate factors on emergency visits for childhood asthma attack. *Pediatrics International, 46*, 48–52.

Hashizume, M., Wagatsuma, Y., Hayashi, T., Saha, S. K., Streatfield, K., & Yunus, M. (2009). The effect of temperature on mortality in rural Bangladesh—a population-based time-series study. *International Journal of Epidemiology, 38*(6), 1689–1697.

Ho, H. C., & Wong, M. S. (2019). Urban environmental influences on the temperature-mortality relationship associated mental disorders and cardiorespiratory diseases during normal summer days in a subtropical city. *Environmental Science and Pollution Research, 26*(23), 24272–24285.

Intergovernmental Panel on Climate Change (IPCC). Special report on climate change and land. Available at: https://www.ipcc.ch/srccl/chapter/summary-for-policymakers/

Keatinge, W. R., Coleshaw, S. R. K., Easton, J. C., Cotter, F., Mattock, M. B., & Chelliah, R. (1986). Increased platelet and red cell counts, blood viscosity, and plasma cholesterol levels during heat stress, and mortality from coronary and cerebral thrombosis. *American Journal of Medicine, 81*(5), 795–800.

Kenney, W. L., Craighead, D. H., & Alexander, L. M. (2014). Heat waves, aging, and human cardiovascular health. *Medicine and Science in Sports Exercise, 46*(10), 1891–1899.

Kim, K. H., Kabir, E., & Ara Jahan, S. (2014). A review of the consequences of global climate change on human health. *Journal of Environmental Science and Health, Part C, 32*(3), 299–318.

Klepeis, N. E., Nelson, W. C., Ott, W. R., et al. (2001). The national human activity pattern survey (NHAPS): A resource for assessing exposure to environmental pollutants. *Journal of Exposure Science & Environmental Epidemiology, 11*, 231–252.

Knowlton, K., Rotkin-Ellman, M., King, G., et al. (2009). The 2006 California heat wave: Impacts on hospitalizations and emergency department visits. *Environmental Health Perspectives, 117*(1), 61–67.

Koken, P. J., Piver, W. T., Ye, F., Elixhauser, A., Olsen, L. M., & Portier, C. J. (2003). Temperature, air pollution, and hospitalization for cardiovascular diseases among elderly people in Denver. *Environmental Health Perspectives, 111*(10), 1312–1317.

Kovats, R. S., Hajat, S., & Wilkinson, P. (2004). Contrasting patterns of mortality and hospital admissions during hot weather and heat waves in Greater London, UK. *Occupational and Environmental Medicine, 61*(11), 893–898.

Lam, L. T. (2007). The association between climatic factors and childhood illnesses presented to hospital emergency among young children. *International Journal of Environmental Health Research, 17*(1), 1–8.

Lavigne, E., Gasparrini, A., Wang, X., et al. (2014). Extreme ambient temperatures and cardiorespiratory emergency room visits: Assessing risk by comorbid health conditions in a time series study. *Environmental Health, 13*(1), 5.

Lee, S., Lee, E., Park, M. S., et al. (2014). Short-term effect of temperature on daily emergency visits for acute myocardial infarction with threshold temperatures. *PLoS One, 9*(4), e94070.

Leyva, E. W. A., Beaman, A., & Davidson, P. M. (2017). Health impact of climate change in older people: An integrative review and implications for nursing. *Journal of Nursing Scholarship, 49*(6), 670–678.

Liang, W. M., Liu, W. P., Chou, S. Y., & Kuo, H. W. (2008). Ambient temperature and emergency room admissions for acute coronary syndrome in Taiwan. *International Journal of Biometeorology, 52*(3), 223–229.

Liao, S.-Y., Tseng, W.-C., Chen, P.-Y., Chen, C.-C., & Wu, W.-M. (2010). Estimating the economic impact of climate change on cardiovascular diseases—evidence from Taiwan. *International Journal of Environmental Research and Public Health,* 7(12), 4250–4266.

Lim, Y. H., Hong, Y. C., & Kim, H. (February 15, 2012). Effects of diurnal temperature range on cardiovascular and respiratory hospital admissions in Korea. *Science of The Total Environment, 417–418,* 55–60.

Linares, C., & Diaz, J. (2008). Impact of high temperatures on hospital admissions: Comparative analysis with previous studies about mortality (Madrid). *European Journal of Public Health, 18*(3), 317–322.

Lin, S., Luo, M., Walker, R. J., Liu, X., Hwang, S. A., & Chinery, R. (2009). Extreme high temperatures and hospital admissions for respiratory and cardiovascular diseases. *Epidemiology, 20*(5), 738–746.

Loh, T. P., Lai, F. Y., Tan, E. S., et al. (2011). Correlations between clinical illness, respiratory virus infections and climate factors in a tropical paediatric population. *Epidemiology & Infection, 139*(12), 1884–1894.

Luo, Q., Li, S., Guo, Y., Han, X., & Jaakkola, J. J. K. (2019). A systematic review and meta-analysis of the association between daily mean temperature and mortality in China. *Environmental Research, 173,* 281–299.

Ma, W., Xu, X., Peng, L., & Kan, H. (2011). Impact of extreme temperature on hospital admission in Shanghai, China. *Science of The Total Environment, 409*(19), 3634–3637.

McCormack, M. C., Belli, A. J., Waugh, D., et al. (2016). Respiratory effects of indoor heat and the interaction with air pollution in chronic obstructive pulmonary disease. *Annals of the American Thoracic Society, 13*(12), 2125–2131.

Michelozzi, P., Accetta, G., De Sario, M., et al. (2009). High temperature and hospitalizations for cardiovascular and respiratory causes in 12 European cities. *American Journal of Respiratory and Critical Care Medicine, 179*(5), 383–389.

Miller, M. E., Shapiro, N. L., & Bhattacharyya, N. (2012). Annual temperature and the prevalence of frequent ear infections in childhood. *American Journal of Otolaryngology, 33*(1), 51–55.

Misailidou, M., Pitsavos, C., Panagiotakos, D. B., Chrysohoou, C., & Stefanadis, C. (2006). Short-term effects of atmospheric temperature and humidity on morbidity from acute coronary syndromes in free of air pollution rural Greece. *European Journal of Cardiovascular Prevention and Rehabilitation, 13*(5), 846–848.

Mitchell, D., Heaviside, C., Vardoulakis, S., et al. (2016). Attributing human mortality during extreme heat waves to anthropogenic climate change. *Environmental Research Letters, 11*(7), 074006.

Nastos, P. T., Paliatsos, A. G., Papadopoulos, M., Bakoula, C., & Priftis, K. N. (2008). The effect of weather variability on pediatric asthma admissions in Athens, Greece. *Journal of Asthma, 45*(1), 59–65.

Nayha, S., Rintamaki, H., Donaldson, G., et al. (2017). The prevalence of heat-related cardiorespiratory symptoms: The vulnerable groups identified from the national FIN-RISK 2007 study. *International Journal of Biometeorology, 61*(4), 657–668.

Ng, C. F. S., Boeckmann, M., Ueda, K., et al. (2016). Heat-related mortality: Effect modification and adaptation in Japan from 1972 to 2010. *Global Environmental Change, 39,* 234–243.

Nitschke, M., Tucker, G. R., & Bi, P. (2007). Morbidity and mortality during heatwaves in metropolitan Adelaide. *Medical Journal of Australia, 187*(11–12), 662–665.

Odame, E. A., Li, Y., Zheng, S., Vaidyanathan, A., & Silver, K. (July 27, 2018). Assessing heat-related mortality risks among rural populations: A systematic review and meta-analysis of epidemiological evidence. *International Journal of Environmental Research and Public Health, 15*(8).

Phung, D., Thai, P. K., Guo, Y., Morawska, L., Rutherford, S., & Chu, C. (April 15, 2016). Ambient temperature and risk of cardiovascular hospitalization: An updated systematic review and meta-analysis. *Science of The Total Environment, 550*, 1084–1102.

Robine, J.-M., Cheung, S. L. K., Le Roy, S., et al. (2008). Death toll exceeded 70,000 in Europe during the summer of 2003. *Comptes Rendus Biologies, 331*(2), 171–178.

Schneider, A., Schuh, A., Maetzel, F. K., Ruckerl, R., Breitner, S., & Peters, A. (July 2008). Weather-induced ischemia and arrhythmia in patients undergoing cardiac rehabilitation: Another difference between men and women. *International Journal of Biometeorology, 52*(6), 535–547.

Schweizer, C., Edwards, R. D., Bayer-Oglesby, L., et al. (2007). Indoor time-microenvironment-activity patterns in seven regions of Europe. *Journal of Exposure Science & Environmental Epidemiology, 17*(2), 170–181.

Semenza, J. C., McCullough, J. E., Flanders, W. D., McGeehin, M. A., & Lumpkin, J. R. (1999). Excess hospital admissions during the July 1995 heat wave in Chicago. *American Journal of Preventive Medicine, 16*(4), 269–277.

Shaposhnikov, D., Revich, B., Gurfinkel, Y., & Naumova, E. (2014). The influence of meteorological and geomagnetic factors on acute myocardial infarction and brain stroke in Moscow, Russia. *International Journal of Biometeorology, 58*(5), 799–808.

Sheridan, S. C., & Lin, S. (2014). Assessing variability in the impacts of heat on health outcomes in New York City over time, season, and heat-wave duration. *Ecohealth, 11*(4), 512–525.

Silverberg, J. I., Braunstein, M., & Lee-Wong, M. (2015). Association between climate factors, pollen counts, and childhood hay fever prevalence in the United States. *Journal of Allergy and Clinical Immunology, 135*(2), 463–469.

Tchidjou, H. K., Vescio, F., Boros, S., et al. (2010). Seasonal pattern of hospitalization from acute respiratory infections in Yaounde, Cameroon. *Journal of Tropical Pediatrics, 56*(5), 317–320.

Tham, S., Thompson, R., Landeg, O., Murray, K. A., & Waite, T. (2020). Indoor temperature and health: A global systematic review. *Public Health, 179*, 9–17.

Tian, Z., Li, S., Zhang, J., & Guo, Y. (2013). The characteristic of heat wave effects on coronary heart disease mortality in Beijing, China: A time series study. *PLoS One, 8*(9), e77321.

Tian, Z., Li, S., Zhang, J., Jaakkola, J. J. K., & Guo, Y. (2012). Ambient temperature and coronary heart disease mortality in Beijing, China: A time series study. *Environmental Health, 11*, 56–62.

Turner, L. R., Barnett, A. G., Connell, D., & Tong, S. (2012). Ambient temperature and cardiorespiratory morbidity: A systematic review and meta-analysis. *Epidemiology, 23*(4), 594–606.

Turner, L. R., Connell, D., & Tong, S. (October 2013). The effect of heat waves on ambulance attendances in Brisbane, Australia. *Prehospital and Disaster Medicine, 28*(5), 482–487.

Uejio, C. K., Tamerius, J. D., Vredenburg, J., et al. (2016). Summer indoor heat exposure and respiratory and cardiovascular distress calls in New York City, NY, U.S. *Indoor Air, 26*(4), 594–604.

Urban, A., & Kyselý, J. (2014). Comparison of UTCI with other thermal indices in the assessment of heat and cold effects on cardiovascular mortality in the Czech Republic. *International Journal of Environmental Research and Public Health, 11*(1), 952–967.

Vaneckova, P., & Bambrick, H. (2013). Cause-specific hospital admissions on hot days in Sydney, Australia. *PLoS One, 8*(2), e55459.

Webb, L., Bambrick, H., Tait, P., Green, D., & Alexander, L. (February 13, 2014). Effect of ambient temperature on Australian northern territory public hospital admissions for cardiovascular disease among indigenous and non-indigenous populations. *International Journal of Environmental Research and Public Health, 11*(2), 1942–1959.

Welsskopf, M. G., Anderson, H. A., Foldy, S., et al. (2002). Heat wave morbidity and mortality, Milwaukee, Wis, 1999 vs 1995: An improved response? *American Journal of Public Health, 92*(5), 830–833.

Wilker, E. H., Yeh, G., Wellenius, G. A., Davis, R. B., Phillips, R. S., & Mittleman, M. A. (2012). Ambient temperature and biomarkers of heart failure: A repeated measures analysis. *Environmental Health Perspectives, 120*(8), 1083–1087.

Xu, Z., Etzel, R. A., Su, H., Huang, C., Guo, Y., & Tong, S. (2012). Impact of ambient temperature on children's health: A systematic review. *Environmental Research, 117*, 120–131.

Xu, Z., FitzGerald, G., Guo, Y., Jalaludin, B., & Tong, S. (2016). Impact of heatwave on mortality under different heatwave definitions: A systematic review and meta-analysis. *Environment International, 89–90*, 193–203.

Xu, Z., Sheffield, P. E., Su, H., Wang, X., Bi, Y., & Tong, S. (2014). The impact of heat waves on children's health: A systematic review. *International Journal of Biometeorology, 58*(2), 239–247.

Yin, Q., & Wang, J. (2017). The association between consecutive days' heat wave and cardiovascular disease mortality in Beijing, China. *BMC Public Health, 17*(1), 223.

Zanobetti, A., & O'Neill, M. S. (2018). Longer-Term outdoor temperatures and health effects: A review. *Current Epidemiology Reports, 5*(2), 125–139.

Zanobetti, A., O'Neill, M. S., Gronlund, C. J., & Schwartz, J. D. (2012). Summer temperature variability and long-term survival among elderly people with chronic disease. *Proceedings of the National Academy of Sciences, 109*(17), 6608–6613.

Zhang, X. H., Lu, Z. L., & Liu, L. (September 2008). Coronary heart disease in China. *Heart, 94*(9), 1126–1131.

Zhao, Q., Li, S., Coelho, M. S. Z. S., et al. (2019). Ambient heat and hospitalisation for COPD in Brazil: A nationwide case-crossover study. *Thorax, 74*, 1031–1036.

CHAPTER 8

Heat exposure and mental health in the context of climate change

Hyewon Lee[1], Ho Kim[2] and Nazife Pehlivan[2]
[1]Department of Health Administration and Management, Soonchunhyang University, Republic of Korea; [2]Department of Public Health Science, Graduate School of Public Health, Seoul National University, Republic of Korea

1. Introduction

Rapid climate change by human activities and its corresponding negative impacts on human health is inevitable, thus demanding comprehensive and thorough assessment of negative health outcomes, and establishment of appropriate health interventions test caused by climate change in near future are urgent issues. Despite increasing attention and public awareness about this climate change–health issue, the mental health impacts of climate change have received less attention compared with physical health impacts.

A better understanding of the link between our environment and mental well-being is critical for a nuanced apprehension of the societal burden of poor mental health, as well as for recognizing and effectively planning for the ways in which variation in weather and climate conditions will impact population's mental health in the future. This investigation contributes to a better understanding of the link between the environment and mental well-being.

Climate change is an emerging interest in mental health specifically. However, mental health needs funding of services (Nordon et al., 2009) and research (Hansen et al., 2008) are not being adequately addressed. The same neglect is reflected in research investment in the mental health implications of climate change (Wang et al., 2012). For instance, a Scopus search yielded 9672 publications for 2007–16 matching the terms "climate change *and* health," but only 208 were returned for "climate change *and* mental health." Only 29 of these critically evaluated mental health, and the rate of publication on this topic domain has remained flat. Mental health, including its relationship to climate change, urgently needs greater priority. Here, we outline current thinking on this topic, describe relationships to modern epidemiology, and propose the adoption of a systems approach to

Heat Exposure and Human Health in the Context of Climate Change
ISBN 978-0-12-819080-7
https://doi.org/10.1016/B978-0-12-819080-7.00008-2
© 2023 Elsevier Inc.
All rights reserved.

remedy present limitations. We finish by introducing a top-level causal process diagram, some notes about research methods and approaches that could be applied, a future research agenda, and a comment on the style of scientific leadership that will be required.

According to World Health Organization, health is "a state of complete physical, mental and social well-being and not merely the absence of disease or infirmity," and thus a comprehensive understanding of health impacts across all aspects of physical, mental, and social well-being is important to prevent various diseases and promote public health in the future.

Global burden of mental diseases is staggering to both society and individuals; therefore, it is essential to address the causes of mental diseases and the number of people suffering from them. The burden should be even heavier if suicides and violence attributable to mental disorders are included. Yet even these figures likely underestimate the full toll of mental disorders, which also increase the risk of and harm from physical illnesses (Prince et al., 2007).

One of the critical issues related to current climate change is the increase of global average temperature. According to Intergovernmental Panel on Climate Change (IPCC), ambient temperature is steeply increasing, and heatwaves have been frequently occurring. Furthermore, although it has recently been increasing, focus on heat effects by climate change is still limited. Therefore, this review is conducted to focus on the heat effect on mental health.

Grasping the effects of climate change on mental health is necessary for prevention in both physical and psychological aspects. There are various changes to climate and consequences to those changes, which may be linked to mental diseases with a limited literature exploring that link, but this chapter focuses solely on the heat effect on mental health.

An accumulating body of research studies has presented effects of climate change on human health. The current state of climate change supports the notion of elevated incidences of heatwaves in future. The understanding of role of increasing temperatures and heatwaves on altered mental health conditions stands crucial, in light of wide spectrum of effects warming climate imparts on mental health and well-being status.

The burden of mental illness is immense. Nearly 45,000 people in the United States and 800,000 globally die by suicide each year (National Institute of Mental Health, 2018; World Health Organization, 2018). In 2017, 18.1% of American adults reported struggling with a diagnosable mental, behavioral, or emotional disorder (National Institute of Mental Health, 2018). The direct cost of treating mental disorders in the United States is growing at 5% annually and surpassed $200 billion in 2013 (Roehrig, 2016). Direct spending on the treatment of mental disorders now outstrips that for any other single category of illness. It is also widely acknowledged that the indirect costs of mental disorders—realized through lost wages, reduced productivity, disability, and mortality—are likely even larger (Trautmann et al., 2016). Total direct and indirect costs imposed on society have been estimated at nearly $450 billion per year in the United States and some $2.5 trillion globally, with these amounts expected to double by 2030 (Trautmann et al., 2016). Yet even these figures likely underestimate the full toll of mental disorders, which also increase the risk of and harm from physical illnesses (Prince et al., 2007). The relationship between environmental conditions and mental well-being has long been acknowledged and has recently garnered additional attention in the face of climate change (Berry et al., 2010; McMichael et al., 2006). A better understanding of the link between our environment and mental well-being is critical for decreasing the direct and indirect costs of mental disorders on society and individuals.

Here, our primary objective of this review was to determine the mental health impacts of heat exposure by thoroughly reviewing previous original research and systematic review articles on the association of ambient temperature and heatwave with various mental health outcomes, to contribute to establishing appropriate policies and interventions for mental health in warming near future.

2. Findings

We classified mental health outcomes into five categories: (1) diagnosed mental disorders (based on ICD-9 codes 290–319 or ICD-10 codes F00–F99) and mental illnesses, (2) attempted and completed suicides, (3) violence/assault, (4) subjective well-being, and (5) other outcomes.

2.1 Mental disorders and illnesses

2.1.1 Mortality

Several studies have found positive associations between deaths with diagnosed mental disorders and illnesses and heatwave or high ambient temperature.

2.1.1.1 Heatwave

The interest on mental health impacts of heat exposure started from evaluating individuals who died during heatwave events. After deaths of two psychiatric patients following heatstroke during a heatwave, Bark (1998) studied if psychiatric patients were at increased risk of death during heatwaves compared with other periods and the general population in New York City, during 1950–84. He defined a heatwave as a period of three or more consecutive days with a maximum temperature of ≥89° (31.7°C) and included one extra day in the case period considering a lag effect. Compared with the control period selected as the closest days before or after the heat wave with a maximum temperature of <85°, higher significant risk of death was observed both in the patients hospitalized in psychiatric hospitals and in the general population on the heatwave periods, but the risk was much larger in the psychiatric patients (relative risk [RR], 1.38) than in the general population (RR, 1.17). Notably, the risk of death of psychiatric patients was highest in the 1970s, when higher dosage of antipsychotic medications was used (RR, 1.49) in the period-specific analysis, suggesting that both mental illness and antipsychotic medications could be risk factors for heat-related deaths.

In July 1995, there was a record-breaking heatwave lasted for 4 days in Chicago, and at least 700 heat-related deaths occurred (Semenza et al., 1996). Semenza et al. (1996) conducted a case–control study with 339 deaths and 339 matched controls to identify risk factors associated with heat-related deaths and found that preexisting mental problem was associated with an increased risk of heat-related deaths (odds ratio [OR], 3.5; 95% CI, 1.7–7.3). There was a Chicago's second deadliest heatwave lasted for 3 days during the summer of 1999, and a case–control study with 63 deaths and 77 controls was conducted to determine related factors (Naughton et al., 2002). The study confirmed that the risk of death during the heatwave was associated with underlying depression (OR, 4.1; 95% CI, 1.3–12.5) and other mental problem including developmental delay and schizophrenia (OR, 11.7; 95% CI, 1.5–92.2). In the same period of 1999, Cincinnati also experienced a heat wave, and Kaiser et al. (2001) examined 17 heat-related deaths by focusing on preexisting mental illness through a 1:2 matched case–control study. Mental illness was confirmed as a possible risk factor for heat-related deaths (OR, 14.0; 95% CI, 1.8–633).

After the heatwaves during 1990s in the Unites States, the dreadful heatwave hit the European countries (France, Spain, Portugal, Italy, England and Wales, Switzerland, the Netherlands, Germany and Belgium) in

2003. Especially, 14,802 people died during the heatwave period of 4–13 August in France (García-Herrera et al., 2010). To identify the risk factors related to death of the victims of the heatwave of August 2003 in France, Davido et al. (2006) compared 134 survivors and 31 nonsurvivors who visited an emergency department (ED) during the heatwave by thorough chart review. Altered mental status was observed among 96.8% of deaths and 36.6% of nonsurvivors ($P < .005$). Moreover, while the use of diuretics and cardiotropes was not significantly different between two groups, the use of psychotropic drugs was higher in nonsurvivors (68.2% vs. 46.1; $P < .05$). Fouillet et al. (2006) compared the number of deaths (O) during the heatwave period of August–November 2003 to those observed from 2000 to 2002 and the estimated number (E) in 2003. There were 15,000 excess deaths during the 2003 heatwave, and 748 excess deaths (E-O) occurred due to mental disorders (ICD-10 codes F00–F99 as causes of death) (mortality ratio [O/E]. 1.8; 95% CI, 1.7–1.9). Comparison was also made between the observed number of deaths and the expected number during the heatwaves occurring between 1971 and 2003 in France (Rey et al., 2007). Heatwaves were defined as periods of at least three consecutive days with both the maximum and the minimum temperature of exceeding the 95th percentile of each temperature. There were 1330–13,734 excess deaths (mortality ratio [O/E], 1.17–1.78) during the six heatwave events, and mental disorders (ICD-10 codes F00–F99) were among the greatest risk factors for deaths (mortality ratio, 1.02–1.25), with heat-related causes including dehydration and hyperthermia, respiratory diseases, nervous system diseases, infectious diseases, and endocrine and nutritional diseases. Moreover, a study focused on a possible negative effect of psychotropic drug use during the heatwave period of August 2003 in France and conducted a population-based case–control study with cases who died before (N = 2093) and during (N = 9513) the heatwave and their 1:1 matched controls (Nordon et al., 2009). The study found that any psychotropic drug was related to a 30% increased risk of death only during the heatwave period (adjusted OR [aOR], 1.29; 95% CI, 1.22–1.37), with a significant dose–response tendency (aOR for linear trend, 1.25; 95% CI, 1.21–1.29). In drug-specific analyses, antidepressants (aOR, 1.71; 95% CI, 1.21–1.29) and antipsychotics (aOR, 2.09; 95% CI, 1.89–2.35) showed a significant association with a risk of death, contrary to anxiolytics/hypnotics use (aOR, 0.85; 95%CI, 0.79–0.91). A metaanalysis combining six case–control studies conducted between 1966 and 2006 reported that preexisting psychiatric illness increased a risk of death during heatwaves by three times

(OR, 3.61; 95% CI, 1.3−9.8), followed by cardiovascular (OR, 2.48; 95% CI, 1.3−4.8) and pulmonary (OR, 1.61; 95% CI, 1.2−2.1) diseases.

More recent studies assessed the association between mortality due to mental disorders and heatwave with a time series data. Hansen et al. (2008) conducted a time series analysis of daily deaths due to mental disorders (ICD-10 codes F00−F99) and heatwaves defined as ≥3 consecutive days with a daily maximum temperature exceeding 35°C in Adelaide, South Australia, for 1993−2006. There was a significant increase in deaths attributed to all mental disorders in 65−74 years (incidence rate ratios [IRR], 2.40; 95% CI, 1.17−4.92), organic, including symptomatic, mental disorders (F00−F09) in 15−64 years (all: IRR, 5.01; 95% CI, 1.21−21.23; male: IRR, 12.73; 95% CI, 2.06−78.52), disorders due to psychoactive substance use (F10−F19) in females (all ages: IRR, 3.10; 95% CI, 1.43−7.16); 15−64 years: IRR, 3.21; 95% CI, 1.30−7.95), and schizophrenia, schizotypal, and delusional disorders (F20−F29) in all ages (all: IRR, 2.08; 95% CI, 1.05−4.14; male: IRR, 4.05; 95% CI, 1.39−11.84) and ≥75 years (all: IRR, 2.11; 95% CI, 1.02−4.38; male: IRR, 5.26; 95% CI, 1.75−15.76). However, a study conducted in Brisbane, Australia, between 1996 and 2005 did not find a significant increase in death attributable to mental disorders (ICD-9 codes 290−319; ICD-10 codes F00−F99) during heatwaves (≥2 consecutive days with a daily maximum ≥37°C) (Wang et al., 2012).

Several studies have also been conducted in European countries. A study of 651,195 residents aged ≥65 years in Rome, Italy, for summer of 1993−2006, confirmed a nonsignificant higher risk of dying among 65−74 aged individuals with a previous admission for psychiatric disorders (RR, 1.27; 95% CI, 0.91−1.77) than those without a previous admission (RR, 1.08; 95% CI, 1.01−1.15) during heatwave episodes (daily maximum apparent temperature exceeding the monthly threshold [May, 28.5; June, 32.5°C; July, 33.5°C; August/September, 34.5°C]) (Schifano et al., 2009). A time series study of daily number of deaths occurring in the summer (15th May and 15th September) of 2000−08 in Rome and Stockholm found a significant increase in death attributable to psychiatric disorders both in Rome (RR, 1.21; 95% CI, 1.06−1.38) and in Stockholm (RR, 1.33; 95% CI, 1.10−1.61) during heatwave days (≥2 consecutive days exceeding the city-specific 95th percentile of maximum apparent temperature) (Åström et al., 2015). Basagaña et al. (2011) evaluated the association between cause-specific mortality and extremely hot days (maximum temperature > the 95th percentile) in the Catalonia region of Spain for the warm seasons of 1983–2006 using a time-stratified case-crossover design and confirmed a significant increase in death due to mental disorders (ICD-9 codes 290–319; ICD-10 codes F00–F99) (RR, 1.3; 95% CI, 1.2–1.4). Rocklöv et al. (2014) examined susceptibility to mortality associated with heatwave duration in the residents of Stockholm County during 1990–2002 and found a significantly increased risk of death among over-65 aged individuals who ever experienced an admission for mental disorders (ICD–9 codes 290–319; ICD-10 codes F00–F99) (OR, 1.099; 95% CI, 1.027–1.175) per 1-day increase of heatwave duration (days in sequence above the 98th percentile of minimum temperature [17.4°C]; value 1 was given for the second day of heatwave).

A time series analysis of daily number of cause-specific deaths in King County, Washington, during May to September 1980–2010, found a significant increase of death due to mental disorders (ICD-9 codes 290–316; ICD-10 codes F01–F69) among individuals aged 65–84 years (RR, 1.43; 95% CI, 1.02–2.01) during heat days (maximum humidex [temperature, relative humidity, and vapor pressure are combined] > the 99th percentile) (Isaksen et al., 2016). Zanobetti et al. (2013) investigated susceptibility to mortality during extreme weather events in Medicare participants in 135 US cities between 1985 and 2006 and found that people with preexisting dementia (ICD–9 codes 290) were more susceptible to extreme heat (days with a minimum temperature ≥ the 99th percentile), with an additional increase in the risk of death (6%; 95% CI, 3%–9%).

2.1.1.2 High ambient temperature

Perhaps the first study on the association between high ambient temperature and mortality was conducted in four Italian cities (Bologna, Milan, Rome, and Turin), with a total of 205,019 residents aged ≥35 years who died during 1997–2003 (Stafoggia et al., 2006). Compared with apparent temperature (lag0–1) of 20°C, the risk of death significantly increased at 30°C among residents with a previous admission for psychoses (ICD–9 codes 290–299) (OR, 1.70; 95% CI, 1.39–2.09) or depression (ICD–9 codes 300.4, 301.1, 309.0, 309.1, 311) (OR, 1.71; 95% CI, 1.23–2.38) in the case-crossover analysis. Stafoggia et al. (2008) conducted a same analysis for residents aged ≥65 years who died during 1997–2004 in four Italian cities. The risk of death of residents who experienced a previous hospitalization for psychoses (OR, 2.05; 95% CI, 1.44–2.93) significantly increased at apparent temperature (lag0–1) of 30°C compared with 20°C, but not for depression (OR, 2.05; 95% CI, 1.44–2.93).

Rocklöv et al. (2014) also investigated susceptibility to mortality associated with temperature in the residents of Stockholm County during 1990–2002 and confirmed a significantly increased risk of death among over-65 aged individuals who ever experienced an admission for mental disorders (ICD–9 codes 290–319; ICD-10 codes F00–F99) (OR, 1.011; 95% CI, 1.002–1.019) per 1°C increase above the median maximum temperature for the summer period (20.7°C). A time series study of high temperature and cause-specific mortality in 10 regions in England and Wales, during the summers (June–September) of 1993–2006, found a significant increase in mortality attributed to mental diseases (ICD–9 codes 290–319; ICD-10 codes F00–F99) (3.1%; 95% CI, 1.7%–4.6%) per 1°C increase above the region-specific thresholds (from 20.9°C in North East to 24.7°C in Greater London) (Gasparrini et al., 2012). In subtype-specific analyses, organic mental disorders (F00–F09) (3.5%) and psychoactive substance drug use (F10–F19) (5.3% for $\leq$65 years) showed a significant association, whereas schizophrenia (F20–F29) exhibited no association. Page et al. (2012) examined the risk of temperature-related death among patients with mental illness who died between 1998 and 2007 in England. There was a 4.9% (95% CI 2.0%–7.8%; RR, 1.04) increase in risk of death per 1°C increase above the 93rd percentile of the annual temperature distribution. Dementia (RR, 1.03), alcohol misuse (RR, 1.08), and other substance misuse (RR, 1.20) exhibited a significant increase in deaths in subtype-specific analyses, and antipsychotics (RR, 1.07) and hypnotics/anxiolytics (RR, 1.08) showed a significant increase in deaths in prescribed drug-specific analyses.

A few studies have been conducted in Asian countries. Kim et al. (2015) assessed the association of cause-specific mortality with high ambient temperature in Seoul, South Korea, for 1992–2009 and confirmed a significantly increased risk of death attributable to mental disorders (ICD-10 codes F00–F99) (RR, 1.04; 95% CI, 1.01–1.07) per 1°C increase above the 90th percentile of maximum temperature (29.5°C). Disorders due to psychoactive substance use (F10–F19) (RR, 1.07; 95% CI, 1.02–1.13) showed a significant association in subtype-specific analyses. Ho and Wong (2019) analyzed the temperature–mortality relationship in Hong Kong between 2007 and 2014 and found a significant increase in deaths due to mental disorders (IRR, 1.033; 95% CI, 1.004–1.062), but not due to cardiovascular or respiratory diseases, per 1°C increase above the 50th percentile of mean temperature (24.51°C).

2.1.2 Hospitalizations

2.1.2.1 Heatwave

As a morbidity outcome, hospital admissions related to heatwave have been investigated in several study areas. During the July 1995 heatwave in Chicago, there were 1072 (11%) more hospital admissions compared with averages, and admissions of diagnose for mental conditions marginally increased (excess of 323; $P = .061$) (Semenza et al., 1999).

Several studies have been conducted in cities in Australia. In the summers of 2008 and 2009, extreme heatwaves lasting 15 and 13 days, respectively, occurred in Adelaide, South Australia. Nitschke et al. (2011) examined the impact of the 2008 and 2009 heatwaves (≥35°C of maximum daily temperature for three or more days) on cause-specific mortality and morbidity and found no significant impact on hospital admissions due to mental disorders (ICD-9, 290−294-9; ICD-10, F00−F99.9). Wang et al. (2012) also did not find a significant increase in emergency hospital admissions due to mental health (ICD-9, 290−319; ICD-10, F00−F99) during the heatwaves (daily maximum ≥37°C for two or more consecutive days) in Brisbane between 1996 and 2005. However, Vaneckova and Bambrick (2013) observed a significant increase in hospital admissions due to total mental disorders (OR, 1.02; 95% CI, 1.00−1.03) and psychoses (OR, 1.03; 95% CI, 1.02−1.05) during hot days (≥95th percentile of daily mean temperature) compared with control days in Sydney between 1991 and 2009. Khalaj et al. (2010) also evaluated the impacts of extreme heat events (≥99th percentile of daily maximum temperature) on emergency hospitalizations in five regions of New South Wales and found a significant increase in admissions for mental disorders as a primary diagnosis (OR, 1.07; 95% CI, 1.00−1.15 for 3-day moving averages) and as an underlying condition (OR, 1.07; 95% CI, 1.03−1.11 for the dame day; OR, 1.05; 95% CI, 1.01−1.09 for the day after; OR, 1.11; 95% CI, 1.06−1.17 for 3-day moving average).

A study focusing on the association of heatwaves with hospital admissions for mental disorders has been conducted in Northern Vietnam for 2008−12 (Trang et al., 2016). The study found that heatwaves (≥35°C of daily maximum temperature for three or seven consecutive days) significantly increased the risk of admissions for total mental disorders (ICD-10, F00−F79) (OR, 1.15; 95% CI, 1.01−1.31 for 3 days; OR, 1.36; 95% CI, 1.00−1.90 for 7 days), mental retardation (F70-79) (OR, 1.68; 95% CI, 1.08−2.62 for 3 days), and organic mental disorders (F00-09) (OR, 3.62; 95% CI, 1.70−7.42 for 7 days).

2.1.2.2 High ambient temperature

In European region, the first study linking mental disorders and temperature was conducted in England, Wales, and Scotland for 1964–74 (Myers & Davies, 1978). The study investigated factors inducing a seasonal variation in the admission rate for mania (ICD-7, 301.0; ICD-8, 296.1 and 296.3) and observed the highest correlation between monthly hospital admission rates and current month's mean temperature ($r = 0.71$). McWilliams et al. (2013) conducted a time series analysis in Ireland between 1971 and 2002 to examine the association between admissions for schizophrenia, schizoaffective disorders, and other psychoses and meteorological variables and did not find a significant association with any meteorological variables. In the same study area, McWilliams et al. (2014) also did not observe the association between meteorological variables and admissions for mania and depression. Similarly, Medici et al. (2016) found no link between the seasonal pattern in admissions for mania and weather variables in Denmark for 1995–2012.

However, a recent study applying an advanced statistical model (distributed lag nonlinear model [DLNM]) found a significant increase in hospital admissions for mental disorders at the 99th percentile of mean temperature (lag0–1) compared with the 50th percentile (16.8°C) (RR, 1.20; 95% CI, 1.03–1.40) in the Lisbon Metropolitan Area, Portugal, between 2008 and 2014 (Almendra et al., 2019). Another recent study conducted in Turin, Italy, from September 2013 to August 2015 examined the association between meteorological variables and admissions for bipolar disorder (Aguglia et al., 2019). The study found an increased risk of admission of bipolar patients when the maximum temperature was high (OR, 1.018; 95% CI, 1.006–1.030). Linares et al. (2017) evaluated the association between daily dementia-related hospital admissions (ICD-9, 290–294 excluding 291.0 and 292.0) and environmental variables in Madrid, Spain, for 2001–09 and observed a significantly increased risk of the admission per 1°C increase over a threshold (34°C) of maximum temperature (RR, 1.19; 95% CI, 1.09–1.30).

Several studies investigating the correlation between mental disorders and temperature have been performed in Middle East countries. Shapira et al. (2004) observed a significant correlation with mean maximum monthly temperature among 4117 bipolar depressed patients (ICD-9, 296.5) ($r = 0.27$), but not in 1036 unipolar depressed patients (296.2–296.3) ($r = 0.02$) in Tel Aviv, Israel, for 1981–91. In the same study area, Shiloh et al. (2005) also found a significant correlation between mean maximum monthly temperature and monthly admission rates of schizophrenia patients (r = 0.35). Amr and Volpe (2012) assessed the correlation between monthly climatic variables and monthly number of admissions for mania, depression, and schizophrenia (based on DSM-IV-TR criteria) in Mansoura, Egypt, between 2003 and 2007. The study found a significant correlation between monthly mean temperature and admissions for mania (r = 0.48) and depression (r = −0.51) but not for schizophrenia. Talaei et al. (2014) also observed a significant correlation between daily number of psychiatric admissions and daily mean temperature in Mashhad, Iran, from March 2009 to February 2010.

In Adelaide, South Australia, a study examined heat thresholds for daily mortality and morbidity outcomes for 1993−2009 and found a significant increase in hospital admissions for mental disorders (ICD-9, 290−294.9; ICD-10, F00−F99) per 1°C increase over threshold temperature (IRR, 1.003; 95% CI, 1.001−1.005 for maximum temperature [26°C]; IRR, 1.004; 95% CI, 1.001−1.007 for minimum temperature [16°C]) (Williams et al., 2012).

In North America region, Sherbakov et al. (2018) found a significant association between daily hospitalizations for mental health (ICD-9, 290−391) and high mean temperature (RR, 1.04; 95% CI, 1.01−1.07 for 26°C compared with 20°C) in California, United States, between 1999 and 2009.

The relationship between mental disorders and high temperature has also been assessed in Asian countries. A study conducted in Taiwan for 1999−2003 examined the association between monthly admission rates for manic (ICD-9, 296.0, 296.1, and 209.4), depressive (296.5), and mixed/unspecified episodes (296.7−296.9) of bipolar disorder and climatic variables (Lee et al., 2007). The study observed a significant association of monthly mean temperature with depressive (β = 0.0044) and mixed/unspecified episodes (β = 0.0025). Another study conducted in Taiwan between 1996 and 2007 also found a significant relationship between daily hospital admissions for bipolar disorder (ICD-9, 296.0−296.1, and 296.4−296.8) and daily mean temperature (RR, 1.51; 95% CI, 1.23−1.84 for 30.7°C [≥99th percentile] compared with 25th−49th percentile) (Sung et al., 2013).

Three studies conducted in Hefei, China, found a significant association of daily admissions for schizophrenia (ICD-10, F20−F29) with high ambient temperature (RR, 1.07; 95% CI, 1.04−1.11 at lag0−4 for 28°C [75th percentile] vs. 24.8°C) during warm season in 2005−14 (Wang et al., 2018), high apparent temperature (RR, 1.062; 95% CI, 1.019−1.106 at lag0 for 28.3°C [90th percentile] vs. 3.3°C) for 2005–14 (Yi et al., 2019), and attributable fraction by hot temperature (7.41%; 95% empirical CI, 3.97%–10.86% over 19°C) for warm season in 2005–14 (Pan et al., 2019).

In Shanghai, China, Peng et al. (2017) found a significant positive association between daily hospital admissions for mental disorders (ICD-10, F00–F99) and hot temperature (RR, 1.266; 95% CI, 1.074–1.493 at lag0–1 for 33.1°C [99th percentile] vs. 24.6°C) during 2008–15.

Chan et al. (2018) assessed the association between daily mental disorder hospitalizations (ICD-9, 209–319) and mean temperature in Hong Kong between 2002 and 2011. A significant increase in hospitalizations was observed for total mental disorders (RR, 1.09; 95% CI, 1.03–1.15), episodic mood disorders (ICD-9, 296; RR, 1.34; 95% CI, 1.05–1.71), transient mental disorders (293; RR, 1.51; 95% CI, 1.00–2.27), and drug-related mental disorders (292 and 305; RR, 1.13; 95% CI, 1.00–1.27) at 28°C relative to 19.4°C over 0–2 lag days.

Lee et al. (2018) also evaluated the effect of mean temperature on mental disease-related emergency hospital admissions in six metropolitan cities in South Korea for 2003–13. The study found a significantly increased risk of admissions for total mental disorders (ICD-10, F00–F99; RR, 1.158; 95% CI, 1.092–1.227), anxiety (F40–F41; RR, 1.375; 95% CI, 1.173–1.613), dementia (F00–F03 and G30-G31; RR, 1.213; 95% CI, 1.040–1.414), and depression (F32–F33; RR, 1.229; 95% CI, 1.013–1.491) at extreme hot temperature (29.4°C, 99th percentile) compared with 15.2°C (median temperature) over 0–4 lag days.

2.1.3 *Visits of emergency departments or hospitals*

Visits of emergency departments or hospitals have also been studied as the morbidity outcome to assess the acute impacts of intensive heatwave and high ambient temperature.

2.1.3.1 Heatwave

A study conducted in Barcelona, Spain, examined the impact of the dreadful heatwave lasting 14 days in 2003 on psychiatric emergencies (Merino et al., 2009). During the heatwave days, there were more psychiatric patients (diagnosis of DSM-IV) with alcohol abuse (OR, 2.12; 95% CI, 1.12–4.01) compared with nonheatwave days. When the psychiatric disease severity scale was compared, the mean score of the item "dangerousness toward others" was higher on patients who visited EDs during heatwave days.

In the summer of 2012, heatwave lasted from July 22 to August 11. To collate data for patients with heat-related illnesses, the Korea Centers for Disease Control and Prevention operated a heatwave surveillance system during June 1–August 31. Using these data, a study conducted a case–control study to investigate characteristics of patients who visited EDs with heat stroke (Kim et al., 2014). The study found that among various pre-existing medical conditions, only neuropsychiatric disorder (ICD-10, F00–F48, F70–F79, G20–G26) was significantly related to heat stroke (OR, 7.69; 95% CI, 4.06–14.54) during hot weather period.

A study by Liu et al. (2019) examined the association between daily hospital visit for mental illness and four heatwave events during the summer of 2010 in Jinan, China. The HW events significantly increased the risk of hospital visits for mental illness (OR, 2.23; 95% CI, 1.44–3.47 for first heatwave; OR, 2.84; 95% CI, 1.78–4.53 for second heatwave; OR. 3.18; 95% CI, 2.00–5.06 for third heatwave; OR, 2.99; 95% CI, 2.16–4.14 for fourth heatwave).

2.1.3.2 High ambient temperature

In Adelaide, South Australia, Williams et al. (2012) assessed not only hospitalizations described before but also the association between high temperature and ED presentations during 1993–2009. The risk of ED presentations for mental disorders (ICD-9, 290–294.9; ICD-10, F00–F99) was significantly associated with both high daily maximum (IRR, 1.011; 95% CI, 1.006–1.016 over 32°C) and minimum temperature (IRR, 1.012; 95% CI, 1.005–1.019 over 16°C) per 1°C increase over threshold temperature.

In three regions of Québec, Canada, during the warm season (May–September) of 1995–2007, a study examined the relationship between daily ambient temperature and the number of ED visits for mental and psychological problems (Vida et al., 2012). The risk of the ED visits significantly increased at 25°C (IRR ranged from 1.03 to 1.09 for <65 and ≥65 years in three areas) in comparison with overall warm season mean temperature. Another study conducted in Toronto, Canada, between 2002 and 2010 investigated the association between hot and cold temperatures and ED visits for mental disorders (ICD-10, F00–F99) (Wang et al., 2014). At the 99th percentile of the daily men temperature distribution compared with the 50th percentile, the risk of ED visits significantly increased for total disorders (RR, 1.29; 95% CI, 1.03–1.09), psychoactive substance abuse (F10–F19; RR, 1.14; 95% CI, 1.03–1.32), schizophrenia and schizotypal and delusional disorders (F20–F29; RR, 2.49; 95% CI, 1.69–3.69), mood disorders (F30–F39; RR, 1.68; 95% CI, 1.17–2.40), and neurotic disorders (F40–F49; RR, 1.12; 95% CI, 1.00–1.27) over 0–7 days. However, no association was found at cold temperature (first percentile).

In King county, Washington, United States, for 2007–12, a study evaluated the impact of extreme heat (maximum daily humidex) on emergency medical service calls and observed a significant association between basic life support calls for psychological problems among the 15–44 aged group and extreme heat exceeding the 95th percentile humidex (29.7°C; RR, 1.07; 95% CI, 1.01–1.14) (Calkins et al., 2016). A study conducted in California, United States, between 2005 and 2013 examined the 16 climate zone-level associations between heat and mental health-related ED visits (Basu et al., 2018). The metaanalysis revealed a significantly increased risk of ED visits for total mental disorders, neurotic disorders, self-injury/suicide, and inflicted injury/homicide both during heatwave (≥2 days in the 95th percentile for climate zone) and at high mean apparent temperature (over 90th percentile).

In European region, a study was conducted to investigate the link between ED visits for psychiatric emergencies and daily mean temperature in Parma, the northern Italy between 2002 and 2010 (Cervellin et al., 2014). The significant positive association was observed through a multiple linear regression analysis (R, 0.82; $P < .001$). Another study conducted in Gothenburg, Sweden, from July 1, 2012 to December 31, 2017 assessed the association of ambient temperature with psychiatric emergency visits (Carlsen et al., 2019). The study found that high temperature was significantly associated with 14% (95% CI, 2%–28% at lag 0–3) and 22% (95% CI, 6%–40% at lag 0–14) increases in psychiatric emergency visits at 95th percentile temperature for the warm season, relative to the seasonal minimum effect temperature. However, the impact of cold temperature was not significant.

2.1.4 Other mental health outcomes

2.1.4.1 Aggravation of symptoms in patients with mental disorders

There have been several studies to investigate the effect of heat exposure on symptom aggravations of patients with mental disorders. Especially due to their seasonal patterns, mood disorders and schizophrenia have been studied as major disorders related to climatic factors. A study investigated whether a seasonal pattern–peak in summer of seasonal affective disorder (SAD) was observed in Queensland, Australia (Morrissey et al., 1996). From the main survey, it was found that 9% of the respondents exhibited summer SAD, whereas only 1.7% revealed winter SAD. It has been reported that excessive heat is one of the most influential factors affecting mood and behavior in summer SAD. A study conducted in regional Victoria, Australia, examined the association between self-reported mood change in patients with bipolar disorder and meteorological factors to determine a factor affecting a seasonal pattern of manic and depressive symptoms (Bullock et al., 2017). The study found that only daily maximum temperature was significantly associated with greater odds of transition into manic mood states (OR, 1.19; 95% CI, 1.04–1.37). Using data of schizophrenia patients who met the DSM-IV criteria and admitted in a mental health center between January 1 and August 6, 2003 in Israel, Shiloh et al. (2007) assessed the effect of environmental (ward) temperature on mental status of schizophrenia inpatients. The mental status measured with the positive and negative syndrome scale of schizophrenia patients was significantly associated with high environmental temperature ($r = 0.52–0.64$), but nonschizophrenia psychotic patients did not show deterioration in mental status. A study reported that schizophrenia patients maintained with antipsychotic drugs exhibited the impaired heat tolerance (Hermesh et al., 2000); thus, they are likely to be more vulnerable to heat exposure.

2.1.4.2 Incidence of mental disorders

Moreover, a cohort study was conducted to assess risk factors for incidence of depression including climatic factors in 13,938 participants from the SUN cohort in Spain (Henríquez-Sánchez et al., 2014). There was a higher risk of depression in participants living in areas with more daily light hours (hazard ratio [HR], 1.29; 95% CI, 1.08–1.54 for first tertile vs. third tertile) and higher mean temperature (HR, 1.31; 95% CI, 1.09–1.56 for first tertile vs. third tertile). Another study focused on the relationship between ambient temperature and risk of developing schizophrenia using data from two multicenter studies by the World Health Organization for 1979 and 1987 (Gupta & Murray, 1992). The study found a strong correlation between mean overall outcome score for schizophrenia and mean temperature ($r = 0.75$).

2.1.4.3 Dementia

Although dementia has primarily been regarded as a neurological–neurodegenerative disease, dementia is also considered a psychiatric disorder because major symptoms of dementia include behavioral and psychological symptoms. A study conducted in Italy by Cornali et al. (2004) examined six patients discharged from an Alzheimer rehabilitation unit during very hot days (30.4°C [86.7°F], June 14–21, 2002) and compared characteristics of the cases with those of two control groups discharged previous/following 5 weeks. The study found that behavioral and psychological symptoms of dementia measured with neuropsychiatry inventory were only deteriorated in the case group due to the high climate temperature. Alzheimer's disease (AD) accounting for large portion of dementia has also been linked to high temperature. A study conducted in Madrid, Spain, between 2001 and 2009 examined the short-term association between emergency hospital admissions for AD and high temperature (Culqui et al., 2017). AD hospital admissions significantly increased when daily maximum temperature surpasses a threshold of 34°C at lag 3 (RR, 1.30; 95% CI, 1.12–1.52). Another study focusing on AD investigated the heatwave impacts on hospitalizations and postdischarge deaths in Brisbane, Australia, for 2005–13 (Xu et al., 2019). During intensive heatwave events (two or more consecutive days with a mean temperature ≥95th percentile), both hospitalizations and post-discharge deaths for AD significantly increased by 51% (95% CI, 2%–126%) and 269% (95% CI: 76%–665%), respectively.

2.1.4.4 Mental health and well-being

For protecting mental health from the climate change, it is important not only to evaluate the impacts of heat exposure on objectively defined mental disorders but to evaluate the impacts on mental health and well-being. Several studies have been performed to examine the effect of heat exposure on mental health and well-being, which were defined variously. A study conducted in Austria investigated the effects of extreme heat on individual-level mental health and well-being that were evaluated using four domains (mental health, vitality, social functioning, and emotional role) in the Household, Income and Labour Dynamics in Australia (HILDA) Survey (Ding et al., 2015). With 12,891 participants and 22,476 observations, the study found that extreme heat (≥30°C) occurring on the same day led to significant deterioration in total mental health score equivalent to the impact of unemployment, and the heat effects remained up to 3 months. Using data of the Social, Economic and Environmental Factors (SEEF) project in the southeast Australian state, a study examined the association between mental health defined using the Kessler 10-item measure for symptoms of nonspecific psychological distress during the past 4 weeks and heat exposure between 2006 and 2008 (Ding et al., 2016). The higher average daily maximum temperature in the past 4 weeks was significantly associated with an increase in high or very high distress by 0.2% (99% CI: 0.1%–0.3% per 1°C increase). Another study conducted across Australia during 2008–14 explored the relationship between hotter climate and children's mental health using a national survey of children aged 6–11 years old (Xu et al., 2018). The mental health was evaluated using the Strengths and Difficulties Questionnaire targeted at children 4–16 years old. Through a mediation analysis, the study observed that an increase in annual average daily maximum temperature significantly worsened children's mental health due to a direct and indirect effect via reduced participation in organized physical activities. This association was only observed among boys in the sex-specific analysis.

A study conducted in Australia investigated the relationship between subjective well-being (SWB) and self-reported heat stress due to high temperature using data from an online survey with 1101 respondents between May 9 and 15, 2017 (Zander et al., 2019). The SWB contained life satisfaction, social state, happiness, purpose in life, uncertainty, and capability. The study found that heat stress was significantly associated with uncertainty about future and mentioned that people worry about the impact of increased heat on their future well-being.

A study by Connolly (2013) examined the weather effect on life satisfaction using a sample of 4000 adults surveyed in the United States. The study observed that rain and high temperature had a negative significant impact on life satisfaction in female respondents. Using a survey data obtained from the 75 students of Osaka University in Japan from November 2006 to March 2008, a study assessed the association between daily changes in weather and individual happiness (Tsutsui, 2013). Subjective happiness was only related to temperature among meteorological variables; happiness was maximized at 13.9°C. A large study including 1.9 million Americans aged 18+ investigated whether ambient temperature affects emotional well-being and whether adaptation to different climate could weaken the heat impact on well-being by including large US areas with different climates for 2008–13 (Noelke et al., 2016). Using survey responses from the study participants, the study found that high temperature significantly reduced emotional well-being. Compared with daily mean temperature of 50–60°F (10–16°C), ≥70°F (21°C) reduced positive emotions (joy and happiness), increased negative emotions (stress and anger), and increased fatigue (feeling tired and low energy). The study also observed that individuals living in regions with different climates did not differently respond to heat exposure, suggesting that heat adaptation had a limited role in mitigating the heat impact on well-being in the United States. In India, a study examined whether extreme temperature affects psychological well-being in developing countries with agriculture-based economies using two survey data (the World Health Survey and the Study on global AGEing and adult health) (Pailler & Tsaneva, 2018). The psychological well-being was measured using following items: depressive symptoms, cognitive difficulties, sleep difficulties, inability to cope, and inability to control. The additional month of unusually high temperature during the past year adversely affected cognition, sleep, ability to cope, and ability to control important things in life. Notably, the temperature shocks during the agricultural season significantly increased depressive symptoms. The association between hot temperature and psychological well-being was only observed in rural areas, but not in urban areas.

2.2 Suicide

Suicide is the top 10 leading causes of death in most countries, increasing by 6.7% (95% CI 0.4%–15.6%) between 1990 and 2016 globally (Naghavi, 2019, p. 364). Previous studies have shown the suicidal behavior could be affected by environment characteristics as well as several socioeconomic and biological factors (Goldney, 2000; Yang et al., 2011). In particular, the study investigating the association between temperature and suicide has been conducted in diverse areas and periods. Suicide was primarily defined as E950-E959 and X60-X84 in ninth and tenth revision of the International Classification of Disease (ICD), respectively. Burke et al. showed 0.7% and 2.1% increase of suicide rates in the United States from 1968 to 2004 and Mexico 1990 to 2010, respectively, for a 1°C increase in monthly average temperature (Burke et al., 2018). Furthermore, they projected 9–40 thousand additional suicides across the United States and Mexico by 2050 under RCP 8.5 scenario. In addition, a study conducted in nine countries of the United States between 1975 and 2010 examined the late peak of suicide rates in spring and summer, confirming the average-above temperature was associated with risk of suicide in area located in midlatitude and subtropical climate types (Dixon & Kalkstein, 2018). Dixon et al. also confirmed the warmer week was more likely to be associated with frequency increasement of suicide in distributed lag nonlinear model (DLNM) using maximum and minimum daily temperature (Dixon et al., 2014).

However, research investigating the temperature variation on suicide in 1991–2001 argued the seasonality of suicide was not associated with temperature (Dixon et al., 2007).

In South America, researches on suicide have been conducted actively in Brazil and Mexico. Gaxiola-Robles et al. assessed the association between the ambient maximum temperature and suicide rate in generalized linear model with Poisson distribution for 23 years (Díaz-Castro & Zenteno-Savín, 2013). They confirmed the annual increasement of the number of suicides correlated with temperature in both hot (May–October) and cold (November–April) season ($P < .01$). Positive correlation was also estimated in association between summer and suicidality in Belo Horizonte, Brazil ($r = 0.28$; $P < .05$) (Volpe et al., 2008). In particular, male Mexican completed suicide by hanging in temperature between 30 and 40°C and days without rain when using the database of 1357 patients from 2005 to 2012 (Fernández-Arteaga et al., 2016). In Brazil, a 2.37% increase (95% CI, 0.82–3.96) in male suicide at lag0 was also associated with 1°C increase of temperature in GAM adjusted by cumulative weekly hours of insolation, averages of minimum air temperature, and cumulative weekly rainfall in lag0 to lag3 (Bando et al., 2017).

Long-term research studies on suicide have been investigated mainly in Europe. One study from 1877 through 2000 found nonsignificant understanding of temperature and similar meteorological characteristics on the seasonality of suicide (Ajdacic-Gross et al., 2007). However, most research studies have shown the effects of high temperature on suicide. Page et al. in England and Wales assessed 1°C rise of mean temperature was positively associated with 3.8% and 5% increase for suicide and violent suicide, respectively, without peak in spring and summer between 1993 and 2003 (Page et al., 2007). Study in Austria also showed the effects of 10°C increase on temperature associated with 12% increase of suicide risk in period 1995–2000 (Deisenhammer et al., 2003), suggesting the most significant predictor on suicide rate was higher ambient temperature (Maes et al., 1994). In addition, it has been found that not only hot temperatures but also cold temperatures are associated with suicide (Marion et al., 1999). A study conducted in Finland from 1974 to 2010 showed cold temperature was associated with lower suicide rates in men (Hiltunen et al., 2014).

Likewise, study groups in Asia have investigated the effects of temperature on suicide. Likhvar et al. observed the positive relationship between suicide mortality in Japan and temperature in GAM adjusted by the time trend, season, selected meteorological parameters, day of the week, and holiday, using the historical data for period 1972–95 (Likhvar et al., 2011). A study in Taiwan from 1997 to 2003 confirmed the seasonality of suicide rate in spring (Lee et al., 2006), particularly not showing peak seasonal trend of nonviolent suicide but violent suicide in early to late spring (Lin et al., 2008). In case of China, the pooled relative risk of high temperature defined as 95th percentile of temperature on suicide in multivariate metaanalysis was 1.37 (95% CI, 0.96–2.57) based on 31 capital cities-specific estimation during 2008–13 (Luan et al., 2019). Their findings were greater in male and age below 65 years than female and age over 65 years. A study in Korea from 2001 to 2005 showed similar results that 1°C increase of daily mean temperature was associated with 1.4% (95% CI, 1.0%–1.7%) increase of suicide (Kim et al., 2011). In a study of comparing results in East Asia, pooled country-level estimation in Taiwan (1994–2007), Korea (1992–2010), and Japan (1972–2010) was 7.8% (95% CI, 5%–10.8%) for 2.3°C increase, 6.8% (95% CI, 5.4%–8.2%) for 4.7°C increase, and 4.5% (95% CI, 3.3%–5.7%) for 4.2°C increase, respectively (Kim et al., 2016). Several studies have investigated the temperature effect on the emergency admission caused by suicide in Israel and Iran. The study conducted in a desert area observed patients with suicide attempts over prior 2 days in Israel were positively associated with the cumulative effect of 5°C increment (Yarza et al., 2020) while no significant relationships between temperature and suicide rate were shown in Iran (Talaei et al., 2014).

2.3 Violence

Violence is one of significant public health problems accounting for approximately 1.4 million deaths globally each year and for millions of violence-related injuries (WHO, 2017). Among deaths due to violence, approximately 33% of deaths occur by homicides and assaults by other person and 11% occur by war or other collective violence. The link between heat exposure and violence has been studied for a very long time. It has been consistently observed that high ambient temperature caused aggressive or violent behaviors of individuals towards others (Aguglia et al., 2019).

2.3.1 Heat and collective or intergroup violence

A study using archival data with 102 cases of collective violence examined the relationship between ambient temperature and the occurrence of collective violence in America between 1967 and 1971 (Baron & Ransberger, 1978). The study found that the temperature-collective violence association was curvilinear; the frequency of riots increased when temperature increased up through about 85°F, but shortly decreased after above 90°F. Against this previous study, Carlsmith and Anderson (1979) suggested that the curvilinear relationship was observed not because extreme temperature lowered the probability of collective violence, but because fewer riots occurred on days with temperatures over 90°F. After controlling for the bias, the study observed that relationship between temperature and collective violence became linear.

In human history, armed conflict within countries has caused dreadful consequences all over the world. Burke et al. (2009) firstly investigated the impact of global climate change characterized by global warming on armed conflict in sub-Saharan Africa. There was a significantly positive association between annual temperature and civil war; the 1°C increase in temperature was associated with 4.5% and 0.9% increase in civil war in the same year and the next year, respectively. Moreover, when the estimated historical response was applied to projected future climate, an approximately 54% increase in armed conflict incidence was expected to occur by 2030, which corresponds to an additional 393,000 battle deaths. Another study conducted in East Africa for 1990–2009 evaluated the relationship between climate and conflict by using a geographically disaggregated approach (O'Loughlin et al., 2012). Precipitation and temperature effects were significant; increase in precipitation decreased the risk of conflict (30.3% decrease per 2 standard deviation increase above the long-term mean precipitation), whereas increase in ambient temperature increased the conflict risk (29.6% increase per 2 standard deviation increase above the long-term mean temperature).

A study conducted in the United States assessed the relationship between temperature and retaliation during 57,293 Major League Baseball games between 1952 and 2009 (Larrick et al., 2011). Higher temperature significantly interacted with a larger number of teammates being hit by a pitch and increased the probability of a pitcher subsequently hitting an opposing batter. Increased anger and arousal by heat was suggested as a possible explanation.

A study examined the causal association between climate chocks and collective violence in India between 1954 and 2006 (Gangopadhyay & Nilakantan, 2018). The study found that rather than precipitation, increasing maximum temperature had long- and short-run effects on collective violence (55% increase per 1 standard deviation increase).

Hsiang et al. (2011) focused not on local random weather events but on global climate to evaluate if weather changes are responsible for collective violence and civil conflicts. The study examined the association between the El Niño/Southern Oscillation (ENSO), the dominant interannual mode of the modern climate, and global patterns of civil conflicts for 1950–2004. The risk of new civil conflicts arising across the tropics doubled during El Niño years compared with La Niña years. The ENSO may have accounted for 21% of all civil conflicts since 1950.

2.3.2 Heat and interpersonal violence and crimes

Substantial studies have been conducted in the United States to explore the impact of heat exposure on violence and crimes. Feldman and Jarmon (1979) examined factors affecting criminal behaviors in Newark, New jersey, the United States, from June 1, 1971 through May 31, 1972. Among various weather variables, temperature was only positively correlated with assaults ($r = 0.33$), and the most important prognostic factor of total crime and assault rates in a multiple regression analysis. A study conducted in Dallas, Texas, the United States, from March through October 1980 investigated the relationship between ambient temperature and assaults using a discomfort index (calculated with temperature and humidity) (Harries & Stadler, 1983). The study observed that the daily maximum discomfort index was significantly related to a higher number of incident events of aggravated assaults. By using data of logged calls to police for aid in assaults and other family disturbances, Rotton and Frey (1985) conducted a time series analysis to assess the relationship between assaults and environmental factors including air pollutants and weather variables in Dayton, Ohio, the United States. Daily mean temperature and wind speed were significantly associated with assaults and disturbances. Especially, high temperature and low wind preceded violent events.

A study assessed seasonal changes in assaults, rapes, robberies, and murders and their associations with temperature in 16 locations in the United States for 2–4 years between 1975 and 1979 (Michael & Zumpe, 1983). Using monthly level data, the study observed that mean temperature was highly correlated with mean numbers of assaults ($r = 0.58–0.97$) and rapes ($r = 0.64–0.97$) in most locations. The results of the ordinary least-squares regression analyses also showed significant associations of temperatures with assaults amplitudes and rape amplitudes. Another study examined the association between monthly number of offenses of murder, rape, and aggravated assault and environmental variables in Raleigh, North Carolina, the United States for 1972–81 (Perry & Simpson, 1987). The study also found positive significant correlations between minimum temperature and rape ($r = 0.26$) and aggravated assault ($r = 0.28$). The significant association of temperature with rape and assault was also observed in the stepwise regression analysis. However, among crimes, robbery or property crime were not related to ambient temperature (Anderson et al., 1997; Michael & Zumpe, 1983). It was suggested that property crime was motivated mostly by economic need and thus not an aggressive crime (Anderson & Anderson, 1984).

Recent studies have also consistently found a positive association between heat and various types of violence. A study conducted in Baltimore, MD, for 2008–13 observed that maximum daily temperature was associated with the increased risk of gunshot wounds (IRR, 1.0165, 95% CI, 1.0042–1.0289) along with total crime (1.0069, 1.0056–1.0081), violent crime (1.0128, 1.0096–1.0160), homicides (1.010, 0.9973–1.0228), and ratio of shootings and firearm (RR1.0124, 1.0014–1.0236) in Baltimore City (Michel et al., 2016). Another study conducted in Philadelphia, PA, during 2006–15 found that the rate of violent crimes was 9% (95% CI 6%–12%) higher when the mean daily heat index was at the 99th percentile of the distribution, compared with the median of the distribution of mean daily heat index values (combined temperature and dew point). A Korean study conducted for 2014 also observed a positive and significant linear relationship between temperature and assaults (RR, 1.114; 95% CI, 1.10–1.13).

3. Discussions

To date, a number of studies have been conducted to elucidate the mental health impacts of climate change by focusing on heat exposure globally. The mental health outcomes affected by heatwave and high temperature included mortality due to mental disorders, hospitalizations, ED or outpatient visits, aggravation of symptoms, incidence of mental disorders, dementia, suicide, and violence including assault and crimes.

3.1 Biological mechanisms

Subjects with neurological disorders and dementia are more vulnerable to extreme heat due to physiological, behavioral, and social factors(Lee et al., 2018), but also because psychiatric medications can increase vulnerability to heat-related morbidity by altering the body's ability to thermoregulate (Kim et al., 2014; Merino et al., 2009) due to pharmacologic effects on the parasympathetic pathway (Liu et al., 2019). Our study therefore confirms and supports previous findings of increased risk of exposure to warm weather in persons with mental illness, dementia, and neurological disorders (Isaksen et al., 2016; Kaiser et al., 2001; Khalaj et al., 2010; McWilliams et al., 2013; Myers & Davies, 1978; Rocklöv et al., 2014; Trang et al., 2016; Vida et al., 2012; Wang et al., 2014).

Rising temperatures and heatwaves pose a substantial negative influence on human health (Campbell et al., 2018), and in view of this mass awareness campaigns planning and implementation, health facilities access enhancement, and heatwaves-related early warning schemes development are suggested (Bakhsh et al., 2018; Li et al., 2015).

The current gaps in knowledge regarding high temperature influence on mental health (Thompson et al., 2018) should be filled by carrying out studies that uncover rising temperatures impact on psychological, mental, and behavioral outcomes, to contribute to better understanding of their association and help strategize concrete plans for tackling mental health burden in face of predicted increase in heatwaves in the future. Moreover, efforts should be made to uncover role of elevating temperatures and heatwaves on mental health outcomes of vulnerable populations (Bi et al., 2011).

Information from heatwave events in the United States and Europe suggests that vulnerability to heat can be modified by intrinsic factors such as age (Berry et al., 2010), sex (Stafoggia et al., 2006, 2008), and possibly race (Stafoggia et al., 2006). Other "acquired"' characteristics may also predispose individuals to heightened susceptibility to environmental exposures (Stafoggia et al., 2008). These include preexisting medical conditions such as cardiorespiratory diseases, neurological diseases, and mental illness. For instance, discomfort caused by high temperatures may aggravate behavioral and psychological symptoms in demented patients due to their diminished coping ability to manage stressful environmental conditions and their brain being more sensitive to the increase in temperature, which results in poorer adaptability to heat stress (Cornali et al., 2004). Also, in many studies, heat has found to be associated with escalated aggression resulting in increased likelihood of intergroup conflicts and use of force by police officers (Hsiang et al., 2013).

4. Heat and provocation

In their general aggression model, Anderson and Bushman (2002) proposed three routes by which temperature can lead to aggression: affect, arousal, and cognition (Anderson & Bushman, 2002). Laboratory studies have shown that high temperatures influence behavior through affective aggression and arousal (Anderson & Bushman, 2002; Anderson et al., 1995, 1996, 2000). Conversely, aggression cues (e.g., images of weapons) influence behavior through cognition. Anderson et al. (1995) proposed that the internal states of affect, arousal, and cognition are likely to have interactive effects if they are all triggered at once (Anderson et al., 1995).

5. Future directions

Increasing temperatures, which is a phenomenon of climate change, presents threat to physical, as well as mental health. Keeping in view the effects of warming climate on mental health outcomes, it is important to promote research aiming to decipher environmental heat-induced mental health alterations, to further contribute to build mental health response plans, and reduce heat-mediated mental illness–related morbidity and mortality. High ambient temperatures have a range of mental health effects. The strongest evidence was found for increased suicide risk. Limited evidence was found for an increase in heat-related morbidity and mortality among people with known mental health problems. Knowledge gaps exist on the impact of high temperatures on many common mental health disorders. Mental health impacts should be fully incorporated into plans for the public health response to high temperatures.

References

Aguglia, A., Serafini, G., Escelsior, A., Canepa, G., Amore, M., & Maina, G. (2019). Maximum temperature and solar radiation as predictors of bipolar patient admission in an emergency psychiatric ward. *International Journal of Environmental Research and Public Health, 16*(7), 1140.

Ajdacic-Gross, V., Lauber, C., Sansossio, R., Bopp, M., Eich, D., Gostynski, M., et al. (2007). Seasonal associations between weather conditions and suicide—evidence against a classic hypothesis. *American Journal of Epidemiology, 165*(5), 561–569.

Almendra, R., Loureiro, A., Silva, G., Vasconcelos, J., & Santana, P. (2019). Short-term impacts of air temperature on hospitalizations for mental disorders in Lisbon. *Science of the Total Environment, 647*, 127–133.

Amr, M., & Volpe, F. M. (2012). Seasonal influences on admissions for mood disorders and schizophrenia in a teaching psychiatric hospital in Egypt. *Journal of Affective Disorders, 137*(1–3), 56–60.

Anderson, C. A., & Anderson, D. C. (1984). Ambient temperature and violent crime: Tests of the linear and curvilinear hypotheses. *Journal of Personality and Social Psychology, 46*(1), 91.

Anderson, C. A., Anderson, K. B., & Deuser, W. E. (1996). Examining an affective aggression framework weapon and temperature effects on aggressive thoughts, affect, and attitudes. *Personality and Social Psychology Bulletin, 22*(4), 366–376.

Anderson, C. A., Anderson, K. B., Dorr, N., DeNeve, K. M., & Flanagan, M. (2000). Temperature and aggression. *Advances in Experimental Social Psychology, 32*, 63–133. Elsevier.

Anderson, C. A., & Bushman, B. J. (2002). Media violence and the American public revisited. *American Psychologist, 57*(6–7), 448–450. Epub 2002/07/04. PubMed PMID: 12094445.

Anderson, C. A., Bushman, B. J., & Groom, R. W. (1997). Hot years and serious and deadly assault: Empirical tests of the heat hypothesis. *Journal of Personality and Social Psychology, 73*(6), 1213.

Anderson, C. A., Deuser, W. E., & DeNeve, K. M. (1995). Hot temperatures, hostile affect, hostile cognition, and arousal: Tests of a general model of affective aggression. *Personality and Social Psychology Bulletin, 21*(5), 434–448.

Åström, D. O., Schifano, P., Asta, F., Lallo, A., Michelozzi, P., Rocklöv, J., et al. (2015). The effect of heat waves on mortality in susceptible groups: A cohort study of a mediterranean and a northern European city. *Environmental Health, 14*(1), 30.

Bakhsh, K., Rauf, S., & Zulfiqar, F. (2018). Adaptation strategies for minimizing heat wave induced morbidity and its determinants. *Sustainable Cities and Society, 41*, 95–103.

Bando, D. H., Teng, C. T., Volpe, F. M., Masi, E., Pereira, L. A., & Braga, A. L. (2017). Suicide and meteorological factors in Sao Paulo, Brazil, 1996-2011: A time series analysis. *Brazilian Journal of Psychiatry, 39*(3), 220–227.

Bark, N. (1998). Deaths of psychiatric patients during heat waves. *Psychiatric Services, 49*(8), 1088–1090.

Baron, R. A., & Ransberger, V. M. (1978). Ambient temperature and the occurrence of collective violence: The "long, hot summer" revisited. *Journal of Personality and Social Psychology, 36*(4), 351.

Basagaña, X., Sartini, C., Barrera-Gómez, J., Dadvand, P., Cunillera, J., Ostro, B., et al. (2011). Heat waves and cause-specific mortality at all ages. *Epidemiology*, 765–772.

Basu, R., Gavin, L., Pearson, D., Ebisu, K., & Malig, B. (2018). Examining the association between apparent temperature and mental health-related emergency room visits in California. *American Journal of Epidemiology, 187*(4), 726–735.

Berry, H. L., Bowen, K., & Kjellstrom, T. (2010). Climate change and mental health: A causal pathways framework. *International Journal of Public Health, 55*(2), 123–132. https://doi.org/10.1007/s00038-009-0112-0

Bi, P., Williams, S., Loughnan, M., Lloyd, G., Hansen, A., Kjellstrom, T., et al. (2011). The effects of extreme heat on human mortality and morbidity in Australia: Implications for public health. *Asia Pacific Journal of Public Health, 23*(2_Suppl. l), 27S–36S.

Bullock, B., Murray, G., & Meyer, D. (2017). Highs and lows, ups and downs: Meteorology and mood in bipolar disorder. *PloS One, 12*(3).

Burke, M., González, F., Baylis, P., Heft-Neal, S., Baysan, C., Basu, S., et al. (2018). Higher temperatures increase suicide rates in the United States and Mexico. *Nature Climate Change, 8*(8), 723–729.

Burke, M. B., Miguel, E., Satyanath, S., Dykema, J. A., & Lobell, D. B. (2009). Warming increases the risk of civil war in Africa. *Proceedings of the National Academy of Sciences, 106*(49), 20670–20674.

Calkins, M. M., Isaksen, T. B., Stubbs, B. A., Yost, M. G., & Fenske, R. A. (2016). Impacts of extreme heat on emergency medical service calls in king county, Washington,

2007–2012: Relative risk and time series analyses of basic and advanced life support. *Environmental Health, 15*(1), 13.

Campbell, S., Remenyi, T. A., White, C. J., & Johnston, F. H. (2018). Heatwave and health impact research: A global review. *Health Place, 53*, 210–218. https://doi.org/10.1016/j.healthplace.2018.08.017

Carlsen, H. K., Oudin, A., Steingrimsson, S., & Oudin Åström, D. (2019). Ambient temperature and associations with daily visits to a psychiatric emergency unit in Sweden. *International Journal of Environmental Research and Public Health, 16*(2), 286.

Carlsmith, J. M., & Anderson, C. A. (1979). Ambient temperature and the occurrence of collective violence: A new analysis. *Journal of Personality and Social Psychology, 37*(3), 337.

Cervellin, G., Comelli, I., Lippi, G., Comelli, D., Rastelli, G., Ossola, P., et al. (2014). The number of emergency department visits for psychiatric emergencies is strongly associated with mean temperature and humidity variations. Results of a nine year survey. *Emergency Care Journal, 10*(1).

Chan, E. Y., Lam, H. C., So, S. H., Goggins, W. B., Ho, J. Y., Liu, S., et al. (2018). Association between ambient temperatures and mental disorder hospitalizations in a subtropical city: A time-series study of Hong Kong special administrative region. *International Journal of Environmental Research and Public Health, 15*(4), 754.

Connolly, M. (2013). Some like it mild and not too wet: The influence of weather on subjective well-being. *Journal of Happiness Studies, 14*(2), 457–473.

Cornali, C., Franzoni, S., Riello, R., Ghianda, D., Frisoni, G. B., & Trabucchi, M. (2004). Effect of high climate temperature on the behavioral and psychological symptoms of dementia. *Journal of the American Medical Directors Association, 5*(3), 161–166.

Culqui, D., Linares, C., Ortiz, C., Carmona, R., & Diaz, J. (2017). Association between environmental factors and emergency hospital admissions due to Alzheimer's disease in Madrid. *Science of the Total Environment, 592*, 451–457.

Davido, A., Patzak, A., Dart, T., Sadier, M., Méraud, P., Masmoudi, R., et al. (2006). Risk factors for heat related death during the August 2003 heat wave in Paris, France, in patients evaluated at the emergency department of the Hôpital Européen Georges Pompidou. *Emergency Medicine Journal, 23*(7), 515–518.

Deisenhammer, E., Kemmler, G., & Parson, P. (2003). Association of meteorological factors with suicide. *Acta Psychiatrica Scandinavica, 108*(6), 455–459.

Díaz-Castro, S. C., & Zenteno-Savín, T. (2013). Ambient temperature increase and its possible association with suicide in the Population of Baja California Sur (BCS) 1985–2008. *Salud Mental, 36*, 389–395.

Ding, N., Berry, H. L., & Bennett, C. M. (2016). The importance of humidity in the relationship between heat and population mental health: Evidence from Australia. *PloS One, 11*(10).

Ding, N., BERRY, H., & O'Brien, L. (2015) The effect of extreme heat on mental health—evidence from Australia. *International Journal of Epidemiology, 44*(Suppl. 1). i64.

Dixon, P. G., & Kalkstein, A. J. (2018). Where are weather-suicide associations valid? An examination of nine US counties with varying seasonality. *International Journal of Biometeorology, 62*(5), 685–697.

Dixon, P., McDonald, A., Scheitlin, K., Stapleton, J., Allen, J., Carter, W., et al. (2007). Effects of temperature variation on suicide in five US counties, 1991–2001. *International Journal of Biometeorology, 51*(5), 395–403.

Dixon, P. G., Sinyor, M., Schaffer, A., Levitt, A., Haney, C. R., Ellis, K. N., et al. (2014). Association of weekly suicide rates with temperature anomalies in two different climate types. *International Journal of Environmental Research and Public Health, 11*(11), 11627–11644.

Feldman, H., & Jarmon, R. (1979). Factors influencing criminal behavior in Newark: A local study in forensic psychiatry. *Journal of Forensic Science, 24*(1), 234–239.

Fernández-Arteaga, V., Tovilla-Zárate, C. A., Fresán, A., González-Castro, T. B., Juárez-Rojop, I. E., López-Narváez, L., et al. (2016). Association between completed suicide and environmental temperature in a Mexican population, using the Knowledge Discovery in Database approach. *Computer Methods and Programs in Biomedicine, 135*, 219–224.

Fouillet, A., Rey, G., Laurent, F., Pavillon, G., Bellec, S., Guihenneuc-Jouyaux, C., et al. (2006). Excess mortality related to the August 2003 heat wave in France. *International Archives of Occupational and Environmental Health, 80*(1), 16–24.

Gangopadhyay, P., & Nilakantan, R. (2018). Estimating the effects of climate shocks on collective violence: ARDL evidence from India. *The Journal of Development Studies, 54*(3), 441–456.

García-Herrera, R., Díaz, J., Trigo, R. M., Luterbacher, J., & Fischer, E. M. (2010). A review of the European summer heat wave of 2003. *Critical Reviews in Environmental Science and Technology, 40*(4), 267–306.

Gasparrini, A., Armstrong, B., Kovats, S., & Wilkinson, P. (2012). The effect of high temperatures on cause-specific mortality in England and Wales. *Occupational and Environmental Medicine, 69*(1), 56–61.

Goldney, R. D. (2000). The privilege and responsibility of suicide prevention. *Crisis: The Journal of Crisis Intervention and Suicide Prevention, 21*(1), 8.

Gupta, S., & Murray, R. M. (1992). The relationship of environmental temperature to the incidence and outcome of schizophrenia. *The British Journal of Psychiatry, 160*(6), 788–792.

Hansen, A., Bi, P., Nitschke, M., Ryan, P., Pisaniello, D., & Tucker, G. (2008). The effect of heat waves on mental health in a temperate Australian city. *Environmental Health Perspectives, 116*(10), 1369–1375.

Harries, K. D., & Stadler, S. J. (1983). Determinism revisited: Assault and heat stress in Dallas, 1980. *Environment and Behavior, 15*(2), 235–256.

Henríquez-Sánchez, P., Doreste-Alonso, J., Martínez-González, M. A., Bes-Rastrollo, M., Gea, A., & Sanchez-Villegas, A. (2014). Geographical and climatic factors and depression risk in the SUN project. *The European Journal of Public Health, 24*(4), 626–631.

Hermesh, H., Shiloh, R., Epstein, Y., Manaim, H., Weizman, A., & Munitz, H. (2000). Heat intolerance in patients with chronic schizophrenia maintained with antipsychotic drugs. *American Journal of Psychiatry, 157*(8), 1327–1329.

Hiltunen, L., Haukka, J., Ruuhela, R., Suominen, K., & Partonen, T. (2014). Local daily temperatures, thermal seasons, and suicide rates in Finland from 1974 to 2010. *Environmental Health and Preventive Medicine, 19*(4), 286–294.

Ho, H. C., & Wong, M. S. (2019). Urban environmental influences on the temperature–mortality relationship associated mental disorders and cardiorespiratory diseases during normal summer days in a subtropical city. *Environmental Science and Pollution Research, 26*(23), 24272–24285.

Hsiang, S. M., Burke, M., & Miguel, E. (2013). Quantifying the influence of climate on human conflict. *Science, 341*(6151). https://doi.org/10.1126/science.1235367, 1235367.

Hsiang, S. M., Meng, K. C., & Cane, M. A. (2011). Civil conflicts are associated with the global climate. *Nature, 476*(7361), 438–441.

Isaksen, T. B., Fenske, R. A., Hom, E. K., Ren, Y., Lyons, H., & Yost, M. G. (2016). Increased mortality associated with extreme-heat exposure in King County, Washington, 1980–2010. *International Journal of Biometeorology, 60*(1), 85–98.

Kaiser, R., Rubin, C. H., Henderson, A. K., Wolfe, M. I., Kieszak, S., Parrott, C. L., et al. (2001). Heat-related death and mental illness during the 1999 Cincinnati heat wave. *The American Journal of Forensic Medicine and Pathology, 22*(3), 303–307.

Khalaj, B., Lloyd, G., Sheppeard, V., & Dear, K. (2010). The health impacts of heat waves in five regions of New South Wales, Australia: A case-only analysis. *International Archives of Occupational and Environmental Health, 83*(7), 833–842.

Kim, S.-H., Jo, S.-N., Myung, H.-N., & Jang, J.-Y. (2014). The effect of pre-existing medical conditions on heat stroke during hot weather in South Korea. *Environmental Research, 133*, 246–252.

Kim, Y., Kim, H., Honda, Y., Guo, Y. L., Chen, B.-Y., Woo, J.-M., et al. (2016). Suicide and ambient temperature in East Asian countries: A time-stratified case-crossover analysis. *Environmental Health Perspectives, 124*(1), 75–80.

Kim, Y., Kim, H., & Kim, D.-S. (2011). Association between daily environmental temperature and suicide mortality in Korea (2001–2005). *Psychiatry Research, 186*(2–3), 390–396.

Kim, C. T., Lim, Y.-H., Woodward, A., & Kim, H. (2015). Heat-attributable deaths between 1992 and 2009 in Seoul, South Korea. *PloS One, 10*(2).

Larrick, R. P., Timmerman, T. A., Carton, A. M., & Abrevaya, J. (2011). Temper, temperature, and temptation: Heat-related retaliation in baseball. *Psychological Science, 22*(4), 423–428.

Lee, S., Lee, H., Myung, W., Kim, E. J., & Kim, H. (2018). Mental disease-related emergency admissions attributable to hot temperatures. *Science of the Total Environment, 616*, 688–694.

Lee, H.-C., Lin, H.-C., Tsai, S.-Y., Li, C.-Y., Chen, C.-C., & Huang, C.-C. (2006). Suicide rates and the association with climate: A population-based study. *Journal of Affective Disorders, 92*(2–3), 221–226.

Lee, H.-C., Tsai, S.-Y., & Lin, H.-C. (2007). Seasonal variations in bipolar disorder admissions and the association with climate: A population-based study. *Journal of Affective Disorders, 97*(1–3), 61–69.

Li, M., Gu, S., Bi, P., Yang, J., & Liu, Q. (2015). Heat waves and morbidity: Current knowledge and further direction-a comprehensive literature review. *International Journal of Environmental Research and Public Health, 12*(5), 5256–5283.

Likhvar, V., Honda, Y., & Ono, M. (2011). Relation between temperature and suicide mortality in Japan in the presence of other confounding factors using time-series analysis with a semiparametric approach. *Environmental Health and Preventive Medicine, 16*(1), 36–43.

Linares, C., Culqui, D., Carmona, R., Ortiz, C., & Díaz, J. (2017). Short-term association between environmental factors and hospital admissions due to dementia in Madrid. *Environmental Research, 152*, 214–220.

Lin, H.-C., Chen, C.-S., Xirasagar, S., & Lee, H.-C. (2008). Seasonality and climatic associations with violent and nonviolent suicide: A population-based study. *Neuropsychobiology, 57*(1–2), 32–37.

Liu, X., Liu, H., Fan, H., Liu, Y., & Ding, G. (2019). Influence of heat waves on daily hospital visits for mental illness in Jinan, China—a case-crossover study. *International Journal of Environmental Research and Public Health, 16*(1), 87.

Luan, G., Yin, P., Wang, L., & Zhou, M. (2019). Associations between ambient high temperatures and suicide mortality: A multi-city time-series study in China. *Environmental Science and Pollution Research, 26*(20), 20377–20385.

Maes, M., De Meyer, F., Thompson, P., Peeters, D., & Cosyns, P. (1994). Synchronized annual rhythms in violent suicide rate, ambient temperature and the light-dark span. *Acta Psychiatrica Scandinavica, 90*(5), 391–396.

Marion, S. A., Agbayewa, M. O., & Wiggins, S. (1999). The effect of season and weather on suicide rates in the elderly in British Columbia. *Canadian Journal of Public Health, 90*(6), 418–422.

McMichael, A. J., Woodruff, R. E., & Hales, S. (2006). Climate change and human health: Present and future risks. *Lancet, 367*(9513), 859–869. https://doi.org/10.1016/S0140-6736(06)68079-3

McWilliams, S., Kinsella, A., & O'Callaghan, E. (2013). The effects of daily weather variables on psychosis admissions to psychiatric hospitals. *International Journal of Biometeorology, 57*(4), 497–508.

McWilliams, S., Kinsella, A., & O'Callaghan, E. (2014). Daily weather variables and affective disorder admissions to psychiatric hospitals. *International Journal of Biometeorology, 58*(10), 2045–2057.

Medici, C. R., Vestergaard, C. H., Hadzi-Pavlovic, D., Munk-Jørgensen, P., & Parker, G. (2016). Seasonal variations in hospital admissions for mania: Examining for associations with weather variables over time. *Journal of Affective Disorders, 205*, 81–86.

Merino, A., Mateu, G., Torrens, M., San Gil, J., & Cunillera, J. (2009). Impact of the summer 2003 heat wave on the activity of two psychiatric emergency departments. *Actas Esp Psiquiatr, 37*(3), 158–165.

Michael, R. P., & Zumpe, D. (1983). Annual rhythms in human violence and sexual aggression in the United States and the role of temperature. *Social Biology, 30*(3), 263–278.

Michel, S. J., Wang, H., Selvarajah, S., Canner, J. K., Murrill, M., Chi, A., et al. (2016). Investigating the relationship between weather and violence in Baltimore, Maryland, USA. *Injury, 47*(1), 272–276.

Morrissey, S. A., Raggatt, P. T., James, B., & Rogers, J. (1996). Seasonal affective disorder: Some epidemiological findings from a tropical climate. *Australian and New Zealand Journal of Psychiatry, 30*(5), 579–586.

Myers, D., & Davies, P. (1978). The seasonal incidence of mania and its relationship to climatic variables. *Psychological Medicine, 8*(3), 433–440.

Naghavi, M. (2019). Global, regional, and national burden of suicide mortality 1990 to 2016: Systematic analysis for the global burden of disease study 2016. *Bmj, 364*.

National Institute of Mental Health. (2018). Available from: https://www.nimh.nih.gov/health/statistics/suicide.shtml.

Naughton, M. P., Henderson, A., Mirabelli, M. C., Kaiser, R., Wilhelm, J. L., Kieszak, S. M., et al. (2002). Heat-related mortality during a 1999 heat wave in Chicago. *American Journal of Preventive Medicine, 22*(4), 221–227.

Nitschke, M., Tucker, G. R., Hansen, A. L., Williams, S., Zhang, Y., & Bi, P. (2011). Impact of two recent extreme heat episodes on morbidity and mortality in Adelaide, south Australia: A case-series analysis. *Environmental Health, 10*(1), 42.

Noelke, C., McGovern, M., Corsi, D. J., Jimenez, M. P., Stern, A., Wing, I. S., et al. (2016). Increasing ambient temperature reduces emotional well-being. *Environmental Research, 151*, 124–129.

Nordon, C., Martin-Latry, K., de Roquefeuil, L., Latry, P., Bégaud, B., Falissard, B., et al. (2009). Risk of death related to psychotropic drug use in older people during the European 2003 heatwave: A population-based case–control study. *The American Journal of Geriatric Psychiatry, 17*(12), 1059–1067.

O'Loughlin, J., Witmer, F. D., Linke, A. M., Laing, A., Gettelman, A., & Dudhia, J. (2012). Climate variability and conflict risk in East Africa, 1990–2009. *Proceedings of the National Academy of Sciences, 109*(45), 18344–18349.

Page, L. A., Hajat, S., & Kovats, R. S. (2007). Relationship between daily suicide counts and temperature in England and Wales. *The British Journal of Psychiatry, 191*(2), 106–112.

Page, L. A., Hajat, S., Kovats, R. S., & Howard, L. M. (2012). Temperature-related deaths in people with psychosis, dementia and substance misuse. *The British Journal of Psychiatry, 200*(6), 485–490.

Pailler, S., & Tsaneva, M. (2018). The effects of climate variability on psychological well-being in India. *World Development, 106*, 15–26.

Pan, R., Zhang, X., Gao, J., Yi, W., Wei, Q., Xu, Z., et al. (2019). Impacts of heat and cold on hospitalizations for schizophrenia in Hefei, China: An assessment of disease burden. *Science of the Total Environment, 694*, 133582.

Peng, Z., Wang, Q., Kan, H., Chen, R., & Wang, W. (2017). Effects of ambient temperature on daily hospital admissions for mental disorders in Shanghai, China: A time-series analysis. *Science of the Total Environment, 590*, 281–286.

Perry, J. D., & Simpson, M. E. (1987). Violent crimes in a city: Environmental determinants. *Environment and Behavior, 19*(1), 77–90.

Prince, M., Patel, V., Saxena, S., Maj, M., Maselko, J., Phillips, M. R., et al. (2007). No health without mental health. *The Lancet, 370*(9590), 859–877.

Rey, G., Jougla, E., Fouillet, A., Pavillon, G., Bessemoulin, P., Frayssinet, P., et al. (2007). The impact of major heat waves on all-cause and cause-specific mortality in France from 1971 to 2003. *International Archives of Occupational and Environmental Health, 80*(7), 615–626.

Rocklöv, J., Forsberg, B., Ebi, K., & Bellander, T. (2014). Susceptibility to mortality related to temperature and heat and cold wave duration in the population of Stockholm County, Sweden. *Global Health Action,* 7(1), 22737.

Roehrig, C. (2016). Mental disorders top the list of the most costly conditions in the United States: $201 billion. *Health Affairs, 35*(6), 1130–1135.

Rotton, J., & Frey, J. (1985). Air pollution, weather, and violent crimes: Concomitant time-series analysis of archival data. *Journal of Personality and Social Psychology, 49*(5), 1207.

Schifano, P., Cappai, G., De Sario, M., Michelozzi, P., Marino, C., Bargagli, A. M., et al. (2009). Susceptibility to heat wave-related mortality: A follow-up study of a cohort of elderly in Rome. *Environmental Health, 8*(1), 50.

Semenza, J. C., McCullough, J. E., Flanders, W. D., McGeehin, M. A., & Lumpkin, J. R. (1999). Excess hospital admissions during the July 1995 heat wave in Chicago. *American Journal of Preventive Medicine, 16*(4), 269–277.

Semenza, J. C., Rubin, C. H., Falter, K. H., Selanikio, J. D., Flanders, W. D., Howe, H. L., et al. (1996). Heat-related deaths during the July 1995 heat wave in Chicago. *New England Journal of Medicine, 335*(2), 84–90.

Shapira, A., Shiloh, R., Potchter, O., Hermesh, H., Popper, M., & Weizman, A. (2004). Admission rates of bipolar depressed patients increase during spring/summer and correlate with maximal environmental temperature. *Bipolar Disorders, 6*(1), 90–93.

Sherbakov, T., Malig, B., Guirguis, K., Gershunov, A., & Basu, R. (2018). Ambient temperature and added heat wave effects on hospitalizations in California from 1999 to 2009. *Environmental Research, 160*, 83–90.

Shiloh, R., Munitz, H., Stryjer, R., & Weizman, A. (2007). A significant correlation between ward temperature and the severity of symptoms in schizophrenia inpatients—a longitudinal study. *European Neuropsychopharmacology, 17*(6–7), 478–482.

Shiloh, R., Shapira, A., Potchter, O., Hermesh, H., Popper, M., & Weizman, A. (2005). Effects of climate on admission rates of schizophrenia patients to psychiatric hospitals. *European Psychiatry, 20*(1), 61–64.

Stafoggia, M., Forastiere, F., Agostini, D., Biggeri, A., Bisanti, L., Cadum, E., et al. (2006). Vulnerability to heat-related mortality: A multicity, population-based, case-crossover analysis. *Epidemiology*, 315–323.

Stafoggia, M., Forastiere, F., Agostini, D., Caranci, N., De'Donato, F., Demaria, M., et al. (2008). Factors affecting in-hospital heat-related mortality: A multi-city case-crossover analysis. *Journal of Epidemiology & Community Health, 62*(3), 209–215.

Sung, T.-I., Chen, M.-J., & Su, H.-J. (2013). A positive relationship between ambient temperature and bipolar disorder identified using a national cohort of psychiatric inpatients. *Social Psychiatry and Psychiatric Epidemiology, 48*(2), 295–302.

Talaei, A., Hedjazi, A., Rezaei Ardani, A., Fayyazi Bordbar, M. R., & Talaei, A. (2014). The relationship between meteorological conditions and homicide, suicide, rage, and psychiatric hospitalization. *Journal of Forensic Sciences, 59*(5), 1397–1402.

Thompson, R., Hornigold, R., Page, L., & Waite, T. (2018). Associations between high ambient temperatures and heat waves with mental health outcomes: A systematic review. *Public Health, 161*, 171–191.

Trang, P. M., Rocklöv, J., Giang, K. B., Kullgren, G., & Nilsson, M. (2016). Heatwaves and hospital admissions for mental disorders in Northern Vietnam. *PloS One, 11*(5).

Trautmann, S., Rehm, J., & Wittchen, H. U. (2016). The economic costs of mental disorders: Do our societies react appropriately to the burden of mental disorders? *EMBO Reports, 17*(9), 1245–1249.

Tsutsui, Y. (2013). Weather and individual happiness. *Weather, Climate, and Society, 5*(1), 70–82.

Vaneckova, P., & Bambrick, H. (2013). Cause-specific hospital admissions on hot days in Sydney, Australia. *PloS One, 8*(2).

Vida, S., Durocher, M., Ouarda, T. B., & Gosselin, P. (2012). Relationship between ambient temperature and humidity and visits to mental health emergency departments in Québec. *Psychiatric Services, 63*(11), 1150–1153.

Volpe, F. M., Tavares, A., & Del Porto, J. A. (2008). Seasonality of three dimensions of mania: Psychosis, aggression and suicidality. *Journal of Affective Disorders, 108*(1–2), 95–100.

Wang, X. Y., Barnett, A. G., Yu, W., FitzGerald, G., Tippett, V., Aitken, P., et al. (2012). The impact of heatwaves on mortality and emergency hospital admissions from non-external causes in Brisbane, Australia. *Occupational and Environmental Medicine, 69*(3), 163–169.

Wang, X., Lavigne, E., Ouellette-kuntz, H., & Chen, B. E. (2014). Acute impacts of extreme temperature exposure on emergency room admissions related to mental and behavior disorders in Toronto, Canada. *Journal of Affective Disorders, 155*, 154–161.

Wang, S., Zhang, X., Xie, M., Zhao, D., Zhang, H., Zhang, Y., et al. (2018). Effect of increasing temperature on daily hospital admissions for schizophrenia in Hefei, China: A time-series analysis. *Public Health, 159*, 70–77.

WHO. (2017). *Violence prevention: World health organization* [cited 2020 May 11]. Available from: https://www.who.int/violence_injury_prevention/violence/en/.

Williams, S., Nitschke, M., Sullivan, T., Tucker, G. R., Weinstein, P., Pisaniello, D. L., et al. (2012). Heat and health in Adelaide, south Australia: Assessment of heat thresholds and temperature relationships. *Science of the Total Environment, 414*, 126–133.

World Health Organization. (2018). *Suicide data*. Available from: https://www.who.int/teams/mental-health-and-substance-use/suicide-data.

Xu, Z., Tong, S., Cheng, J., Zhang, Y., Wang, N., Zhang, Y., et al. (2019). Heatwaves, hospitalizations for Alzheimer's disease, and postdischarge deaths: A population-based cohort study. *Environmental Research, 178*, 108714.

Xu, Y., Wheeler, S. A., & Zuo, A. (2018). Will boys' mental health fare worse under a hotter climate in Australia? *Population and Environment, 40*(2), 158–181.

Yang, A. C., Tsai, S.-J., & Huang, N. E. (2011). Decomposing the association of completed suicide with air pollution, weather, and unemployment data at different time scales. *Journal of Affective Disorders, 129*(1–3), 275–281.

Yarza, S., Vodonos, A., Hassan, L., Shalev, H., Novack, V., & Novack, L. (2020). Suicide behavior and meteorological characteristics in hot and arid climate. *Environmental Research*, 109314.

Yi, W., Zhang, X., Gao, J., Wei, Q., Pan, R., Duan, J., et al. (2019). Examining the association between apparent temperature and admissions for schizophrenia in Hefei, China, 2005–2014: A time-series analysis. *Science of the Total Environment, 672*, 1–6.

Zander, K. K., Moss, S., & Garnett, S. T. (2019). Climate change–related heat stress and subjective well-being in Australia. *Weather, Climate, and Society, 11*(3), 505–520.

Zanobetti, A., O'Neill, M. S., Gronlund, C. J., & Schwartz, J. D. (2013). Susceptibility to mortality in weather extremes: Effect modification by personal and small area characteristics in a multi-city case-only analysis. *Epidemiology (Cambridge, Mass), 24*(6), 809.

Brinnel, H., Nagasaka, T., & Cabanac, M. (1987). Enhanced brain protection during passive hyperthermia in humans. *European Journal of Applied Physiology, 56*(5), 540–545. https://doi.org/10.1007/BF00635367

Buffum, M. D., Miaskowski, C., Sands, L., & Brod, M. (2001). A pilot study of the relationship between discomfort and agitation in patients with dementia. *Geriatric Nursing, 22*(2), 80–85.

Elmståhl, S., Annerstedt, L., & Ahlund, O. (1997). How should a group living unit for demented elderly be designed to decrease psychiatric symptoms? *Alzheimer Disease and Associated Disorders, 11*(1), 47–52.

Huffman, J. C., & Kunik, M. E. (2000). Assessment and understanding of pain in patients with dementia. *Gerontologist, 40*(5), 574–581. https://doi.org/10.1093/geront/40.5.574

Ragneskog, H., Gerdner, L. A., Josefsson, K., & Kihlgren, M. (1998). Probable reasons for expressed agitation in persons with dementia. *Clinical Nursing Research*, 7(2), 189–206.

Vrij, A., Van der Steen, J., & Koppelaar, L. (1994). Aggression of police officers as a function of temperature: An experiment with the fire arms training system. *Journal of Community & Applied Social Psychology, 4*(5), 365–370.

Worfolk, J. B. (2000). Heat waves: Their impact on the health of elders. *Geriatric Nursing, 21*(2), 70–77.

Yaqub, B., & Al Deeb, S. (1998). Heat strokes: Aetiopathogenesis, neurological characteristics, treatment and outcome. *Journal of Neurological Science, 156*(2), 144–151. https://doi.org/10.1016/s0022-510x(98)00037-9

CHAPTER 9

Heat exposure and the transmission of infectious diseases

Paul L.C. Chua[1], Xerxes T. Seposo[2,3,4] and Masahiro Hashizume[1,5]
[1]Department of Global Health Policy, Graduate School of Medicine, University of Tokyo, Bunkyo-ku, Tokyo, Japan; [2]Department of Hygiene, Graduate School of Medicine, Hokkaido University, Hokkaido, Japan; [3]School of Tropical Medicine and Global Health, Nagasaki University, Nagasaki, Japan; [4]Ateneo Center for Research and Innovation, Ateneo de Manila School of Medicine and Public Health, Ateneo de Manila University, Pasig, Philippines; [5]Department of Global Health, School of Tropical Medicine and Global Health, Nagasaki University, Nagasaki, Japan

1. Waterborne diseases and heat

1.1 Background

Pathogens and noninfectious agents can be transmitted to humans through water (Griffiths, 2016). Waterborne transmission pathways are not limited to the ingestion of or exposure to water; rather, they also relate to the availability of clean water for the hand washing and sanitation required to separate human and animal feces from water and food (Neil et al., 2017). Waterborne diseases can also be transmitted via aquatic intermediate hosts or arthropods breeding in or biting near open water sources.

About 400 organisms and infectious agents are known to cause waterborne diseases (Neil et al., 2017). In this section, our discussion of waterborne diseases will be limited to the diarrheal diseases considered to be priority water-related diseases by the World Health Organization. Diarrheal diseases are caused by an array of pathogens; therefore, we have only focused on the major enteropathogens (Table 9.1) listed by the Institute of Health Metrics and Evaluation: Global Burden of Disease (Troeger, Blacker, et al., 2018) and the Global Enteric Multicenter Study (Kotloff et al., 2019) because these pathogens cause most of the global health burden seen in children and other age groups (Troeger, Blacker, et al., 2018).

Diarrheal diseases are mainly transmitted via the fecal—oral route wherein food, water, and other surfaces contaminated with fecal material can orally introduce enteropathogens to a susceptible host (Ugboko et al., 2020). They are generally preventable through proper handwashing, use of sanitary toilets, access to clean, safe drinking water, and proper food storage

Heat Exposure and Human Health in the Context of Climate Change
ISBN 978-0-12-819080-7
https://doi.org/10.1016/B978-0-12-819080-7.00003-3
© 2023 Elsevier Inc.
All rights reserved.

Table 9.1 Characteristics of selected diarrheal diseases derived from the United States Centers for Disease Control and Prevention.

Taxa	Pathogen	Incubation period	Signs and symptoms	Mode of transmission
Bacteria	*Aeromonas* spp.	24–48 h	Watery diarrhea; persistent diarrhea; fever; abdominal cramps	Drinking water; foodborne
	Campylobacter spp.	2–5 days	Diarrhea (can be bloody); abdominal pain; fever; nausea	Raw or undercooked poultry; unpasteurized dairy products; contaminated water and produce; animal fecal contact
	Clostridioides (*Clostridium*) *difficile*	>1 week (no-consensus)	Watery diarrhea; fever; stomach tenderness or pain; loss of appetite; nausea; fulminant colitis	Fecal–oral route
	Shiga toxin *Escherichia coli*	1–10 days	Severe stomach cramps; diarrhea (often bloody); vomiting; fever	Contaminated food; unpasteurized milk; drinking water; animal contact
	Enterotoxigenic *E. coli*	1–3 days	Profuse watery diarrhea; abdominal cramping	Contaminated food and water
	Enteropathogenic *E. coli*	0.5–34 h	Diarrhea; abdominal pain; nausea; chills; headache; fever	Contaminated surfaces; weaning fluids; human carriers
	Non-typhoidal *Salmonella* spp.	12–72 h	Diarrhea; abdominal cramps; fever	Contaminated food and water; animal contact
	Shigella spp.	1–8 days	Diarrhea (sometimes bloody); fever; stomach pain; tenesmus	Contaminated surfaces and water; contact with infected individual
	Vibrio cholerae	From 2 h to 5 days	Severe, watery diarrhea; vomiting; severe dehydration	Drinking water; contaminated food; raw or undercooked shellfish
Protozoa	*Cryptosporidium* spp.	2–10 days	Watery diarrhea; stomach cramps or pain; dehydration;	Contaminated soil, food, and water

			nausea; vomiting; fever; weight loss	
	Entamoeba histolytica	2–4 weeks	Diarrhea; stomach pain; stomach cramping	Contaminated surfaces, food, and water
Virus	Adenovirus	From 2 days to 2 weeks	Acute gastroenteritis; flu-like symptoms; fever; sore throat; conjunctivitis	Close contact with infected individuals (e.g., touching and shaking hands; air contact through coughing and sneezing; contact with an infected person's stool; diaper changing)
	Astrovirus	2–3 days	Diarrhea; abdominal pain; vomiting; fever	Fecal–oral route; contaminated food and water
	Norovirus	12–48 h	Diarrhea; vomiting; nausea; stomach pain	Contaminated food and water; touching contaminated surfaces
	Rotavirus	<48 h	Severe watery diarrhea; vomiting; fever; abdominal pain	Fecal–oral route; contaminated food and surfaces
	Sapovirus	<1–4 days	Diarrhea; vomiting; nausea; stomach/abdominal cramps; chills; headache; myalgia; malaise	Fecal–oral route; contaminated food, water, and surfaces

(Webb & Cabada, 2018). Another strategy in preventing diarrheal diseases is by vaccination with the successful vaccines made for rotavirus (Burnett et al., 2020; Troeger et al., 2018), for example, but numerous other vaccines for other enteropathogens are under development (Mokomane et al., 2018). Good childhood nutrition can reduce the risk of diarrheal diseases (Webb & Cabada, 2018). For cases with severe diarrhea and dehydration, oral rehydration salts and zinc supplements are the main interventions used for preventing mortality from diarrheal diseases (Ugboko et al., 2020). Reductions in global diarrhea mortality between 1980 and 2015 were primarily attributed to the use of oral rehydration salts, a reduction in stunting and vitamin A deficiency, along with water, sanitation, and other hygiene interventions (Black et al., 2019).

However, the gains made in reducing diarrheal burden may be hampered by climate change. This is because exposure to changes in ambient temperature, rainfall, flooding, and drought could worsen via climate change (Collins et al., 2013). Such changes can directly or indirectly affect the transmission and incidence of diarrheal diseases (Cissé, 2019; Funari et al., 2012).

This book section focuses on presenting the available evidence on the effects of ambient temperature, particularly heat, on diarrheal diseases. The relationship between heat and diarrhea has been speculated on as far back as the late 1800–1900s when summer diarrhea was a prominent cause of illness and death in young children (Anderson et al., 2020; Zahorsky, 1913). It was theorized that high summertime temperatures directly cause diarrheal symptoms among young children (Zahorsky, 1913). Since then, more studies have been done on the relationship between temperature and diarrheal diseases using more advanced research tools than the simple observations from time-series plots of diarrhea cases versus temperature.

In this section, we present the findings from epidemiological studies analyzing the associations between temperature and diarrhea. The findings from studies on the effects of temperature on the growth and transmission of diarrhea-causing enteropathogens are also presented. Considering the projected future rises in global temperatures (Fan et al., 2020), the findings from studies projecting temperature-related diarrheal diseases are explored and presented. The last part of this section presents the possible future research that could potentially expand and improve the scientific literature on the effects of heat on diarrheal diseases.

1.2 Epidemiological studies

Past studies have largely relied on observations in seasonal changes in diarrhea occurrences or comparisons between time-series patterns of diarrheal occurrences and ambient temperatures (Dean & Jones, 1972;

Woodward et al., 1974). Some have also applied correlation and linear regression to ambient temperatures and diarrhea cases (Lama et al., 2004; Singh et al., 2001; Victora et al., 1985). The earliest advance in analyzing the associations between temperature and diarrhea was from a Peruvian study published in 2000 wherein linear regression models with autoregressive moving-average errors and regression splines were used (Checkley et al., 2000). The study reported that daily hospital admissions for childhood diarrhea increased by 8% [95% confidence intervals (CI): 7%–9%] per 1°C daily temperature increase in 1993–98.

Since then, numerous epidemiological studies have reported associations between temperature and diarrheal diseases in other locations with various enteropathogens. A comprehensive systematic review of quantitative studies in 2016 found that most of the included studies (i.e., 65% of 82 studies) reported positive associations between temperature and diarrhea, particularly all-cause infections and bacterial infections (Levy et al., 2016). The complementary metaanalysis of the same systematic review reported pooled estimates of 7% (95% CI: 3%–10%) increase in the incidence of all-cause diarrhea per 1°C rise in ambient temperature (Carlton et al., 2016). The study also reported a similar percentage change for bacterial diarrhea (i.e., cholera, campylobacteriosis, salmonellosis, typhoid, and bacillary dysentery) at 7% (95% CI: 4%–10%) per 1°C temperature rise. However, viral diarrhea (i.e., rotavirus and norovirus) was found to decrease with increasing temperatures, and the reductions were up to 18% (Carlton et al., 2016).

With the availability of more recent study data, an updated systematic review and metaanalysis study on the associations between temperature and diarrheal diseases for studies published between 2000 and 2019 was conducted in 2020. This study reported a 3% (95% CI: 1%–5%) increase in all-cause diarrheal diseases per 1°C temperature rise (Chua et al., 2022, pp. 19), four percentages lower than that reported in the earlier metaanalysis study (Carlton et al., 2016). The availability of additional studies has allowed researchers to conduct pathogen-specific metaanalysis rather than taxa-specific ones. One recent study reported that positive associations were generally found for bacterial diarrhea, the strongest of which were for typhoid and shigellosis (Fig. 9.1). Negative associations were reported for rotavirus and norovirus. But the highest pathogen-specific pooled estimates were for pathogens such as *Salmonella typhi* and norovirus where few studies were conducted. Another limitation from that systematic review and metaanalysis is the lack of studies on protozoal diarrheal infections such as cryptosporidiosis and amebiasis, which are quite common in many parts of the developing world. More studies are needed on pathogen-specific diarrhea, and this may be facilitated by the availability of better diagnostic

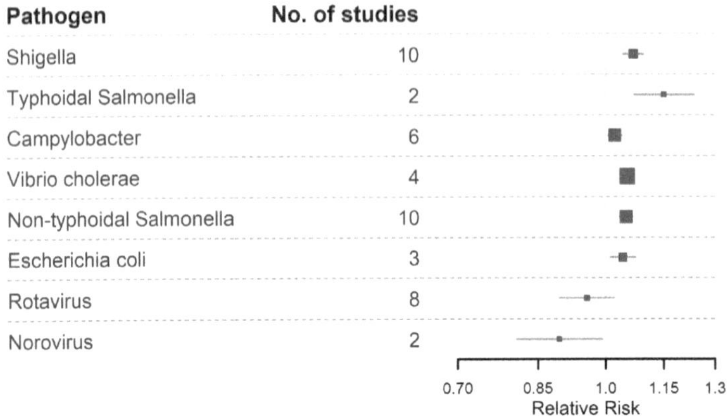

Figure 9.1 Pooled pathogen-specific temperature–diarrhea associations. *(Adopted from Chua, P. L. C., Fook, C., Ng, S., Tobias, A., Seposo, X. T., & Hashizume, M. (2022). Associations between ambient temperature and enteric infections by pathogen: A systematic review and meta-analysis. The Lancet Planetary Health. 6(3): e202-e218.)*

tools than those previously used for routine diagnostics and were often labor-intensive and lacked sensitivity (Ugboko et al., 2020).

The two metaanalysis studies reported high heterogeneity for the pooled estimates despite subgroup analyses on climate and income groups (Carlton et al., 2016; Chua et al., 2022). This is because there were several sources of heterogeneity in the studies that reported percentage increases in the diarrheal incidence per 1°C temperature rise. This heterogeneity includes differences in the exposure and outcome data used, transmission pathways, study years, number of lags, age groups, modeling approaches, and, most importantly, temperature ranges. Therefore, a unified modeling approach may help to address the issue of heterogeneity, and a nonlinear time-series modeling approach should provide an opportunity to resolve this (Gasparrini, 2014).

For the other pathogens in Table 9.1, their associations with ambient temperature varied. *Aeromonas* infections in cirrhotic patients were found to be correlated with high monthly temperatures in southern Taiwan in 2003–13 (Syue et al., 2016). Cryptosporidiosis was positively associated with temperature in two Australian studies citing relatively high percentage increases of 17% (95% CI: 8%–26%) in 1996–2004 (Hu et al., 2010) and 22% (95% CI: 13%–31%) in 2001–09 (Kent et al., 2015), per 1°C rise in weekly maximum and monthly minimum temperatures, respectively. However, a New Zealand study found negative associations between annual temperatures and cryptosporidiosis, and a percentage reduction of 2% (95% CI: 1%–3%) in 1997–2006 (Britton et al., 2010). *Entamoeba histolytica* infections were positively associated with monthly and annual mean and maximum temperatures in a Turkish city in 2001 and in Mexico in 2010 (Erdem et al., 2003; Zavala et al., 2020). Two studies on *Clostridioides difficile* in South Korea and Australia found no associations with ambient temperatures (Furuya-Kanamori et al., 2014). For enteric viruses, a study in the capital city of Burkina Faso found that monthly mean temperatures were negatively correlated with adenovirus and sapovirus cases, whereas astrovirus was not correlated with monthly mean temperatures in 2011–12 (Ouedraogo et al., 2017).

In terms of summer heat and heatwaves, several studies reported increased risks in hospital admissions for diarrheal diseases. In a study in New York State, United States, relatively low positive associations were reported between daily summer maximum temperatures and daily diarrheal hospital admissions in 1991–2004 in lag 0–3 days (Lin et al., 2016). Additionally, extreme heat or the 90th percentile of daily maximum temperatures was positively associated with daily diarrheal hospital admissions in New York State in lag 0–4 days. In Zurich, Switzerland, heatwaves reportedly increased the risk of daily diarrheal hospital admissions by 5% (95% CI: 2%–7%) for every additional day within the heatwave period with lags of up to 14 days in 2001–05 (Manser et al., 2013). This study defined heatwaves as 6 days with maximum temperatures 5°C above the daily average maximum temperature. However, the added effect of heatwaves on diarrheal emergency department visits in Brisbane, Australia, was only found in the 99th percentile of land surface temperatures in 2001–10 (Xu et al., 2014).

1.3 Heat impacts on enteropathogen growth and transmissibility

Heat is expected to affect the growth and survival of enteropathogens (Levy et al., 2016). Warm temperatures provide an ideal growth situation for bacterial enteropathogens, and some of them can survive relatively high temperatures. *Aeromonas* spp. have optimal growth at 35–37°C (Parker & Shaw, 2011), which possibly explains the correlation between *Aeromonas* infections and warm temperatures. Members of this genus produce heat shock proteins or contain polyhydroxyalkanoate synthesis genes, which help them to tolerate heat stress and survive high temperatures (Awan et al., 2018). *Campylobacter* spp. are thermophilic and grow between 37 and 42°C (minimum growth at 30°C) (Bolton, 2015; Rukambile et al., 2019). Their spores, which are the main infection morphotypes and disease transmitters, are generally heat resistant with thermal inactivation only occurring at temperatures exceeding 70°C (Pickering et al., 2019). *C. difficile* infections are yet to be associated with ambient temperatures because most of these infections are nosocomial rather than community-acquired (Czepiel et al., 2019). *Escherichia coli* growth is optimal at 36–40°C in the intestinal tract and 37°C in soil, although its death rate increases over time in warm soil (Jang et al., 2017). Remarkably, foodborne *Salmonella* serovars can grow between 5 and 47°C and optimally at 35–37°C, especially in settings with poor food safety practices or facilities allowing food spoilage (Pui et al., 2011). *Salmonella* serovars have developed heat stress adaption mechanisms and can survive and proliferate in soil and water (Rukambile et al., 2019). *Vibrio cholerae* thrive in water at 30°C temperature, pH 8.5%, and 15% salinity (Chowdhury et al., 2017). Increasing water temperatures, which could be directly affected by warmer ambient temperatures (Lee & Cho, 2015), hasten the growth of *V. cholerae* (Chowdhury et al., 2017).

With protozoal enteropathogens, the impact of warmer temperatures on their survival and infectivity differs. *Cryptosporidium parvum* oocysts reportedly remain infective in water temperatures between 4 and 15°C for over 12 weeks (King et al., 2005). Temperatures beyond 25°C cause increased oocyst die-off rates in water, soil, and feces (Peng et al., 2008), but this contrasts with the positive associations reported between ambient temperature and cryptosporidiosis cases (Hu et al., 2010; Kent et al., 2015). Unlike *C. parvum* oocysts, *E. histolytica* trophozoites can grow between 32 and 40°C and optimally at 35–37°C (Gillin & Diamond, 1980); these findings may support the reported positive associations between ambient temperature and amebiasis cases (Erdem et al., 2003; Zavala et al., 2020).

Viral enteropathogens generally cannot survive warm temperatures. In a systematic review and metaanalysis of experiments reporting the decay rate constants for rotavirus, there were positive associations between water temperatures and decay rates on various water surfaces (Boehm et al., 2019). Generally, rotavirus survives best in relatively low temperatures of 4–20°C and rapidly loses its infectivity at 37°C (D'Souza et al., 2008). Similarly, adenovirus and astrovirus survive best in low temperatures of 4–20°C on various water surfaces and bodies (Bosch et al., 1997; Rigotto et al., 2011). Surrogates for human norovirus can survive well at 4 and 15°C on oysters and peppers but rapidly inactivate at 25 and 40°C (Lee et al., 2015). Another study also reported the rapid inactivation of surrogate human norovirus at 37°C and beyond (Seo et al., 2012). The greater inactivation of viral enteropathogens under warm temperatures may explain why a negative association exists between ambient temperature and viral diarrhea.

The experimental studies conducted on the effects of temperature on the growth and survival of enteropathogens have limitations that may not truly explain the associations between ambient temperature and diarrheal diseases (Goddard et al., 2020). First, these studies were conducted in laboratories using highly specific conditions for culturing the various enteropathogens, and the different media they used would not truly represent the natural environment outside a human host. Second, enteropathogen growth and survival depend on more than ambient temperatures; rather, a full set of environmental conditions (e.g., nutrients, oxygen, light, and moisture) is necessary (Levy et al., 2016; Rzezutka & Cook, 2004).

For some zoonotic enteropathogens, warm seasons can affect their presence and proliferation in nonhuman animals because these animals are reservoirs for human enteric infections. Two studies reported that *Campylobacter*-positive broilers peak in summer and their incidence was positively correlated with monthly mean temperatures at lag 0 and 1 month in some European countries (Hofshagen & Kruse, 2005; Jore et al., 2010). A slightly older study also observed high *Campylobacter* isolation rates in dairy cows in summer in New Zealand and England (Stanley & Jones, 2003). In a study of Dutch and Argentine farms, *E. coli* isolates in cattle were detected more in summer than in winter (Fernández et al., 2009; Schouten et al., 2005). *C. parvum* oocysts in fecal samples from preweaned calves were relatively higher in number compared with postweaned and adult dairy cattle; they were also more prevalent during summer than winter in New York City, United States (Szonyi et al., 2010). However, other studies have reported contradictory or no clear seasonality of enteropathogen detection in animals (Brankston et al., 2018; Johnsen et al., 2006), thereby showing variability in the geographical and other factors involved in seasonality.

Animals play a major role in transmitting enteric infections to humans by contaminating water, soil, food, and other surfaces with their manure. This is because unlike human feces, which is treated before environmental release, animal manure is not usually treated (Suresh et al., 2012). Households with livestock are more likely to have animal manure present. A study in rural Bangladesh showed that households with livestock or animal feces in their surroundings had relatively high *E. coli* levels in soil, stored water, and food; in turn, they had an increased presence of *E. coli* on flies and on their children's hands (Ercumen et al., 2017). An increased risk of *E. coli* contamination of household drinking water at the point of consumption was also reportedly associated with ownership of large numbers of livestock in Ghana, Nepal, and Bangladesh (Wardrop et al., 2018). To simulate diarrhea in children caused by contaminated drinking water, a study in an Indian city used an agent-based model that incorporated recorded ambient temperatures and rainfall, and the model estimated an increased diarrhea incidence of 6% per 1°C temperature rise (Mellor et al., 2016). However, a Peruvian study did not find any reduced risk in clinic visits for diarrhea in locations with high access to piped water (Delahoy et al., 2021), although water safety and quality are not assured from piped water sources. In Minnesota, United States, summer farm day camps or farm visitations (where the guests bottle fed the calves) were thought to have caused two diarrheal illness outbreaks from *E. coli* and *C. parvum* (Smith et al., 2004). Apart from direct contact with calves, paddock soils and forage could have been the source of *C. parvum* oocysts as demonstrated on a farm in Virginia, United States (Boyer & Kuczynska, 2010).

Summer-related recreational activities can also lead to enteric infections. One is recreational swimming during summer, especially in untreated water bodies such as rivers and lakes where water ingestion contaminated with protozoal oocysts can occur (Ikiroma & Pollock, 2020). In Belgium, France, and the Netherlands, relatively low numbers of *Cryptosporidium* oocysts were detected in recreational lakes (Coupe et al., 2006; Ehsan et al., 2015; Schets et al., 2008), whereas in Northern Spain and Western Australia, recreational water bodies were found to have relatively high numbers of *Cryptosporidium* oocysts (Castro-Hermida et al., 2010; Loganthan et al., 2012). Even treated waterbodies can be the sources of diarrheal diseases. Treated public swimming pools were also associated with 76 *Cryptosporidium*-related diarrhea outbreaks in the United States between 2015 and 2019 (Hlavsa et al., 2021). Summer barbequing was another recreational activity associated with salmonellosis in a community in Ontario, Canada (Ravel et al., 2010). The study authors found that monthly salmonellosis cases were associated with the monthly mean temperature, but not from the monthly proportion of *Salmonella*-positive retail chicken breasts. Rather, the summer salmonellosis cases were associated with attending barbeque events, as compared with the nonsummer cases where no barbeque events usually occurred.

1.4 Climate change impact projections

Given the associations between temperature and diarrheal diseases, the several studies that have conducted projections on future temperature-related diarrhea (Kovats & Lloyd, 2014; Liu et al., 2020; Mellor, Levy, et al., 2016; Onozuka et al., 2019) were based on projected future changes in temperature, which is expected to increase over time primarily from anthropogenic greenhouse gas emissions (Rosenzweig et al., 2008). Projection studies on the impact of climate change on future diarrhea episodes are valuable for guiding the development of adaptation strategies for governments and policymakers (Levy et al., 2018). Moreover, depending on their results, projection studies may also support the mitigation of the anthropogenic greenhouse gas emissions that heavily contribute to climate change (Nissan & Conway, 2018).

Old global-level projections have reported increasing risks for temperature-related all-cause diarrhea incidence in geographical regions containing developing nations such as Africa and South Asia by 2030, compared with the risks in 2000 (Campbell-Lendrum et al., 2003; McMichael et al., 2004). Fewer risks were assumed for developed countries in future with mitigated emissions. However, these assertions were only based on two studies that quantified the associations between temperature and all-cause diarrhea cases in Lima, Peru, and Fiji (Checkley et al., 2000; Singh et al., 2001). This is a major limitation because the modeling approach in projections relies more on the empirical climate—health impacts than the variability of global climate models (Kolstad & Johansson, 2011). Using a wide array of linear regression coefficients of temperature—diarrhea associations, together with the temperature changes from 19 global climate models, one study estimated the global diarrhea risk at 22%—29% in 2070—2099 compared with only 8%—11% in 2010—2039 per 1°C temperature rise (Kolstad & Johansson, 2011). The authors explained that the temperature—diarrhea associations contributed more in the projections rather than the variability of temperature change from the 19 global climate models (Kolstad & Johansson, 2011). Hence, with more studies available, a global projection of diarrheal deaths in children was conducted in 2014, which used the temperature—diarrhea associations from six studies but only a single emission scenario (Kovats & Lloyd, 2014). The study highlighted the importance of economic status as the authors estimated the excess temperature-related all-cause diarrheal deaths to be up to 81,000 under the low economic growth scenario compared with the high economic growth scenario of only 30,000 in 2050. The estimated diarrhea incidence increase from temperature changes is expected to increase the costs of treating diarrheal symptoms (Ebi, 2008; Trærup et al., 2011).

A major shortcoming of the all-cause diarrhea approach in the global projections is the failure to take into consideration the negative associations between ambient temperature and viral diarrhea. In a projection study in Japan, the risks of diarrhea were greater in cold temperatures than in hot temperatures in 47 prefectures, and this resulted in projected overall reductions of temperature-related diarrhea cases of up to 15% under a high temperature rise by 2090–99 relative to 2010–19 (Onozuka et al., 2019). The authors associated the overall reductions from the viral enteropathogens that were assumed to comprise most diarrhea cases in Japan (Thongprachum et al., 2016). To consider the temperature associations of viral diarrhea, a study used a pathogen-specific approach in an Indian city to account for diarrhea infections caused by *E. coli*, *Cryptosporidium*, and rotavirus in children from contaminated drinking water (Mellor, Levy, et al., 2016). The authors used agent-based modeling with quantitative microbial risk assessment to analyze the dose–response relationships from the existing studies. They reported an overall increase of 18% (range: 9%–26%) in temperature-related all-cause diarrhea given the positive associations for *E. coli* and *Cryptosporidium;* this was despite projected reductions of about 10% in rotavirus infections in 2080–2099. The pathogen-specific approach was also used in a recent global projection study that included 10 different enteropathogens (Chua et al., 2022). Similarly, an overall increase in global excess temperature-related enteric infection-related deaths was estimated; this was despite reductions from viral enteropathogens in 2080–2095 against the baseline comparison of 1990–2005 (Fig. 9.2). Most of the excess deaths from enteric infections were from sub-Saharan Africa and South Asia, whereas many of the other regions such as East Asia and the Pacific and Latin America and the Caribbean had overall reductions (Chua et al., 2022).

1.5 Further research

There is an abundance of studies quantifying the historical associations between ambient temperature and diarrheal diseases, especially in the years since data and modeling tools became more widely available. But some aspects of this research field remain poorly understood. One is the temperature association related to diarrhea mortality. Few studies have

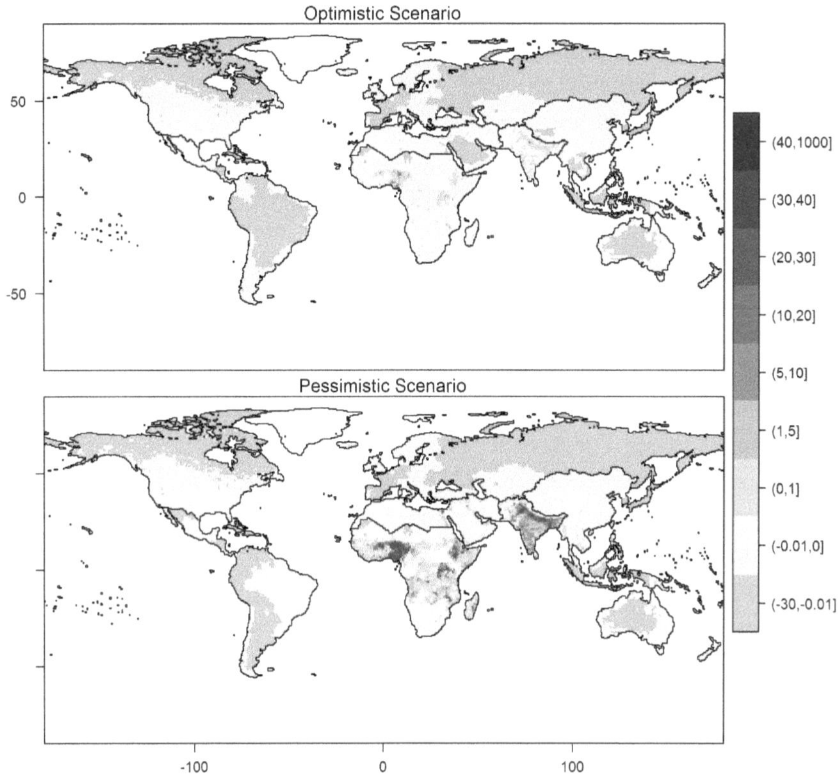

Figure 9.2 Projected number of excess temperature-related enteric infection deaths in 2080–2095 by 0.5 × 0.5 degrees grids. The most optimistic scenario is the combination of Representative Concentration Pathway (RCP) 2.6, Shared Socioeconomic Pathway (SSP) 1, and additional health investments. The most pessimistic scenario is the combination of RCP 3, SSP 3, and low health investments. *(Adopted from Chua, P.L., Huber, V., Ng, C.F., Seposo, X.T., Madaniyazi, L., Hales, S., Woodward, A., Hashizume, M. Global projections of temperature-attributable mortality due to enteric infections: a modelling study. The Lancet Planetary Health. 2021 Jul 1;5(7):e436-45.)*

established that diarrhea mortality is positively associated with temperature (in Mexico and the Netherlands) (Alonso et al., 2012; Ekamper et al., 2009), and only one study quantified such an association in a region of the Philippines (Chua et al., 2022). This type of research does not materialize because data are lacking for diarrheal mortality, particularly in low- to middle-income countries. Therefore, more empirical work is needed to determine whether temperature and diarrheal mortality are associated. The advent of new data sets like the ones from MDS (Million Death Study) (Farrar et al., 2019) and CHAMPS (Child Health and Mortality Prevention

Surveillance) (Taylor et al., 2020) could help to provide quality data for modeling temperature associations.

Apart from mortality, another poorly understood aspect is the modifying impact of social and ecological factors on the associations between temperature and diarrhea. Studies in this area are valuable because they help to provide information on vulnerability and opportunities for adaptive capacity (Levy et al., 2018). For example, two metaanalysis studies conducted subgroup analyses for country income groups and reported slightly higher temperature—diarrhea associations for the studies conducted in low- to middle-income countries compared with high-income countries (Carlton et al., 2016; Chua et al., 2020,). On a smaller-scale, temperature—diarrhea associations are reportedly higher in rural areas than in urban areas (Phung et al., 2015; Thiam et al., 2017). This may explain why temperature associations could be attenuated by better socioeconomic status because more resources would be available for diarrhea prevention and treatment.

Furthermore, other modeling approaches apart from time-series regressions could be explored. One promising approach is mechanistic modeling wherein specific transmission processes are modeled between temperature and diarrheal infections (Mellor, Levy, et al., 2016). This approach could measure the impacts of specific adaptation measures; however, only a few studies have employed this approach. Agent-based modeling has been used to mechanistically model diarrheal disease transmission from drinking water by considering the actual impacts of ambient temperature and rainfall on water quality and diarrhea aversion through the use of water filters and disinfectants (Mellor, Levy, et al., 2016). Apart from that one study, compartmental models have been used to mechanistically model the water transmission of cholera and rotavirus, although both studies focused on the impacts of rainfall (Kraay et al., 2018; Rinaldo et al., 2012). Such modeling approaches could enable better understanding of the specific transmission pathways and related adaptation strategies for specific pathogens rather than all-cause infections.

Projection studies remain limited in a number for diarrheal diseases. Most of the studies thus far were conducted at city or country levels (Liu et al., 2020; Mellor, Levy, et al., 2016; Moors et al., 2013; Onozuka et al., 2019; Trærup et al., 2011); only two were on a global level (Chua et al., 2022; Kovats & Lloyd, 2014) wherein future temperature-related excess diarrhea cases were estimated. Levy et al. (2018) argued against more research to project the impacts of climate change alone; instead projections that consider the impacts of future social and environmental factors related to diarrheal diseases should be considered (Levy et al., 2018). Numerous limitations, however, still affect projection studies, especially those on a global scale. One limitation is the use of temperature—diarrhea associations. Most studies have neglected the spatial variation of risks given that the variable geographical characteristics across the world could affect the local transmission of diarrheal diseases. Some evidence of variations is apparent for regional climate and economic status, but empirical studies remain limited in describing them (Carlton et al., 2016; Chua et al., 2022, pp. 1—28).

Another limitation is the incorporation of interventions in the projections. One global-scale projection considered the impacts of water, sanitation, and nutrition in reducing diarrheal deaths (Chua et al., 2022); however, no studies have properly projected these factors according to their impacts on future temperature and socioeconomic changes. This is because access to water and sanitation and food availability could be directly affected by temperature rises and increased drought occurrences, thereby reducing their capabilities as interventions in future scenarios involving significant climate change. Oral rehydration salts and zinc supplements are very effective interventions for preventing childhood mortality from diarrheal diseases (Black et al., 2019), but they have yet to be incorporated into projections of diarrheal deaths. Vaccinations are also important interventions that could reduce diarrhea by preventing infections in the first place. A good example is the rotaviral vaccine, which when used in very young children prevents global diarrheal deaths (Troeger, Blacker, et al., 2018). The challenge in incorporating vaccines into projections is that not all countries have introduced them in their population and their delivery relies on the stability of health infrastructure and compliance in the targeted population. There are numerous vaccines for other enteropathogens being developed that could potentially address major infections such as shigellosis, but incorporating them into projections would be very challenging given that none have been used in national vaccination programs.

Ideally, diarrhea projections should not be limited to considering ambient temperatures or heat because other climatic factors such as rainfall will impact diarrheal disease transmission, as demonstrated by various studies (Kraay et al., 2020; Levy et al., 2016). Only one study applied the aforementioned factors to projections on a city, but this has yet to be done at country or global levels because of the inherent intricacies behind the quantification and patterns of rainfall associations for diarrhea, unlike temperature associations.

Alongside projection studies, cobenefits studies could be conducted that favor supporting mitigation policies and efforts for greenhouse gas emissions. These studies quantify the positive health effects of policies and measures in reducing emissions of climate warming pollutants. In a systematic review of studies quantifying the public health cobenefits of reducing greenhouse gas emissions, none investigated diarrheal diseases and all of them referred to the air pollution—related and nutrition-related health outcomes arising from the energy, transportation, industrial, and agricultural sectors (Gao et al., 2018). The difficulty in initiating cobenefits studies for infectious illnesses such as diarrhea is the indirect relationship of greenhouse gases to such diseases. The modeling should first consider the impacts of the emissions to ambient temperatures and other climatic factors before the diarrheal diseases are estimated. Nonetheless, studies aimed at quantifying the averted future diarrhea cases from greenhouse gas emission mitigation have potential to contribute to the growing literature on health cobenefits.

2. Vector-borne diseases and heat

2.1 Vector-borne diseases and mosquito-borne diseases

Vector-borne diseases (VBDs) cause substantial global morbidity and mortality. They account for 17% of all infectious diseases and over 700,000 attributable deaths, and 80% of the world's population are at risk of one or more of them (WHO, 2020). Generally, vectors can be any living organism capable of transmitting infectious pathogens (i.e., viruses, parasites, and bacteria) to humans. The typical vectors of VBDs are mosquitoes, ticks, snails, and fleas, among others. Among the VBDs, mosquito-borne diseases (MBDs) have been the largest contributors to the human VBD disease burden. Malaria tops the list of high-morbidity MBDs. It causes approximately 212 million (range, 148—304 million) cases per annum and is followed by dengue (96 million, range 67—136 million) and lymphatic filariasis (96 million, range 67—136 million) (Franklinos et al., 2019). While the magnitude of their cases is thousands to tens of thousands lower, Chikungunya virus, zika virus, yellow fever, Japanese encephalitis, and West Nile fever are equally important, inflicting a substantial burden on healthcare systems (WHO, 2017). Most VBDs, particularly MBDs, occur in tropical and subtropical climates because these environments are conducive to vector growth and viability. With most of the VBD health burden apportioned to MBDs, sizing up the latter's impact on climate change would be more direct than for other vectors. The following sections of this chapter will focus on mosquito dynamics and transmission pathways and on how climate change might affect MBDs. We will present the relevant, up-to-date modeling techniques employed, the historical and future projections of MBD-related health burden, and finally the future directions of MBD-related research under the wider scope of VBD-related research. While several MBDs exist, this chapter particularly focuses on dengue for illustrative and discussion purposes.

2.2 Mosquito life cycle and disease transmission

An adult female *Anopheles* mosquito can be anatomically simplified into three distinct major parts: head, thorax, and abdomen. The four stages in a mosquito's life cycle are egg, larva, pupa, and imago ("adult"). The adult female mosquito needs blood meals to internally develop its eggs. These eggs will then be laid in water as a raft-like structure. In the aquatic stage, the larvae feed on microorganisms in the water. The larvae go through various developmental stages whereby they molt three times (from the first to the fourth instar), before they pupate (Gomes Ade et al., 1995). After the pupal stage, they further metamorphosize into adults, leaving the aquatic stage behind. Along these stages, there are several conjunctions with climate-related factors (i.e., temperature, humidity, and rainfall) that can affect future disease transmission. These conjunctions have been keystones in examining the role of climate change on MBD-related health burden (Hashizume et al., 2012; Iguchi et al., 2018; Lowe et al., 2021).

Disease transmission depends on a variety of factors that may (at least in part) include the physical life cycle of the mosquitoes and that of the (pathogenic) agent's pathogenesis. In most cases, it will involve the interaction between the mosquitoes and humans. Under certain circumstances, it may require an intermediate host (i.e., snails) for effective transmission. In the case of dengue virus, the transmission pathway may include a dengue-infected mosquito inoculating a completely susceptible human host. Inside the host, an elaborate and complex pathway governs the pathogenesis of the dengue virus, and this is fully discussed by Guzman et al. (2016).

2.3 How climate change pathways affect mosquito-borne diseases

Several weather and climate factors related to climate change are documented to affect MBDs. In the Philippines, Iguchi et al. (2018) observed an inverted U-shaped risk curve for rainfall and dengue incidence. In particular, the authors noted that the highest dengue incidence risk was at 32 millimeters (mm) of rainfall, with decreasing risk running alongside increasing volumes of rain. This risk reduction is loosely related to the physical life stage of the mosquito when the eggs and larval stages are aquatic. High volumes of rain will eventually lead to flushing out of the mosquito eggs/larvae, thereby rendering a reduction in vectors and ultimately lowering the risk of dengue transmission. In Brazil, Lowe et al. (2021) noted that the overall dengue risk increased 0–3 months after extremely wet conditions and 3–5 months after drought conditions. The possible role played by meteorological variables on the risk of MBD transmission has been largely documented in tropical and subtropical countries. To be precise, these roles have been largely linked to the local weather patterns and regional climate trends.

➢ *Local weather pathways*

Local weather pathways (e.g., rainfall, temperature, and relative humidity) are known to affect vector viability. In Kenya, Nosrat et al. (2021) noted that the effect of climate extremes on mosquito vector abundance depended on the particular life stage of *Aedes aegypti*. The authors particularly observed an abundance of *A. aegypti* eggs 1 month following the floods. Elsewhere, in the subtropical city of Porto Alegre, Brazil, adult mosquito abundance was observed to follow a strongly seasonal pattern, with low infestation indices during the winters and high infestation during the summers (da Cruz Ferreira et al., 2017).

- Regional climate pathways

Looking at the bigger climatic picture, several studies have documented the role of regional climate variables such as El Nino southern oscillation (ENSO) and Indian Ocean Dipole on mosquitoes and MBDs (Pramanik et al., 2020; Tipayamongkholgul et al., 2009; Vincenti-Gonzalez et al., 2018). These regional climate variables change on a medium- to long-term basis, ranging from weeks, months, years, to decades. Briefly, regional climate variables such as ENSO and Indian Ocean Dipole are large-scale atmospheric variability patterns capable of remotely affecting distant local weather variables (e.g., ambient temperature and rainfall) (McCabe and Muller, 2002). Several studies have noted that greenhouse warming induces the intensification of these regional climate variables, which then leads to an increase in extreme weather events in local areas (Cai et al., 2018).

Several studies have documented how these regional climate variables are associated with dengue incidence. In Venezuela, Vincenti-Gonzalez et al. (2018) noted that the long-term periodicity of ENSO was associated with local changes in weather factors and played a subsequent role in dengue incidence trends. On the other side of the world, in Australia, Hu et al. (2010) observed that the reduction in the average Southern Oscillation Index during the previous 3–12 months was statistically associated with increased monthly dengue fever cases.

2.4 Modeling associations between climate change and mosquito-borne diseases

The health risks associated with exposure to weather/climate factors are largely driven by the underlying physiological mechanisms usually illustrated through exposure–response (E-R) functions. Several studies have employed a linear fit that assumes a uniform risk across the exposure range (Goto et al., 2013; Thammapalo et al., 2005). Recent advances in the field, however, have allowed a relaxation of the linearity assumption, and this has paved the way for more nonlinear modeling techniques to examine the E-R association between weather/climate factors and dengue incidence. In particular, well-grounded work has been done on the use of temporal and spatial modeling techniques in elucidating these E-R functions.

- Temporal modeling techniques

Traditional temporal modeling techniques that estimate the E-R in environmental epidemiology include time-series and case-crossover analyses. However, for the weather/climate factor and dengue association estimation, time-series modeling is popularly used. In most cases, we use the temporal trends between the exposure of interest (i.e., temperature) and the outcome of interest (i.e., dengue) to first visually surmise whether there are interestingly noteworthy patterns. Such patterns can either be further explored or adjusted for (especially for potential outlier data). In Fig. 9.3, we observe that for Singapore the maxima weekly dengue trend visually corresponds with the maxima of the weekly temperature trends, albeit with some additional lag or lead period.

Time-series models take the basic form of the parameterizations in Eq. (9.1) (Peng et al., 2006):

$$Y_t \sim Poisson(\mu_t) \qquad (9.\ 1)$$
$$\log(\mu_t) = \beta_0 + \beta x_t + f(time) + covariates + \varepsilon$$

The health outcome of interest, Y_t, follows a Poisson distribution with mean μ_t. To account for the non-normal distribution of the (i.e. dengue) data, link functions, in this case, the log form, is employed. The intercept is represented by β_0 and the error term ε, whereas $f(time)$ is the smooth

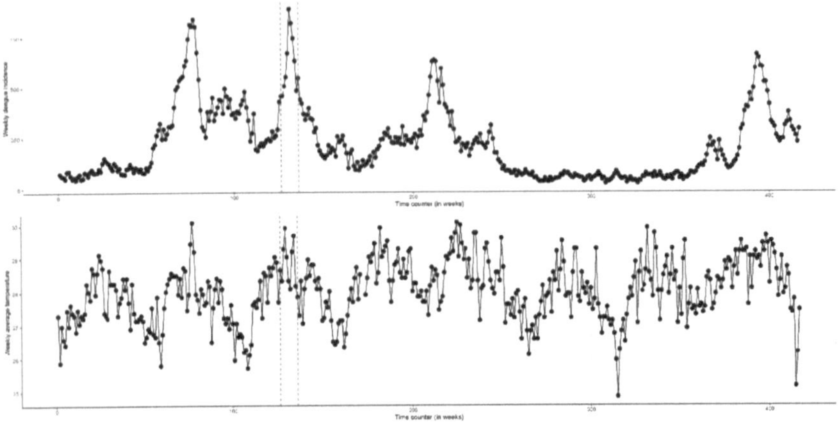

Figure 9.3 Weekly dengue incidence (*upper panel*) and average temperature (*lower panel*) in Singapore, 2012–19. Weekly dengue incidence and average temperatures are shown as black dots. The red (gray in print version) vertical lines correspond to ±5 weeks from the dengue trend maximum. For consistency, we used the same vertical line demarcation for weekly average temperatures.

function of the time variable. Several covariates, which may be potential confounders to the association, may be included, and their adjustments may vary from linear to nonlinear, depending on the existing literature. The focus though is to estimate the β parameter, which represents the linear association between the exposure of interest x_t, and the health outcome of interest, Y_t. Adjustment to account for the nonlinearity of the exposure of interest would entail the substitution of βx_t by a nonlinear function , $f(x_t)$.

Using the publicly available weekly dengue data from Singapore and the simple parameterization in Eq. (9.1), it is possible to derive the E-R function shown in Fig. 9.4. Here, we simplified the estimation using very crude assumptions for illustrative purposes.

- *Spatial modeling techniques*

Apart from temporal associations, several studies have explored how geographically diverse exposures and covariates affect dengue incidence (Aswi et al., 2019; Eisen and Lozano-Fuentes 2009; Karl et al., 2014). In a recent review, Aswi et al. (2019) highlight the popular usage of a Bayesian statistical approach to dengue modeling. Most of the studies included in their review used Bayesian modeling and generalized linear mixed modeling. These models employed a spatially structured random effect using a conditional autoregressive (CAR) prior structure to investigate associations in the risk of dengue and the various exposures of interest. They

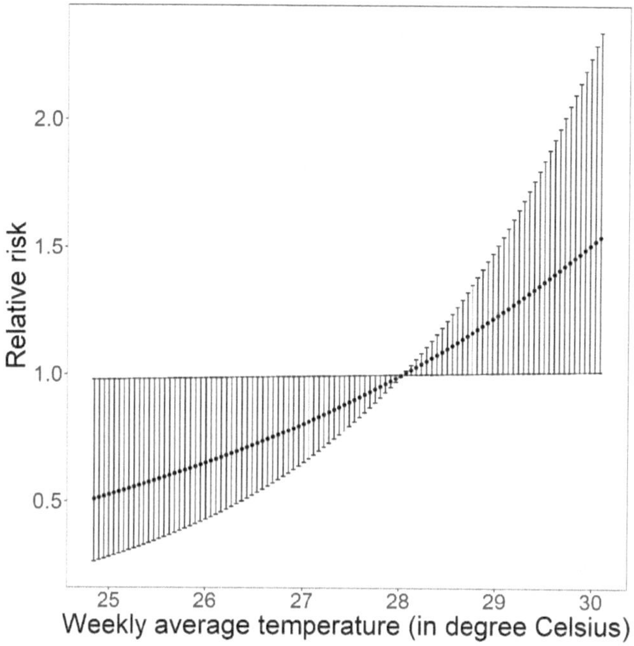

Figure 9.4 Exposure–response function plot for weekly dengue and average temperature in Singapore, 2012–19.

then adjusted for the relevant covariates sourced out from the existing literature. Notably, the spatial modeling trends mostly lean toward the use of empirical Bayes techniques and the full Bayes generalized linear mixed model (GLMM) with a spatial CAR prior.

The empirical Bayes methodology employs the Bayesian principle, wherein the posterior distribution of the outcome of interest is parameterized by combining the prior information and likelihood of the observed data. Contrastingly, a full Bayes model's prior distribution is specified prior to observing the data (Saravana Kumar et al., 2017). The simplified specification of the Bayes generalized linear mixed model with spatial random effects is shown in Eq. (9.2) (Aswi et al., 2019):

$$\begin{aligned} y_i &\sim Poisson(\mu_i) \\ \log(\mu_i) &= \log(e_i) + \theta_i \\ \theta_i &= \alpha + \beta X + \mu_i + v_i \end{aligned} \tag{9.2}$$

wherein y_i is the number of dengue cases in $i = 1, \ldots, I$ areas; e_i and θ_i are, respectively, the expected number of dengue cases in area i and the log relative risk of dengue; α is the overall level of relative risk; $\beta = (\beta_1, \beta_2, \ldots\beta_p)$ represents the coefficient of the covariates; μ_i is a spatially structured random effect with a CAR prior structure, and v_i is a spatially unstructured random effect with mean zero and variance. All authors used an intrinsic Gaussian CAR (ICAR) prior and adopted binary neighborhood weighting. More model specification details are comprehensively discussed by Aswi et al. (2019), Riebler et al. (2016), and Saravana Kumar et al. (2017).

- *Emerging modeling techniques*

Current modeling techniques have been innovated by employing the use of land-use regression, remote sensing techniques, and machine learning algorithms apart from traditional temporal and spatial techniques.

Land-use trends serve as indicators of human activities and the concurrent potential breeding habitats of the mosquito vectors. In Thailand, Vanwambeke et al. (2007) noted that land-use changes have a detectable impact on mosquito populations and on the infections they cause. In Malaysia, Cheong et al. (2014) observed that human settlements, bodies of water, and mixed horticulture lands were associated with dengue patterns in the country.

Remote sensing techniques using satellite imagery data have gained momentum, particularly in their application in human health and especially in locations where exposure measurements have been scarce (Ashby et al., 2017). In Brazil, Lorenz et al. (2020) used remote sensing products such as MODIS (Moderate Resolution Imaging Spectroradiometer), TRMM (Tropical Rainfall Measuring Mission) 3B42, and SRTM (Shuttle Radar Topography Mission) in determining the role of environmental factors in the ecological niche of dengue. In Guangzhou, China, remote sensing variables such as LSTD (land surface temperature of daytime), NDWI (normalized difference water index) land types, and LSTN (land surface temperature of nighttime) were found to be positively correlated with dengue fever outbreaks (Yue et al., 2018).

The increasing complexity in both exposure measurements and outcome measurements has led to the utilization of more flexible modeling techniques to account for these complexities; thus, machine learning and its application to both ecological and health sciences have emerged (Doupe et al., 2019; Olden et al., 2008). In the Philippines, Carvajal et al. (2018) used several machine learning tools (e.g., random forest and gradient boosting) to estimate the effects of meteorological variables on the temporal pattern of dengue incidence in Metropolitan Manila. The authors particularly noted that the random forest model showed the best predictive accuracy among the other statistical techniques.

2.5 Historical and future mosquito-related health burdens

Globally, the total number of dengue cases increased from 23,283,274 (95% uncertainty interval (UI): 453,180.7–51,840,670) in 1990 to 104,771,911 (95% UI: 63,759,019–158,870,031) in 2017. These cases mainly occurred in regions of South Asia, Southeast Asia, and the Caribbean, which have high dengue burdens (Zeng et al., 2021). In Fig. 9.5, a gradual reduction in dengue burden in Indonesia can be seen (*in light green in web version*). Indonesia, however, remains the country with the highest historical dengue burden, whereas the Philippines (*in light blue in web version*) has been experiencing a sustained and increasing dengue burden over the years. In 2017, the Philippines was second only to Indonesia in terms of dengue burden.

Against the backdrop of a changing climate, the current dengue health burden will likely vary in the future; however, this will depend on the relevant and occurring climatic changes, particularly in the exposures of interest. Several geographical areas where disease vectors were previously inviable may become increasingly permissive to vector and viral survival. In modeling future dengue burdens, several studies have used a combination of the historical E-R function, Representative Concentration Pathway (RCP) with Shared Socioeconomic Pathway (SSP), and population

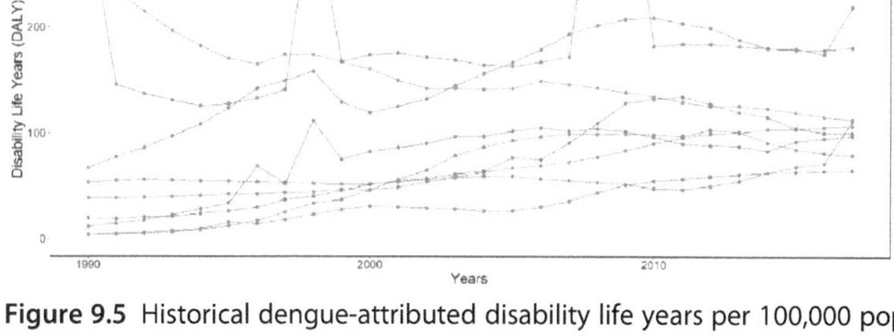

Figure 9.5 Historical dengue-attributed disability life years per 100,000 population for the 10 countries where dengue was highly endemic in 2017. *(Figure generated using the supplementary data from https://doi.org/10.1016/j.eclinm.2020.100712.)*

projections (Messina et al., 2019). Linear E-R functions have been used for future modeling to provide a more practical interpretation of how the changing climate might affect future health burdens.

In brief, the linear assumption implies that the change in risk per unit exposure is homogenous across the concentration range. Usually, the linear health impact assumption for historical burden estimation follows Eq. (9.3):

$$RR = exp^{\beta x} \tag{9.3}$$

$$AF = \frac{(RR - 1)}{RR}$$

$$AN = AF \times Mortality\ rate \times Population$$

Here, the risk estimate, parameter β, which was previously estimated in the linear regression, is exponentiated alongside the concentration, x. The result would yield the relative risk. In basic epidemiology, we assume that the risk from a specific exposure is just one among the sum of the risks from other causes or exposures. Thus, we assume that this risk is just a proportion of the total risk (=1), and we frame this as the attributable fraction. We use the derived attributable fraction to estimate the attributable number of deaths that are the product of the attributable fraction and the baseline mortality. Several studies have used these formulations to estimate both current and future attributable health burden.

This formulation in Eq. (9.3) is further expanded in Eq. (9.4) to account for the future scenarios relative to the RCP, SSP, and the future population structure.

$$RR = exp^{\beta\left(x_{RCP-SSP\ scenario,\ future\ year} - x_{baseline\ year}\right)} \tag{9.4}$$

$$AF = \frac{(RR - 1)}{RR}$$

$$AN = AF \times Mortality\ rate_{baseline\ year} \times Population_{future\ year}$$

Here, we assume that the difference between the future RCP—SSP-derived exposure and baseline year exposure is caused by climate change subject to the RCP—SSP scenarios. Holding the E-R and mortality rate constant in the future projections, we assume that whatever changes we observe are driven by three variabilities, RCP, SSP, and population, which, in principle, may be potentially actionable to a certain extent. In essence, policymakers may tweak how we achieve the exposure levels subject to the RCP and SSP combinations. Likewise, the possibility exists to understand how population structures can change and how we can use this information to reduce the susceptibility of a population to exposures.

2.6 Future directions for climate change and mosquito-borne disease studies

Potential developments are foreseeable for emerging technologies to be used in both health outcome monitoring and exposure measurements. MBD-related health outcomes have heavily relied on weekly incidence counts of syndromic or laboratory-confirmed cases. There are several initiatives in using more individual health outcomes, which may include serological and biomarker data.

Furthermore, as mentioned earlier, the scarcity of exposure measurements in resource-constrained settings limits opportunities to explore how changes in exposure can affect existing health outcomes. Advanced tools such as machine learning technologies, while computationally resource- and time-consuming, allow an omnidirectional examination of an association that is less constrained by the modeler's prior specifications.

Lastly, there is merit in examining how climate change affects MBD-related risks, not just at the global and national scale, but also at the granular scale to provide more accurate information for health managers and policymakers alike.

2.6.1 Code availability

The figures generated in this chapter can be readily reproduced using the codes found in this link: https://github.com/Zorkcis55/climatechangeMBD. All utilized data were sourced from publicly available resources sanctioned under the creative commons license.

References

Alonso, W. J., Acuña-Soto, R., Giglio, R., et al. (2012). Spatio-temporal patterns of diarrhoeal mortality in Mexico. *Epidemiology and Infection, 140*, 91–99.

Anderson, D. M., Rees, D. I., & Wang, T. (2020). The phenomenon of summer diarrhea and its waning, 1910-1930. *Explorations of Economic History, 78*, 101341.

Ashby, J., Moreno-Madriñán, M. J., Yiannoutsos, C. T., & Stanforth, A. (2017). Niche modeling of dengue fever using remotely sensed environmental factors and boosted regression trees. *Remote Sensing, 9*, 328.

Aswi, A., Cramb, S. M., Moraga, P., & Mengersen, K. (2019). Bayesian spatial and spatio-temporal approaches to modelling dengue fever: A systematic review. *Epidemiology and Infection, 147*, e33.

Awan, F., Dong, Y., Wang, N., Liu, J., Ma, K., & Liu, Y. (2018). The fight for invincibility: Environmental stress response mechanisms and Aeromonas hydrophila. *Microbial Pathogenesis, 116*, 135–145.

Black, R., Fontaine, O., Lamberti, L., et al. (2019). Drivers of the reduction in childhood diarrhea mortality 1980-2015 and interventions to eliminate preventable diarrhea deaths by 2030. *Journal of Global Health, 9*, 020801.

Boehm, A. B., Silverman, A. I., Schriewer, A., & Goodwin, K. (2019). Systematic review and meta-analysis of decay rates of waterborne mammalian viruses and coliphages in surface waters. *Water Research, 164*, 114898.

Bolton, D. J. (2015). Campylobacter virulence and survival factors. *Food Microbiology, 48*, 99–108.

Bosch, A., Pintó, R. M., Villena, C., & Abad, F. X. (1997). Persistence of human astrovirus in fresh and marine water. *Water Science and Technology, 35*, 243–247.

Boyer, D. G., & Kuczynska, E. (2010). Prevalence and concentration of cryptosporidium oocysts in beef cattle paddock soils and forage. *Foodborne Pathogens and Diseases, 7*, 893–900.

Brankston, G., Boughen, C., Ng, V., Fisman, D. N., Sargeant, J. M., & Greer, A. L. (2018). Assessing the impact of environmental exposures and Cryptosporidium infection in cattle on human incidence of cryptosporidiosis in Southwestern Ontario, Canada. *PLoS One, 13*, e0196573.

Britton, E., Hales, S., Venugopal, K., & Baker, M. G. (2010). The impact of climate variability and change on cryptosporidiosis and giardiasis rates in New Zealand. *Journal of Water and Health, 8*, 561–571.

Burnett, E., Parashar, U. D., & Tate, J. E. (2020). Real-world effectiveness of rotavirus vaccines, 2006–19: A literature review and meta-analysis. *Lancet Global Health, 8*, e1195–e1202.

Cai, W., Wang, G., Dewitte, B., Wu, L., Santoso, A., Takahashi, K., et al. (2018). Increased variability of eastern pacific el niño under greenhouse warming. *Nature, 564*, 201–206.

Campbell-Lendrum, D. H., Corvalan, C. F., & Pruss-Ustun, A. (2003). How much disease could climate change cause? In A. J. McMichael, Diarmid H. Campbell-Lendrum, C. F. Corvalán, et al. (Eds.), *Climate change and human health: Risks and responses* (pp. 133–159). Geneva: World Health Organization.

Carlton, E. J., Woster, A. P., DeWitt, P., Goldstein, R. S., & Levy, K. (2016). A systematic review and meta-analysis of ambient temperature and diarrhoeal diseases. *International Journal of Epidemiology, 45*, 117–130.

Castro-Hermida, J. A., García-Presedo, I., González-Warleta, M., & Mezo, M. (2010). Cryptosporidium and Giardia detection in water bodies of Galicia, Spain. *Water Research, 44*, 5887–5896.

Carvajal, T. M., Viacrusis, K. M., Hernandez, L. F. T., Ho, H. T., Amalin, D. M., & Watanabe, K. (2018). Machine learning methods reveal the temporal pattern of dengue incidence using meteorological factors in metropolitan manila, Philippines. *BMC Infectious Diseases, 18*, 183.

Checkley, W., Epstein, L. D., Gilman, R. H., et al. (2000). Effects of El Nino and ambient temperature on hospital admissions for diarrhoeal diseases in Peruvian children. *Lancet*. https://doi.org/10.1016/S0140-6736(00)82010-3

Cheong, Y. L., Leitão, P. J., & Lakes, T. (2014). Assessment of land use factors associated with dengue cases in Malaysia using boosted regression trees. *Spatial and Spatio-Temporal Epidemiology, 10*, 75–84.

Chowdhury, F. R., Nur, Z., Hassan, N., Seidlein, L., & Dunachie, S. (2017). Pandemics, pathogenicity and changing molecular epidemiology of cholera in the era of global warming. *Annals of Clinical Microbiology and Antimicrobials, 16*, 10.

Chua, P. L., Ng, C. F., Tobias, A., Seposo, X. T., & Hashizume, M. (2022). Associations between ambient temperature and enteric infections by pathogen: a systematic review and meta-analysis. *The Lancet Planetary health, 6*(3), e202–e218. https://doi.org/10.1016/S2542-5196(22)00003-1

Chua, P. L., Ng, C. F., Madaniyazi, L., Seposo, X., Salazar, M. A., Huber, V., & Hashizume, M. (2022). Projecting Temperature-Attributable mortality and hospital admissions due to enteric infections in the Philippines. *Environmental health perspectives, 130*(2), 027011.

Chua, P. L. C., Ng, C. F. S., Rivera, A. S., et al. (2021). Association between ambient temperature and severe diarrhoea in the national capital region, Philippines. *International Journal of Environmental Research and Public Health, 18*, 8191, 2021; 18: 8191.

Cissé, G. (2019). Food-borne and water-borne diseases under climate change in low- and middle-income countries: Further efforts needed for reducing environmental health exposure risks. *Acta Tropica, 194*, 181–188.

Collins, M., Knutti, R., Arblaster, J., et al. (2013). Long-term climate change: Projections, commitments and irreversibility. In T. F. Stocker, D. Qin, G.-K. Plattner, et al. (Eds.), *Climate change 2013: The physical science basis. Contribution of working group I to the fifth assessment report of the intergovernmental panel on climate change* (pp. 1029–1136). Cambridge, United Kingdom and New York, NY, USA: Cambridge University Press.

Coupe, S., Delabre, K., Pouillot, R., Houdart, S., Santillana-Hayat, M., & Derouin, F. (2006). Detection of *Cryptosporidium*, *Giardia* and *Enterocytozoon bieneusi* in surface water, including recreational areas: A one-year prospective study. *FEMS Immunology and Medical Microbiology, 47*, 351–359.

da Cruz Ferreira, D. A., Degener, C. M., de Almeida Marques-Toledo, C., Bendati, M. M., Fetzer, L. O., Teixeira, C. P., et al. (2017). Meteorological variables and mosquito monitoring are good predictors for infestation trends of aedes aegypti, the vector of dengue, chikungunya and zika. *Parasites & Vectors, 10*, 78.

Czepiel, J., Dróżdż, M., Pituch, H., et al. (2019). *Clostridium difficile* infection: Review. *European Journal of Clinical Microbiology Infectious Diseases, 387*, 1211–1221, 2019; 38.

Dean, A. G., & Jones, T. C. (1972). Seasonal gastroenteritis and malabsorption at an American military based in the Philippines. *American Journal of Epidemiology, 95*, 111–127.

Delahoy, M. J., Cárcamo, C., Huerta, A., et al. (2021). Meteorological factors and childhood diarrhea in Peru, 2005–2015: A time series analysis of historic associations, with implications for climate change. *Environmental Health, 20*, 1–10.

Doupe, P., Faghmous, J., & Basu, S. (2019). Machine learning for health services researchers. *Value in Health, 22*, 808–815.

D'Souza, R. M., Hall, G., & Becker, N. G. (2008). Climatic factors associated with hospitalizations for rotavirus diarrhoea in children under 5 years of age. *Epidemiology and Infection*. https://doi.org/10.1017/S0950268807008229

Ebi, K. L. (2008). Adaptation costs for climate change-related cases of diarrhoeal disease, malnutrition, and malaria in 2030. *Global Health, 4*, 1–9.

Ehsan, A., Casaert, S., Levecke, B., et al. (2015). Cryptosporidium and Giardia in recreational water in Belgium. *Journal of Water Health, 13*, 870–878.

Eisen, L., & Lozano-Fuentes, S. (2009). Use of mapping and spatial and space-time modeling approaches in operational control of aedes aegypti and dengue. *PLOS Neglected Tropical Diseases, 3*, e411.

Ekamper, P., van Poppel, F., van Duin, C., & Garssen, J. (2009). 150 years of temperature-related excess mortality in The Netherlands. *Demographic Research, 21*, 385–426.

Ercumen, A., Pickering, A. J., Kwong, L. H., et al. (2017). Animal feces contribute to domestic fecal contamination: Evidence from *E. coli* measured in water, hands, food, flies, and soil in Bangladesh. *Environmental Science and Technology, 51*, 8725–8734.

Erdem, H., Kiliç, S., Çinar, E., & Pahsa, A. (2003). Symptomatic intestinal amoebiasis and climatic parameters. *Scandinavian Journal of Infectious Diseases, 35*, 186–188.

Fan, X., Duan, Q., Shen, C., Wu, Y., & Xing, C. (2020). Global surface air temperatures in CMIP6: Historical performance and future changes. *Environmental Research Letters, 15*, 104056.

Farrar, D. S., Awasthi, S., Fadel, S. A., et al. (2019). Seasonal variation and etiologic inferences of childhood pneumonia and diarrhea mortality in India. *Elife, 8*. https://doi.org/10.7554/eLife.46202

Fernández, D., Rodríguez, E. M., Arroyo, G. H., Padola, N. L., & Parma, A. E. (2009). Seasonal variation of Shiga toxin-encoding genes (stx) and detection of *E. coli* O157 in dairy cattle from Argentina. *Journal of Applied Microbiology, 106*, 1260–1267.

Franklinos, L. H. V., Jones, K. E., Redding, D. W., & Abubakar, I. (2019). The effect of global change on mosquito-borne disease. *The Lancet Infectious Diseases, 19*, e302–e312.

Funari, E., Manganelli, M., & Sinisi, L. (2012). Impact of climate change on waterborne diseases. *Annali dell'Istituto Superiore di Sanità, 48*, 473–487.

Furuya-Kanamori, L., Robson, J., Soares Magalhães, R. J., et al. (2014). A population-based spatio-temporal analysis of *Clostridium difficile* infection in Queensland, Australia over a 10-year period. *Journal of Infection*. https://doi.org/10.1016/j.jinf.2014.06.014

Gao, J., Kovats, S., Vardoulakis, S., et al. (2018). Public health co-benefits of greenhouse gas emissions reduction: A systematic review. *Science of the Total Environment, 627*, 388–402.

Gasparrini, A. (2014). *Distributed lag linear and non-linear models for time series data. 14310721250.*

Gillin, F. D., & Diamond, L. S. (1980). Entamoeba histolytica and Entamoeba invadens: Effects of temperature and oxygen tension on growth and survival. *Experimental Parasitology, 49*, 328–338.

Goddard, F. G. B., Ban, R., Barr, D. B., et al. (2020). Measuring environmental exposure to enteric pathogens in low-income settings: Review and recommendations of an interdisciplinary working group. *Environmental Science and Technology, 54*, 11673–11691.

Gomes Ade, C., Gotlieb, S. L., Marques, C. C., de Paula, M. B., & Marques, G. R. (1995). Duration of larval and pupal development stages of aedes albopictus in natural and artificial containers. *Revista de Saúde Pública, 29*, 15–19.

Goto, K., Kumarendran, B., Mettananda, S., Gunasekara, D., Fujii, Y., & Kaneko, S. (2013). Analysis of effects of meteorological factors on dengue incidence in Sri Lanka using time series data. *PLOS ONE, 8*, e63717.

Griffiths, J. K. (2016). *Waterborne diseases* (2nd ed.). Elsevier. https://doi.org/10.1016/B978-0-12-803678-5.00490-2

Guzman, M. G., Gubler, D. J., Izquierdo, A., Martinez, E., & Halstead, S. B. (2016). Dengue infection. *Nature Reviews Disease Primers, 2*, 16055.

Hashizume, M., Dewan, A. M., Sunahara, T., Rahman, M. Z., & Yamamoto, T. (2012). Hydroclimatological variability and dengue transmission in dhaka, Bangladesh: A time-series study. *BMC Infectious Diseases, 12*, 98.

Hlavsa, M. C., Aluko, S. K., Miller, A. D., et al. (2021). Outbreaks associated with treated recreational water — United States, 2015–2019. *Morbidity and Mortality Weekly Report, 70*, 733–738.

Hofshagen, M., & Kruse, H. (2005). Reduction in flock prevalence of Campylobacter spp. in broilers in Norway after implementation of an action plan. *Journal of Food Protection, 68*, 2220–2233.

Hu, W., Clements, A., Williams, G., & Tong, S. (2010). Dengue fever and el niño/southern oscillation in queensland, Australia: A time series predictive model. *Occupational and Environmental Medicine, 67*, 307–311.

Hu, W., Mengersen, K., Fu, S. Y., & Tong, S. (2010). The use of ZIP and CART to model cryptosporidiosis in relation to climatic variables. *International Journal of Biometeorology*. https://doi.org/10.1007/s00484-009-0294-4

Iguchi, J. A., Seposo, X. T., & Honda, Y. (2018). Meteorological factors affecting dengue incidence in davao, Philippines. *BMC Public Health, 18*, 629-629.

Ikiroma, I. A., & Pollock, K. G. (2020). Influence of weather and climate on cryptosporidiosis—a review. *Zoonoses Public Health, 68*, 285–298.

Jang, J., Hur, H. G., Sadowsky, M. J., Byappanahalli, M. N., Yan, T., & Ishii, S. (2017). Environmental *Escherichia coli*: Ecology and public health implications—a review. *Journal of Applied Microbiology, 123*, 570–581.

Johnsen, G., Zimmerman, K., Lindstedt, B.-A., Vardund, T., Herikstad, H., & Kapperud, G. (2006). Intestinal carriage of Campylobacter jejuni and Campylobacter coli among cattle from South-western Norway and comparative genotyping of bovine and human isolates by amplified-fragment length polymorphism. *Acta Veterinaria Scandinavica, 48*, 1–6.

Jore, S., Viljugrein, H., Brun, E., et al. (2010). Trends in Campylobacter incidence in broilers and humans in six European countries, 1997–2007. *Preventive Veterinary Medicine, 93*, 33–41.

Karl, S., Halder, N., Kelso, J. K., Ritchie, S. A., & Milne, G. J. (2014). A spatial simulation model for dengue virus infection in urban areas. *BMC Infectious Diseases, 14*, 447.

Kent, L., McPherson, M., & Higgins, N. (2015). A positive association between cryptosporidiosis notifications and ambient temperature, Victoria, Australia, 2001–2009. *Journal of Water Health*. https://doi.org/10.2166/wh.2015.130

King, B. J., Keegan, A. R., Monis, P. T., & Saint, C. P. (2005). Environmental temperature controls Cryptosporidium oocyst metabolic rate and associated retention of infectivity. *Applied and Environmental Microbiology, 71*, 3848–3857.

Kolstad, E. W., & Johansson, K. A. (2011). Uncertainties associated with quantifying climate change impacts on human health: A case study for diarrhea. *Environmental Health Perspectives*. https://doi.org/10.1289/ehp.1002060

Kotloff, K. L., Nasrin, D., Blackwelder, W. C., et al. (2019). The incidence, aetiology, and adverse clinical consequences of less severe diarrhoeal episodes among infants and children residing in low-income and middle-income countries: A 12-month case-control study as a follow-on to the global enteric multicenter study. *Lancet Global Health*, 7, e568–e584.

Kovats, S., & Lloyd, S. (2014). Diarrhoeal disease. In S. Hales, S. Kovats, S. Lloyd, & D. Campbell-Lendrum (Eds.), *Quantitative risk assessment of the effects of climate change on selected causes of death, 2030s and 2050s* (pp. 37–50). World Health Organization.

Kraay, A. N. M., Brouwer, A. F., Lin, N., Collender, P. A., Remais, J. V., & Eisenberg, J. N. S. (2018). Modeling environmentally mediated rotavirus transmission: The role of temperature and hydrologic factors. *Proceedings of the National Academy Sciences*. https://doi.org/10.1073/pnas.1719579115

Kraay, A. N. M., Man, O., Levy, M. C., Levy, K., Ionides, E., & Eisenberg, J. N. S. (2020). Understanding the impact of rainfall on diarrhea: Testing the concentration-dilution hypothesis using a systematic review and meta-analysis. *Environmental Health Perspectives, 128*, 126001-1–126001–16.

Lama, J. R., Seas, C. R., León-Barúa, R., Gotuzzo, E., & Sack, R. B. (2004). Environmental temperature, cholera, and acute diarrhoea in adults in Lima, Peru. *Journal of Health, Population and Nutrition*. https://doi.org/10.3329/jhpn.v22i4.289

Lee, K.-H., & Cho, H.-Y. (2015). Projection of climate-induced future water temperature for the aquatic environment. *Journal of Environmental Engineering, 141*, 06015004.

Lee, S. J., Si, J., Yun, H. S., & Ko, G. P. (2015). Effect of temperature and relative humidity on the survival of foodborne viruses during food storage. *Applied and Environmental Microbiology, 81*, 2075–2081.

Levy, K., Smith, S. M., & Carlton, E. J. (2018). Climate change impacts on waterborne diseases: Moving toward designing interventions. *Current Environmental Health Reports, 5*, 272.

Levy, K., Woster, A. P., Goldstein, R. S., & Carlton, E. J. (2016). Untangling the impacts of climate change on waterborne diseases: A systematic review of relationships between

diarrheal diseases and temperature, rainfall, flooding, and drought. *Environmental Science & Technology, 50*, 4905–4922.
Lin, S., Sun, M., Fitzgerald, E., & Hwang, S. A. (2016). Did summer weather factors affect gastrointestinal infection hospitalizations in New York State? *Science of the Total Environment, 550*, 38–44.
Liu, Z., Tong, M. X., Xiang, J., et al. (2020). Daily temperature and bacillary dysentery: Estimated effects, attributable risks, and future disease burden in 316 Chinese cities. *Environmental Health Perspectives, 128*, 057008.
Loganthan, S., Yang, R., Bath, A., Gordon, C., & Ryan, U. (2012). Prevalence of Cryptosporidium species in recreational versus non-recreational water sources. *Experimental Parasitology, 131*, 399–403.
Lorenz, C., Castro, M. C., Trindade, P. M. P., Nogueira, M. L., de Oliveira Lage, M., Quintanilha, J. A., et al. (2020). Predicting aedes aegypti infestation using landscape and thermal features. *Science Reports, 10*, 21688.
Lowe, R., Lee, S. A., O'Reilly, K. M., Brady, O. J., Bastos, L., Carrasco-Escobar, G., et al. (2021). Combined effects of hydrometeorological hazards and urbanisation on dengue risk in Brazil: A spatiotemporal modelling study. *The Lancet Planetary Health, 5*, e209–e219.
Manser, C. N., Paul, M., Rogler, G., Held, L., & Frei, T. (2013). Heat waves, incidence of infectious gastroenteritis, and relapse rates of inflammatory bowel disease: A retrospective controlled observational study. *American Journal of Gastroenterology, 108*, 1480–1485.
McCabe, G. J., & Muller, R. A. (2002). Effects of enso on weather-type frequencies and properties at new orleans, Louisiana, USA. *Climate Research, 20*, 95–105.
McMichael, A. J., Campbell-Lendrum, D., Kovats, S., et al. (2004). Global climate change. In M. Ezzati, A. D. Lopez, A. Rodgers, & C. J. L. Murray (Eds.), *Comparative quantification of health risks: Global and regional burden of disease attributable to selected major risk factors* (pp. 1543–1650). Geneva, Switzerland: World Health Organization.
Mellor, J., Kumpel, E., Ercumen, A., & Zimmerman, J. (2016). Systems approach to climate, water, and diarrhea in hubli-dharwad, India. *Environmental Science & Technology*. https://doi.org/10.1021/acs.est.6b02092
Mellor, J. E., Levy, K., Zimmerman, J., et al. (2016). Planning for climate change: The need for mechanistic systems-based approaches to study climate change impacts on diarrheal diseases. *Science of the Total Environment, 548–549*, 82–90.
Messina, J. P., Brady, O. J., Golding, N., Kraemer, M. U. G., Wint, G. R. W., Ray, S. E., et al. (2019). The current and future global distribution and population at risk of dengue. *Nature Microbiology, 4*, 1508–1515.
Mokomane, M., Kasvosve, I., Melo, E., Pernica, J. M., & Goldfarb, D. M. (2018). The global problem of childhood diarrhoeal diseases: Emerging strategies in prevention and management. *Therapeutic Advances in Infectious Disease*. https://doi.org/10.1177/2049936117744429
Moors, E., Singh, T., Siderius, C., Balakrishnan, S., & Mishra, A. (2013). Climate change and waterborne diarrhoea in northern India: Impacts and adaptation strategies. *Science of the Total Environment*. https://doi.org/10.1016/j.scitotenv.2013.07.021
Neil, K. P., Yoder, J. S., Hall, A. J., & Bowen, A. (2017). *Enteric diseases transmitted through food, water, and zoonotic exposures (5th ed.)*. Elsevier Inc.. https://doi.org/10.1016/B978-0-323-40181-4.00059-1
Nosrat, C., Altamirano, J., Anyamba, A., Caldwell, J. M., Damoah, R., Mutuku, F., et al. (2021). Impact of recent climate extremes on mosquito-borne disease transmission in Kenya. *PLOS Neglected Tropical Diseases, 15*, e0009182.
Nissan, H., & Conway, D. (2018). From advocacy to action: Projecting the health impacts of climate change. *PLoS Medicine*. https://doi.org/10.1371/journal.pmed.1002624

Olden Julian, D., Lawler Joshua, J., & Poff, N. L. (2008). Machine learning methods without tears: A primer for ecologists. *The Quarterly Review of Biology, 83*, 171–193.

Onozuka, D., Gasparrini, A., Sera, F., Hashizume, M., & Honda, Y. (2019). Modeling future projections of temperature-related excess morbidity due to infectious gastroenteritis under climate change conditions in Japan. *Environmental Health Perspectives, 127*, 077006.

Ouedraogo, N., Ngangas, S. M. T., Bonkoungou, I. J. O., et al. (2017). Temporal distribution of gastroenteritis viruses in Ouagadougou, Burkina Faso: Seasonality of rotavirus. *BMC Public Health, 17*, 1–8.

Parker, J. L., & Shaw, J. G. (2011). Aeromonas spp. clinical microbiology and disease. *Journal of Infection, 62*, 109–118.

Peng, R. D., Dominici, F., & Louis, T. A. (2006). Model choice in time series studies of air pollution and mortality. *Journal of the Royal Statistical Society: Series A (Statistics in Society), 169*, 179–203.

Peng, X., Murphy, T., & Holden, N. M. (2008). Evaluation of the effect of temperature on the die-off rate for Cryptosporidium parvum oocysts in water, soils, and feces. *Applied Environmental Microbiology, 74*, 7101–7107.

Phung, D., Huang, C., Rutherford, S., et al. (2015). Association between climate factors and diarrhoea in a Mekong Delta area. *International Journal of Biometeorology*. https://doi.org/10.1007/s00484-014-0942-1

Pickering, D. S., Vernon, J. J., Freeman, J., Wilcox, M. H., & Chilton, C. H. (2019). Investigating the transient and persistent effects of heat on *Clostridium difficile* spores. *Journal of Medical Microbiology, 68*, 1445–1454.

Pramanik, M., Singh, P., Kumar, G., Ojha, V. P., & Dhiman, R. C. (2020). El niño southern oscillation as an early warning tool for dengue outbreak in India. *BMC Public Health, 20*, 1498.

Pui, C. F., Wong, W. C., Chai, L. C., et al. (2011). Salmonella: A foodborne pathogen. *International Food Research Journal, 18*, 465–473.

Ravel, A., Smolina, E., Sargeant, J. M., et al. (2010). Seasonality in human salmonellosis: Assessment of human activities and chicken contamination as driving factors. *Foodborne Pathogens Diseases, 7*, 785–794.

Riebler, A., Sørbye, S. H., Simpson, D., & Rue, H. (2016). An intuitive bayesian spatial model for disease mapping that accounts for scaling. *Statistical Methods in Medical Research, 25*, 1145–1165.

Rigotto, C., Hanley, K., Rochelle, P. A., De Leon, R., Barardi, C. R. M., & Yates, M. V. (2011). Survival of adenovirus types 2 and 41 in surface and ground waters measured by a plaque assay. *Environmental Science Technology, 45*, 4145–4150.

Rinaldo, A., Bertuzzo, E., Mari, L., et al. (2012). Reassessment of the 2010–2011 Haiti cholera outbreak and rainfall-driven multiseason projections. *Proceedings of the National Academy of Sciences of the United States of America, 109*, 6602–6607.

Rosenzweig, C., Karoly, D., Vicarelli, M., et al. (2008). Attributing physical and biological impacts to anthropogenic climate change. *Nature, 453*, 353–357.

Rukambile, E., Sintchenko, V., Muscatello, G., Kock, R., & Alders, R. (2019). Infection, colonization and shedding of Campylobacter and Salmonella in animals and their contribution to human disease: A review. *Zoonoses Public Health, 66*, 562–578.

Rzezutka, A., & Cook, N. (2004). Survival of human enteric viruses in the environment and food. *FEMS Microbiology Reviews*. https://doi.org/10.1016/j.femsre.2004.02.001

Saravana Kumar, V., Devika, S., George, S., & Jeyaseelan, L. (2017). Spatial mapping of acute diarrheal disease using gis and estimation of relative risk using empirical bayes approach. *Clinical Epidemiology and Global Health, 5*, 87–96.

Schets, F. M., Van Wijnen, J. H., Schijven, J. F., Schoon, H., & De Roda Husman, A. M. (2008). Monitoring of waterborne pathogens in surface waters in Amsterdam, The

Netherlands, and the potential health risk associated with exposure to Cryptosporidium and Giardia in these waters. *Applied and Environmental Microbiology, 74*, 2069–2078.

Schouten, J. M., Graat, E. A. M., Frankena, K., Van De Giessen, A. W., Van Der Zwaluw, W. K., & De Jong, M. C. M. (2005). A longitudinal study of *Escherichia coli* O157 in cattle of a Dutch dairy farm and in the farm environment. *Veterinary Microbiology, 107*, 193–204.

Seo, K., Lee, J. E., Lim, M. Y., & Ko, G. (2012). Effect of temperature, pH, and NaCl on the inactivation kinetics of murine norovirus. *Journal of Food Protection, 75*, 533–540.

Singh, R. B. K., Hales, S., De Wet, N., Raj, R., Hearnden, M., & Weinstein, P. (2001). The influence of climate variation and change on diarrheal disease in the Pacific Islands. *Environmental Health Perspectives*. https://doi.org/10.1289/ehp.01109155

Smith, K. E., Stenzel, S. A., Bender, J. B., et al. (2004). Outbreaks of enteric infections caused by multiple pathogens associated with calves at a farm day camp. *Pediatric Infectious Disease Journal, 23*, 1098–1104.

Stanley, K., & Jones, K. (2003). Cattle and sheep farms as reservoirs of Campylobacter. *Journal of Applied Microbiology, 94*, 104–113.

Suresh, K. G., Toranzos, G. A., Fayer, R., et al. (2012). Assessing the importance of zoonotic waterborne pathogens. In A. Dufour, J. Bartram, R. Bos, & V. Gannon (Eds.), *Animal waste, water quality and human health* (pp. 17–71). London, United Kingdom: IWA Publishing.

Syue, L. S., Chen, P. L., Wu, C. J., et al. (2016). Monomicrobial Aeromonas and Vibrio bacteremia in cirrhotic adults in southern Taiwan: Similarities and differences. *Journal of Microbiology, Immunology and Infection, 49*, 509–515.

Szonyi, B., Bordonaro, R., Wade, S. E., & Mohammed, H. O. (2010). Seasonal variation in the prevalence and molecular epidemiology of Cryptosporidium infection in dairy cattle in the New York City Watershed. *Parasitology Research, 107*, 317–325.

Taylor, A. W., Blau, D. M., Bassat, Q., et al. (2020). Initial findings from a novel population-based child mortality surveillance approach: A descriptive study. *Lancet Global Health, 8*, e909–e919.

Thammapalo, S., Chongsuwiwatwong, V., McNeil, D., & Geater, A. (2005). The climatic factors influencing the occurrence of dengue hemorrhagic fever in Thailand. *Southeast Asian Journal of Tropical Medicine and Public Health, 36*, 191–196.

Thiam, S., Diène, A. N., Sy, I., et al. (2017). Association between childhood diarrhoeal incidence and climatic factors in urban and rural settings in the health district of Mbour, Senegal. *International Journal of Environmental Research and Public Health, 14*, 1049.

Thongprachum, A., Khamrin, P., Maneekarn, N., Hayakawa, S., & Ushijima, H. (2016). Epidemiology of gastroenteritis viruses in Japan: Prevalence, seasonality, and outbreak. *Journal of Medical Virology, 88*, 551–570.

Tipayamongkholgul, M., Fang, C.-T., Klinchan, S., Liu, C.-M., & King, C.-C. (2009). Effects of the el niño-southern oscillation on dengue epidemics in Thailand, 1996–2005. *BMC Public Health, 9*, 422.

Trærup, S. L. M., Ortiz, R. A., & Markandya, A. (2011). The costs of climate change: A study of cholera in Tanzania. *International Journal of Environmental Research and Public Health*. https://doi.org/10.3390/ijerph8124386

Troeger, C., Blacker, B. F., Khalil, I. A., et al. (2018). Estimates of the global, regional, and national morbidity, mortality, and aetiologies of diarrhoea in 195 countries: A systematic analysis for the global burden of disease study 2016. *Lancet Infectious Diseases, 18*, 1211–1228.

Troeger, C., Khalil, I. A., Rao, P. C., et al. (2018). Rotavirus vaccination and the global burden of rotavirus diarrhea among children younger than 5 years. *JAMA Pediatrics*. https://doi.org/10.1001/jamapediatrics.2018.1960

Ugboko, H. U., Nwinyi, O. C., Oranusi, S. U., & Oyewale, J. O. (2020). Childhood diarrhoeal diseases in developing countries. *Heliyon, 6*, e03690.

Vanwambeke, S. O., Lambin, E. F., Eichhorn, M. P., Flasse, S. P., Harbach, R. E., Oskam, L., et al. (2007). Impact of land-use change on dengue and malaria in northern Thailand. *EcoHealth, 4*, 37–51.

Victora, C. G., Vaughan, J. P., & Barros, F. C. (1985). The seasonality of infant deaths due to diarrheal and respiratory diseases in Southern Brazil, 1974–1978. *Bulletin of the Pan American Health Organization, 19*, 29–39.

Vincenti-Gonzalez, M. F., Tami, A., Lizarazo, E. F., & Grillet, M. E. (2018). Enso-driven climate variability promotes periodic major outbreaks of dengue in Venezuela. *Science Reports, 8*, 5727-5727.

Wardrop, N. A., Hill, A. G., Dzodzomenyo, M., Aryeetey, G., & Wright, J. A. (2018). Livestock ownership and microbial contamination of drinking-water: Evidence from nationally representative household surveys in Ghana, Nepal and Bangladesh. *International Journal of Hygiene and Environmental Health, 221*, 33–40.

Webb, C., & Cabada, M. M. (2018). A review on prevention interventions to decrease diarrheal diseases' burden in children. *Current Tropical Medicine Reports, 5*, 31–40.

WHO. (2017). *Global vector control response 2017–2030.*

WHO. (2020). *Vector-borne diseases.* Available: https://www.who.int/news-room/fact-sheets/detail/vector-borne-diseases [accessed June 14, 2021 2021].

Woodward, W. E., Hirschhorn, N., Sack, R. B., et al. (1974). Acute diarrhea on an Apache Indian reservation. *American Journal of Epidemiology, 99*, 281–290.

Xu, Z., Liu, Y., Ma, Z., Toloo, G., Hu, W., & Tong, S. (2014). Assessment of the temperature effect on childhood diarrhea using satellite imagery. *Science Reports.* https://doi.org/10.1038/srep05389

Yue, Y., Sun, J., Liu, X., Ren, D., Liu, Q., Xiao, X., et al. (2018). Spatial analysis of dengue fever and exploration of its environmental and socio-economic risk factors using ordinary least squares: A case study in five districts of guangzhou city, China, 2014. *International Journal of Infectious Disease, 75*, 39–48.

Zeng, Z., Zhan, J., Chen, L., Chen, H., & Cheng, S. (2021). Global, regional, and national dengue burden from 1990 to 2017: A systematic analysis based on the global burden of disease study 2017. *EClinicalMedicine, 32.*

Zahorsky, J. (1913). Heat and summer diarrhea. *American Journal of Diseases of Children, 6*, 289–318.

Zavala, G. A., van Dulm, E., Doak, C. M., García, O. P., Polman, K., & Campos-Ponce, M. (2020). Ascariasis, Amebiasis and giardiasis in Mexican children: Distribution and geographical, environmental and socioeconomic risk factors. *Journal of Parasitic Diseases, 44*, 829–836.

CHAPTER 10

Geographic variation in impacts of heat exposure on human health

Qi Zhao, Yuming Guo and Shanshan Li
Department of Epidemiology and Preventive Medicine, School of Public Health and Preventive Medicine, Monash University, Melbourne, Australia

1. Introduction

The global mean surface temperature has increased by 0.82°C compared with the average level of 1951–80, with 18 of the 19 hottest years occurring since 2001 (Global, 2019). Climate change has been regarded as one of the biggest threats for human health (Intergovernmental Panel on Climate Change, 2015). There is growing amount of evidence showing that exposure to ambient heat may cause a variety of adverse health outcomes without any populations in the world exempted (Cheng et al., 2018; Fletcher et al., 2012; Phung, Guo, et al., 2016). For example, in Vietnam, hospital admissions for intestinal infection were estimated to increase by 5.5% for every 1°C increase in daily mean temperature in the hot Mekong area (Phung, Thai, et al., 2016). In the United States, every 2.78°C increase in daily mean temperature in the hottest months was associated with 9% increase in the risk of hospital admission for acute renal failure (Fletcher et al., 2012). In China, exposure to extreme high temperature was linked to 15% increase in the risk of emergency department visits (Zhao et al., 2017).

Numerous studies have reported that the association between exposure to ambient heat and health outcomes does not equally distribute across the world but may vary geographically (Gasparrini et al., 2015; Guo et al., 2017; Zhao et al., 2019). Generally, individuals living close to the equator may be less sensitive to high temperature than those living in high latitude. The geographic variation in the strength of heat–health association has been widely observed at the global level or within countries/regions, which may be explained as due to the long-term adaptation to the local climatic system (Basu, 2009; Deschenes, 2014; Kovats & Hajat, 2008). Moreover, this variation may be also affected by other factors, resulting in the geographic characteristics of heat–health association more complicated

Heat Exposure and Human Health in the Context of Climate Change
ISBN 978-0-12-819080-7
https://doi.org/10.1016/B978-0-12-819080-7.00002-1
© 2023 Elsevier Inc.
All rights reserved.

than theoretical speculation. In this chapter, we describe the geographic variation in the strength of association between ambient heat and health outcomes at the global level and within countries/regions. In addition, we also discuss the main factors that may modify the geographic variation in heat effect. The overview helps to understand thoroughly the adverse impacts of ambient heat on population health across the world.

Previous evidence on this issue comes from studies with various designs such as descriptive, time-series, cohort, or case-crossover studies (Anderson et al., 2013; Ellis et al., 1980; Fletcher et al., 2012; Guo, 2017; Johnson et al., 2005; Lim et al., 2012; Michelozzi et al., 2009; Phung et al., 2017; Rydman et al., 1999; Schwartz et al., 2004; Wichmann et al., 2011). The heterogeneity in study design may influence the effect estimates of heat exposure and lead to incomparability across findings of different studies. To date, time-series and case-crossover designs are two most commonly used analytic strategies for exploring the adverse impacts of ambient heat. This chapter therefore restricts the discussion to findings from time-series and case-crossover studies. In addition, the definition of heat exposure varies across studies. Some studies have focused on the health effect associated with single day's high temperature (e.g., per unit increase in daily temperature during the hot season or extreme hot days of the year). However, other studies explored the effect of continued extreme heat, i.e., heatwaves, within a few days. In this chapter, we discuss the geographic variations of the two heat events separately.

2. Geographic variation in impacts of high temperature

2.1 At the global level

It has been widely acknowledged that the impacts of exposure to ambient heat on human health vary across different countries. However, it remains a challenge to explore the geographic variation at the global level due to the limited access to necessary heath data in a large number of countries. Although there are studies evaluating the impact of heat on populations in single countries separately, the use of heterogenous models or parameters makes it difficult to compare their findings. Guo et al. explored the association between suboptimal temperature and nonaccidental mortality in 306 cities of 12 countries/regions with upper-middle to high incomes—Australia, Brazil, Thailand, mainland China, Taiwan, South Korea, Japan, Italy, Spain, the United Kingdom, the United States, and Canada

(Fig. 10.1) (Guo et al., 2014). They found that compared with the minimum-mortality temperature percentile (MMP), the cumulative risk of mortality over lags of 0–21 days increased by 30% in Italy, 19% in Spain, and 18% in Taiwan, China, on days with extreme high temperature (the 99th temperature percentile) (Table 10.1). In comparison, the effect estimates of heat exposure were much lower in Australia, the United States, the United States, and Canada (4%–6% increase in mortality). The effect estimates for 99th versus 90th temperature percentiles showed the similar geographic variation, although the strengths were lower.

Using the similar analytic strategy and assuming a causal relationship, Gasparrini et al. (2015) evaluated the fraction of mortality attributable to heat exposure in 13 countries (i.e., all 12 aforementioned countries/regions and Sweden). They found that the heat burden was highest in Italy and Spain, where 1.62% and 1.06% of deaths could be attributable to exposure to ambient heat. In comparison, the fraction of mortality attributable to heat exposure in some of the lowest affected countries such as Sweden, South Korea, and the United States ranged between 0.18% and 0.35%. Although the heat effect (99th vs. MMP) was higher in Taiwan, China, than in Brazil and Thailand, there was no substantial difference in the estimated attributable fraction of mortality. Gasparrini et al. further

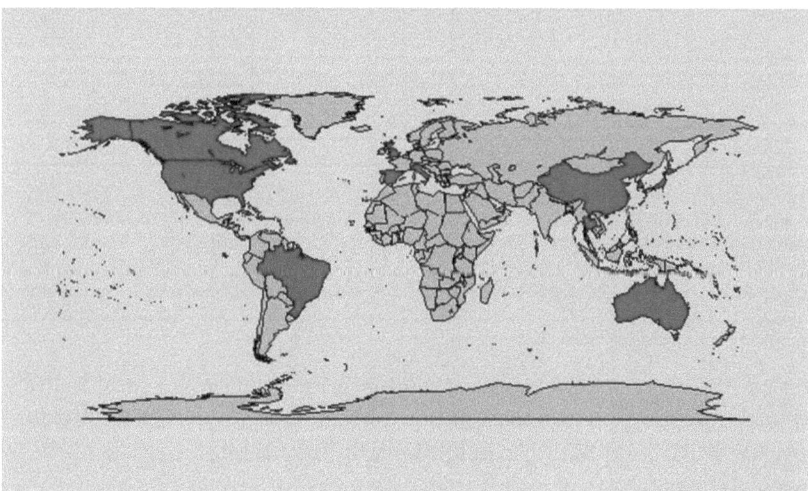

Figure 10.1 The location of the 12 study countries/regions. *(Reproduced with permission from Guo, Y., Gasparrini, A., Armstrong, B., Li, S., Tawatsupa, B., Tobias, A., et al. (2014). Global variation in the effects of ambient temperature on mortality: A systematic evaluation.* Epidemiology, 25*(6), 781. https://www.ncbi.nlm.nih.gov/pubmed/25166878.)*

Table 10.1 The pooled percentage change in risk of mortality associated with high temperature cumulated over lags of 0–21 days in the 12 countries/regions: 99th versus minimum-mortality temperature percentile (MMP) and 99th versus 90th temperature percentile.

		Change in risk of mortality (%, with 95% CI)		
Country/ region	MMP	99th versus MMP	99th versus 90th	Fraction of mortality attributable to ambient heat (%, with 95% CI)[a]
Australia	76th	6 (0, 13)	3 (0, 7)	0.45 (0.20, 0.70)
Brazil	72nd	7 (2, 11)	3 (1, 6)	0.70 (0.45, 0.93)
Thailand	76th	11 (8, 14)	6 (4, 7)	0.76 (0.65, 0.86)
Mainland China	76th	10 (3, 17)	5 (2, 9)	0.64 (0.47, 0.79)
Taiwan, China	66th	18 (12, 25)	7 (5, 10)	0.86 (0.12, 1.50)
South Korea	78th	8 (4, 12)	5 (3, 7)	0.31 (0.15, 0.45)
Japan	76th	7 (4, 11)	4 (3, 6)	0.32 (0.27, 0.36)
Italy	74th	30 (22, 39)	14 (11, 17)	1.62 (1.24, 1.98)
Spain	74th	19 (16, 23)	9 (8, 11)	1.06 (0.96, 1.16)
United Kingdom	80th	5 (4, 6)	3 (3, 4)	0.30 (0.25, 0.36)
United States	76th	4 (2, 6)	2 (1, 3)	0.35 (0.30. 0.39)
Canada	74th	5 (1, 9)	3 (1, 5)	0.54 (0.39, 0.66)
Sweden[b]	—	—	—	0.18 (−0.47, 0.65)

[a]Data on attributable fraction were extracted from Gasparrini et al. (2015), while other data were from Guo et al. (2017).
[b]Sweden was not covered by Guo et al. (2014).
Reproduced with permission from Guo, Y., Gasparrini, A., Armstrong, B., Li, S., Tawatsupa, B., Tobias, A., et al. (2014). Global variation in the effects of ambient temperature on mortality: A systematic evaluation. *Epidemiology, 25*(6), 781. https://www.ncbi.nlm.nih.gov/pubmed/25166878. https://www.ncbi.nlm.nih.gov/pubmed/25166878 and Gasparrini, A., Guo, Y., Hashizume, M., Lavigne, E., Zanobetti, A., Schwartz, J., et al. (2015). Mortality risk attributable to high and low ambient temperature: A multicountry observational study. *The Lancet, 386*(9991), 369–375. https://www.ncbi.nlm.nih.gov/pubmed/26003380.

explored the fractions of mortality attributable to moderate (temperature between MMP and 97.5th percentile) and extreme (temperature ≥97.5th percentile) heat. They found that the attributable burden of moderate heat was higher than extreme heat in Australia, mainland China, Japan, South Korea, Sweden, the United Kingdom, and the United States, while the opposite was also true for Brazil, Italy, Thailand, and Taiwan, China.

McMichael analyzed the association between high temperature and mortality in 12 cities from five continents (Table 10.2) (McMichael et al., 2008). The effect estimate of high temperature was greatest in Monterrey, Mexico, where the risk of mortality increased by 18.8% (95% CI: 13.0%–25.0%) for every 1°C increase in temperature (2 days' average) above the heat threshold. By contrast, the heat–mortality association was insignificant in Chiang Mai, Thailand (2.4 [95% CI: −0.5%–5.4%]) and Cape Town, South Africa (0.5 [95% CI: −0.3%–1.2%]). The risk of mortality associated with heat exposure increased from 0.8% to 5.8% for other cities.

Table 10.2 Percentage change in risk of mortality associated with each 1°C increase in 2-day mean temperature above threshold in the 12 cities.

City, country	Latitude and longitude	Threshold (°C) with 95% CI	Change in risk of mortality (%) with 95% CI
Ljubljana, Slovenia	46° 03′ N 14° 30′ E	17 (7, 20)	3.1 (1.3, 5.0)
Bucharest, Romania	44° 25′ N 26° 07′ E	22 (20, 22)	3.3 (2.4, 4.3)
Sofia, Bulgaria	42° 42′ N 23° 19′ E	16 (15, 17)	2.9 (2.1, 3.7)
Delhi, India	28° 39′ N 77° 13′ E	29 (8, 30)	3.9 (2.8, 5.1)
Monterrey, Mexico	25° 41′ N 100° 18′ W	31 (31, 33)	18.8 (13.0, 25.0)
Mexico City, Mexico	19° 25′ N 99° 08′ W	18 (8, 21)	0.8 (0.1, 1.4)
Chiang Mai, Thailand	18° 47′ N 99° 00′ E	28 (17, [a])	2.4 (−0.5, 5.4)
Bangkok, Thailand	13° 44′ N 100° 30′ E	29 (29, 30)	5.8 (3.5, 8.1)
Salvador, Brazil	12° 58′ S 38° 30′ W	23 (20, 27)	2.5 (0.9, 4.1)
São Paulo, Brazil	23° 32′ S 46° 37′ W	23 (19, 23)	3.5 (2.6, 4.3)
Santiago, Chile	33° 27′ S 70° 38′ W	16 (14, 20)	1.0 (0.3, 1.8)
Cape Town, South Africa	33° 55′ S 18° 26′ E	17 (15, 22)	0.5 (−0.3, 1.2)

[a]CI extends to limit of observed temperature range.

Reproduced with permission from McMichael, A. J., Wilkinson, P., Kovats, R. S., Pattenden, S., Hajat, S., Armstrong, B., et al. (2008). International study of temperature, heat and urban mortality: The 'ISOTHURM' project. *International Journal of Epidemiology*, *37*(5), 1121–1131. https://www.ncbi.nlm.nih.gov/pubmed/18522981.

2.2 At the country/region level

In comparison with the scarce evidence at the global level, reports of the geographic variation at the country or region level are relatively abundant, especially from Europe, America, Asia, and Oceania. Gasparrini et al. assessed the effects of high temperature on mortality across all ten Government Office Regions in England and Wales during the 1993–2006 summers (Armstrong et al., 2011; Gasparrini et al., 2012). The risk of all-cause mortality was estimated to increase by 2.1% at the national level for every 1°C increase in daily maximum temperature above the region-specific threshold. The heat effects varied from 0.8% in the coldest northeast to 3.8% in the hottest London area. Tobías et al. (2014) explored the effects of high summer temperatures on mortality in all 50 provincial capital cities in Spain between 1990 and 2004. They found that the risk of mortality at the national level increased by 14.6% when the heat effect was expressed as 99th versus 90th percentiles of the daily maximum temperature. A clear geographic trend was observed from eastern to western areas, such that the impact of heat exposure was higher for cities in the south and west than the Mediterranean region (Fig. 10.2).

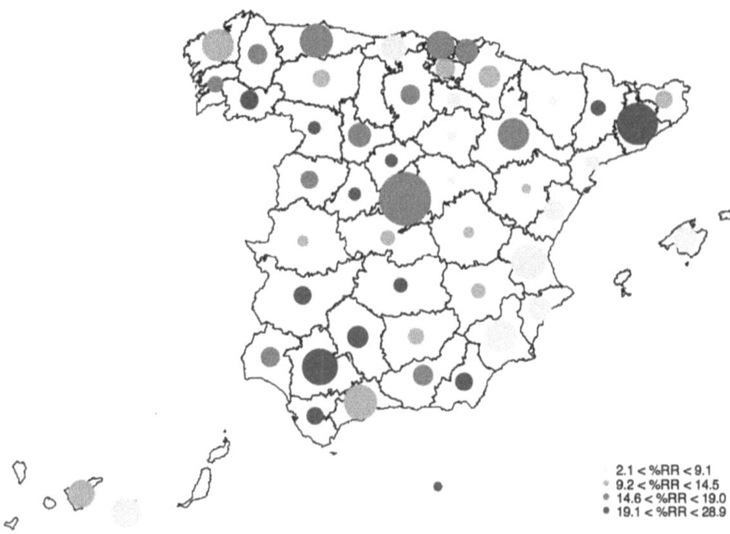

Figure 10.2 Geographic distribution of the city-specific percentage relative risk (%RR) of mortality comparing the 99th to 90th percentiles of maximum temperature. The size corresponds to the inverse of the variance of the risk estimate (i.e., larger circles are more certain). *(Reproduced with permission from Tobías, A., Armstrong, B., Gasparrini, A., Diaz, J. (2014). Effects of high summer temperatures on mortality in 50 Spanish cities.* Environmental Health, 13*(1):48. https://www.ncbi.nlm.nih.gov/pubmed/24912929.)*

Anderson et al. quantified the impact of heat exposure on mortality in 107 US counties between 1987 and 2000, finding the risk at 99th temperature percentile increased by 3.0% comparing with the 90th percentile (Anderson & Bell, 2009). An obvious geographic distribution was found, with the heat effect higher in the cold north and western coast but null in some warm southeastern counties (Fig. 10.3). Vaidyanathan et al. assessed the heat–morbidity association and developed the heat threshold for the early warning system using hospitalization data in 1617 US counties during 2003–12 (Vaidyanathan et al., 2019). Their findings were similar to the study of Anderson et al.: the nine climatic regions with threshold of heat warning from low to high were northwest, west, southwest, northeast, west north central, south, southeast, central, and east north central.

In Brazil, Zhao et al. (2019) explored the association between heat exposure and hospitalizations in 1814 cities (covering 78% of the national population) in the 2000–15 hot seasons. Every 5°C increase in daily mean temperature was associated with 4.0% increase in risk of all-cause hospitalization. The strength of heat effect varied across the five geographic regions in Brazil, which was lowest in the coldest south (2.9%), highest in

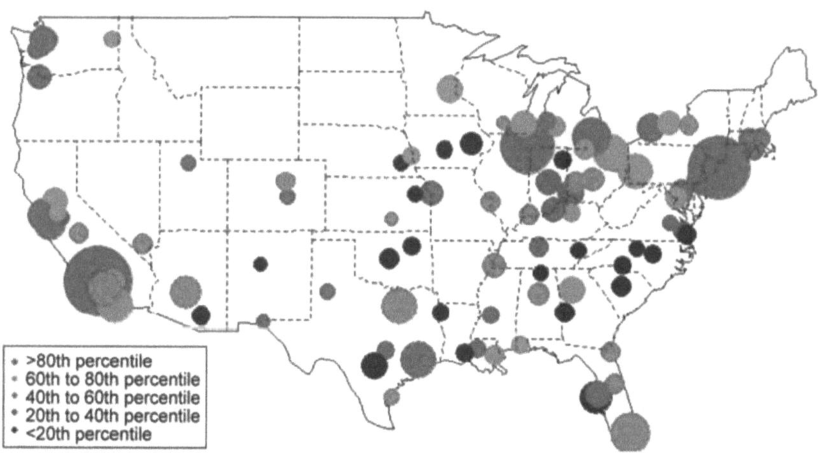

Figure 10.3 Geographic distribution of the county-specific percentage relative risk (%RR) of mortality comparing the 99th to 90th percentiles of temperature. The size corresponds to the inverse of the variance of the risk estimate (i.e., larger circles are more certain). The two noncontinental cities included in the data set, Honolulu HI and Anchorage AK, are not included in this regional analysis. *(Reproduced with permission from Anderson, B. G., & Bell, M. L. (2009). Weather-related mortality: How heat, cold, and heat waves affect mortality in the United States.* Epidemiology, *20(2), 205. https://www.ncbi.nlm.nih.gov/pubmed/19194300.)*

the second hottest northeast (6.3%), and insignificant in the hottest north. This study also estimated that attributable fraction and rate of hospitalizations at the city level (Fig. 10.4). The burden of hospitalization attributable to heat exposure in the hot season was more substantial for cities in the central west and inland of the northeast. In comparison, the lowest attributable fraction and rate occurred for cities in the north and along the eastern coast. Similar geographic trend was also reported by Guo et al. (2014). For example, the risk of mortality increased by 8% in Porto Alegre (a city in the south) for 99th versus 90th temperature percentiles, which however, were insignificant in the Belem (a city in the north).

Numerous studies have explored the geographic variation in heat—health association in mainland China using different grouping strategy, but the findings are largely similar. For example, Ma et al. performed analysis by dividing 66 Chinese cities into seven regions, considering the similarity of cities in geographic, economic, cultural, and other factors (Fig. 10.5) (Ma et al., 2015). The heat effect on mortality (99th vs. threshold) was highest in the coldest northeast. Although the associations were insignificant in the rest regions except for the east, the heat effect was lower in the hot southern

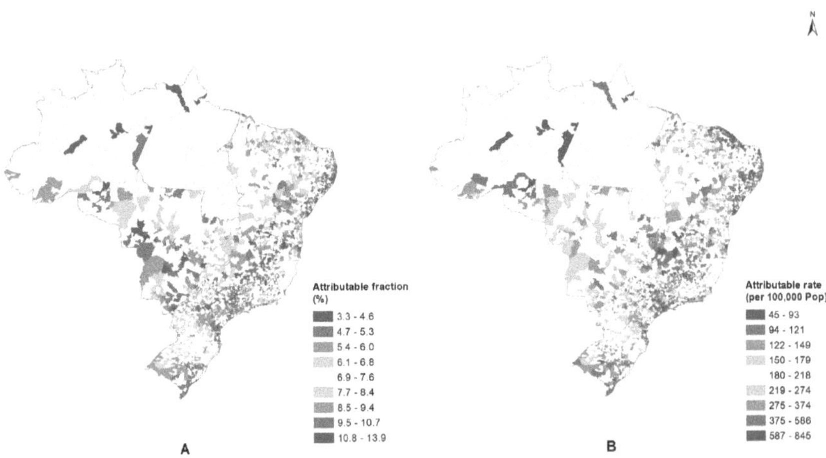

Figure 10.4 The fraction (A) and rate (B) of hospitalizations attributable to heat exposure over lag 0—7 days in the 1814 Brazilian cities during 2000—2015 hot seasons. The best linear unbiased prediction of the cumulative association in each city was used to calculate the attributable burden. *(Reproduced with permission from Zhao, Q., Li, S., Coelho, M. S., Saldiva, P. H., Hu, K., Arblaster, J. M., et al. (2019). Geographic, demographic, and temporal variations in the association between heat exposure and hospitalization in Brazil: A nationwide study between 2000 and 2015.* Environmental Health Perspectives, *127(01), 017001. https://www.ncbi.nlm.nih.gov/pubmed/30620212.)*

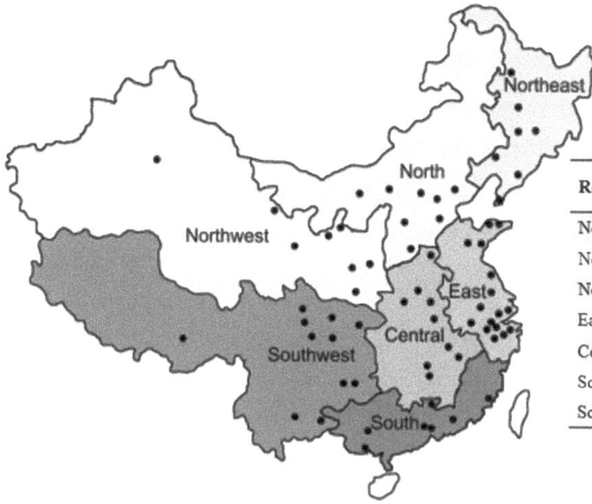

Region	Heat effect (99th vs threshold)
Northeast	36%(17%, 58%)
North	22%(-2%, 53%)
Northwest	26%(-8%, 72%)
East	24%(4%, 50%)
Central	2%(-13%, 20%)
Southwest	3%(-11%, 19%)
South	5% (-18%, 35%)

Figure 10.5 The distribution of 66 Chinese cities in the seven geographic regions, and the heat effects. *(Reproduced with permission from Ma, W., Wang, L., Lin, H., Liu, T., Zhang, Y., Rutherford, S., Luo, Y., Zeng, W., Zhang, Y., Wang, X., Gu, X., Chu, C., Xiao, J., & Zhou, M. (2015). The temperature–mortality relationship in China: An analysis from 66 Chinese communities.* Environmental Research, 137, *72–77. https://pubmed.ncbi.nlm.nih.gov/25490245.)*

areas. Another Chinese study analyzed the heat–mortality association by grouping 122 cities into five temperature zones, i.e., temperate, warm temperate, north subtropical, middle subtropical, and south subtropical (Wang et al., 2017). The heat effect was greatest in the warm temperate zone and minimal in the middle subtropical zone.

In comparison with the abundant evidence from other countries based on data from a large number of communities, few studies have assessed the heat–health association in Australia using data at the national level. Cheng et al. explored this relationship in the five largest Australian cities between 2000 and 2009 (Adelaide, Brisbane, Melbourne, Perth, and Sydney), which comprised over 60% of the national population (Cheng et al., 2019). These cities have different climatic characteristics, with the mean temperature of the hot season ranging from 24.2°C in Brisbane, 22.6°C in Perth, 21.9°C in Sydney, 20.8°C in Adelaide to 18.3°C in Melbourne. However, the heat effects distributed in a different geographic pattern: for per 1°C increase in daily mean temperature in the hot season, the risk of mortality increased by 2.9% in Brisbane and 1.6% in Melbourne, which was insignificant in Perth.

3. Geographic variation in impact of heatwaves

In comparison with heat exposure, few studies have examined the geographic variation in the impact of heatwaves in an international context. Heatwaves are climatic scenarios that the extreme high temperature lasts for a few days. The effects of heatwaves may come from two aspects, i.e., the independent impact of daily high temperature and the added impact due to the consecutive hot days (Anderson & Bell, 2011; Gasparrini & Armstrong, 2011). However, there is no unified definition of the temperature threshold and duration worldwide, making it a challenge to compare the variation in the impact of heatwaves at the global level. Guo et al. (2017) investigated the association between heatwaves and mortality in 18 countries/regions. In this study, 12 types of heatwaves were applied for each city by setting the threshold at the 90th, 92.5th, 95th, or 97.5th percentile of city-specific daily mean temperature range, and the duration as $\geq$ 2, 3, or 4 days. The main finding was that heatwaves had higher impacts in moderate cold and moderate hot countries/regions than in cold and hot areas. The risk of mortality was greater for heatwaves with higher temperature threshold. By contrast, the duration of heatwaves did not add additional effects on mortality in all countries/regions except for Brazil, Moldova, and Taiwan, China. This finding was consistent to several previous studies in the United States (Barnett et al., 2012; Gasparrini & Armstrong, 2011), indicating that for most areas of the world, the impacts of heatwaves could be explained by daily high temperature. Given this, the geographic variation in heatwave effect is not further discussed at the country/region level.

4. Modifiers for the geographic variation in heat effect and the uncertainties

Findings from these multicountry or multicity studies indicate that the geographic variation in heat impacts does not strictly follow the hypothesis that the individuals living in cold areas are more susceptible than individuals in hot areas. Currently, only a few studies have examined potential climatic, demographic, socioeconomic, and infrastructure factors that may modify the geographic variation in the heat—health association (Anderson & Bell, 2009; Tobías et al., 2014). However, the findings were inconsistent.

4.1 Climatic and geographic factors

It is speculated that individuals living in hot areas are less sensitive to heat exposure due to long-term adaptation, while those living in cold areas are more susceptible. Findings from a few studies support this hypothesis. For example, Anderson and Bell (2009) found that per interquartile range (IQR) increases in yearly and summer mean temperature (community-specific), and the absolute effect of heat exposure in 107 US communities declined by 70% and 112%, respectively. Another Chinese study observed an opposite phenomenon in 66 communities (Huang et al., 2015): the relative effect of heat exposure increased by 0.3% per IQR increase in community-specific annual temperature (7.2°C). However, this study also reported a 0.2% decline in heat effect per IQR increase in latitude (7.9°). A Spanish study found that the effect of heat exposure was not affected by the yearly or summer mean temperature or latitude at the city level (Tobías et al., 2014). However, this study found that city-specific longitude could modify the heat effect negatively. Guo et al. (2017) assessed the powers of several country- and community-level factors in explaining the variation in heatwave—mortality association across 400 communities in 18 countries/regions. They found that substantial heterogeneity of heatwave effect could be explained by between-country differences. Although other community-specific factors, e.g., average temperature in the hot season, could slightly modify the strength of association in the single-predictor model, their explainable powers were minimal in the full model when the *country* covariate was included. This finding indicates that between-country differences may play an important role in modifying the geographic variation of heat effect, although the modification effects of climatic and geographic factors such as local mean temperature and latitude should not be overlooked.

4.2 Sociodemographic and infrastructural factors

Another hypothesis of the geographic variation of heat effects is that individuals living in poor countries/regions may be more vulnerable to high temperature than others (IPCC (Intergovernmental Panel on Climate Change), 2014). However, findings from multicity, multicountry studies indicated that some of the most susceptible populations were from more developed areas (Gasparrini et al., 2015; Guo et al., 2014, Guo et al., 2017; McMichael et al., 2008). In addition, several country-level studies also have inconsistent findings. Ma et al. (2014) found that the effect of high temperature was only significant in Chinese cities with the disposable income per capita below 12,000 CNY. However, results from another Chinese study which involved 66 communities indicated that the average income might not modify the heat effects (Huang et al., 2015). The Spanish study has the similar insignificant finding (Tobías et al., 2014). By contrast, the US study found that the absolute effect of heat exposure increased by 57% per IQR increase in median household income ($6538) (Anderson & Bell, 2009). Other demographic, socioeconomic, and infrastructural effect modifiers may include unemployment rate, education level, whether living in an urban setting, the prevalence of air conditioning, population density, and the proportion of the elderly (Huang et al., 2015; Ma et al., 2014; Tobías et al., 2014). These effect modifiers also have supportive theories. For example, numerous studies have shown that the elderly are more vulnerable to high temperature due to their preexisting health conditions and impaired physiological systems (Basu, 2009; Zhao et al., 2017, 2019; Åström et al., 2011). The use of air conditioning is protective to extreme high temperature in China and the United States (Anderson & Bell, 2009; Ma et al., 2014). Similar to the inconsistent findings on economic factors, there have been some other studies reporting opposite results of the modification effects of these socioeconomic, demographic, and infrastructural factors. For example, in Japan, the prevalence of air conditioning shows no protection against heat exposure (Chung et al., 2018).

4.3 Uncertainties

The inconsistent results across studies on the potential effect modifiers are not supportive to the prior hypotheses. However, it would be premature to treat them as strong evidence and then deny these theories due to two reasons. First, most studies exploring the potential effect modifiers of geographic variation in heat effect have used the single-predictor models. This may distort the explanation capacities of a variety of potential factors. Second, different countries/regions may have very different levels of these effect modifiers. The geographic variation in heat effect is speculated as due to the combined effect of these factors. To clarify this issue, it is urgency to explore the potential effect modifiers at the global level.

5. Geographic variation in future heat effect under climate change scenarios

Given the excess emission of greenhouse gases and other reasons, the surface temperature of the earth is projected to increase in the next decades (IPCC (Intergovernmental Panel on Climate Change), 2014). Paralleling with this, the strength and duration of extreme high temperature events are also projected to increase. However, the warming trend will not equally distribute across the world, with some areas predicted to experience more heat exposure than other areas (Pachauri et al., 2014; Rogelj et al., 2012). As a result, the projected heat-related health burden may also exhibit geographic variation. Gasparrini et al., 2017 projected the change in the risk of mortality in 451 locations of 23 countries/regions across nine regions of the world until 2099 under four different climate change scenarios. This study was performed by assuming a constant population size and unchanged adaptive capacity to global warming. Based on five general circulation models, the global average temperature by the end of this century is projected to increase by 0.4°C–0.8°C under representative concentration pathways (RCP) 2.6 and 3.3°C–4.9°C under RCP8.5 (Table 10.3). The greatest and lowest warming will occur in North America and Australia, respectively. Similar to the changing magnitude of ambient temperature, the heat-related excess mortality is also greatest under RCP8.5 and lowest under RCP2.6 by 2090–99. However, the geographic variation in projected heat effects does not strictly follow the pattern of projected temperature change. By the end of this century, the excess mortality associated with heat exposure is projected to increase by 17% in Southeast Asia and 10.5% in Southern Europe under RCP8.5, resulting the two regions most vulnerable to global warming (location-specific results are shown in Fig. 10.6). By contrast, the impact of heat exposure will be lowest in Northern Europe (2.5%), Australia (3.2%), East Asia (3.2%), and North America (3.6%).

Guo et al. (2018) quantified the excess mortality related to heatwaves in 20 countries/regions (data from 412 communities) until 2080 under four different climate change scenarios. For each community, heatwaves were defined as at least 2 days with daily mean temperature exceeding 95th percentile of year-round temperature. This study considered three population change scenarios (low, median, and high) and adaptation (yes/no). The greatest mortality burden appeared under RCP8.5. For the RCP, the heat-related excess mortality was highest under the high-variant population scenario. Without considering thermal adaptation, the excess mortality related to heatwaves will be greatest in tropical and subtropical areas, particularly in Colombia, Philippines, and Brazil (Fig. 10.7). By contrast, the increase in heatwave-related mortality will be much smaller in European countries. The excess mortality related to heatwaves will be lower with considering adaptation to temperature change (Fig. 10.8). The mortality burden is projected to decline by 2080 in Italy, Japan, Moldova, and Spain under the low-variant population scenario. This projected decline remains for Japan and Moldova under the high-variant population scenario.

Table 10.3 Projected temperature change and excess risk of mortality (%) under four different RCP scenarios.

	Current temperature (2010–19)	Projected increase in temperature (2090–99 vs. 2010–19)				Projected heat-related excess mortality (%) in 2090–99			
		RCP2.6	RCP4.5	RCP6.0	RCP8.5	RCP2.6	RCP4.5	RCP6.0	RCP8.5
North America	14.2 (3.4, 26.0)	0.8 (0.5, 1.2)	2.2 (1.3, 3.0)	2.8 (1.8, 3.6)	4.9 (3.2, 6.3)	0.7 (0.0, 1.6)	1.4 (−0.0, 3.0)	1.8 (−0.1, 3.9)	3.6 (−0.5, 7.6)
Central America	19.0 (14.1, 23.5)	0.6 (0.4, 1.0)	1.9 (1.7, 2.3)	2.6 (2.3, 3.3)	4.5 (4.1, 5.4)	0.9 (0.2, 1.6)	2.3 (0.4, 4.9)	3.2 (0.4, 6.8)	7.0 (0.4, 13.7)
South America	22.8 (11.8, 27.8)	0.5 (0.3, 0.7)	1.5 (1.0, 2.0)	1.9 (1.4, 2.6)	3.7 (2.8, 5.1)	1.6 (0.0, 3.6)	2.9 (−1.3, 6.8)	3.5 (−2.5, 8.9)	7.7 (−14.0, 22.1)
Northern Europe	10.2 (6.9, 12.0)	0.5 (0.4, 1.1)	1.4 (1.1, 2.4)	2.1 (1.6, 3.3)	3.4 (2.8, 5.4)	0.5 (0.1, 1.0)	0.9 (0.2, 1.9)	1.4 (0.4, 3.0)	2.5 (0.5, 5.0)
Central Europe	11.8 (8.7, 16.5)	0.7 (0.4, 1.0)	1.8 (1.5, 2.0)	2.4 (2.1, 2.6)	4.3 (3.5, 4.8)	1.6 (0.8, 2.8)	3.0 (1.6, 5.5)	4.0 (2.1, 6.6)	7.1 (2.9, 11.3)
Southern Europe	15.9 (11.3, 21.9)	0.7 (0.6, 0.8)	1.9 (1.3, 2.2)	2.5 (1.8, 2.7)	4.5 (3.0, 5.1)	2.4 (1.3, 4.4)	4.6 (2.3, 7.3)	5.8 (3.5, 9.3)	10.5 (5.6, 17.3)
East Asia	15.6 (7.6, 24.1)	0.7 (0.4, 1.1)	1.9 (1.4, 2.6)	2.5 (1.7, 3.2)	4.3 (3.1, 6.0)	0.8 (0.2, 1.3)	1.3 (0.4, 2.3)	1.8 (0.5, 3.2)	3.2 (0.9, 5.6)
Southeast Asia	27.8 (23.6, 29.6)	0.6 (0.4, 0.8)	1.5 (1.2, 1.7)	2.0 (1.7, 2.3)	3.8 (3.2, 4.3)	2.8 (0.7, 5.3)	5.4 (1.0, 10.1)	1.3 (0.4, 2.7)	16.7 (−1.7, 33.2)
Australia	18.5 (16.1, 20.7)	0.4 (0.2, 0.6)	1.2 (1.1, 1.3)	1.8 (1.6, 1.9)	3.3 (3.2, 3.6)	0.8 (0.1, 1.5)	1.2 (0.2, 2.2)	1.5 (0.3, 2.9)	3.2 (0.6, 6.3)

Reproduced with permission from Gasparrini, A., Guo, Y., Sera, F., Vicedo-Cabrera, A. M., Huber, V., Tong, S., et al. (2017). Projections of temperature-related excess mortality under climate change scenarios. *The Lancet Planetary Health, 1*(9), e360–e367. https://www.ncbi.nlm.nih.gov/pubmed/29276803.

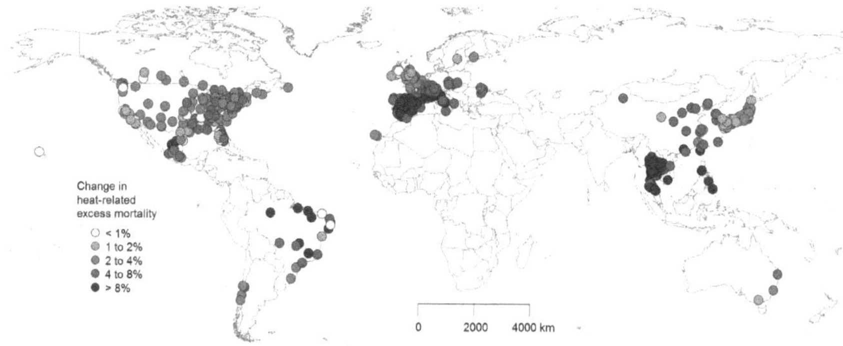

Figure 10.6 The projected change in heat-related excess mortality (%) in 2090–99 compared with 2010–19 under the RCP8.5 in the 451 locations. *(Reproduced with permission from Gasparrini, A., Guo, Y., Sera, F., Vicedo-Cabrera, A. M., Huber, V., Tong, S., et al. (2017). Projections of temperature-related excess mortality under climate change scenarios.* The Lancet Planetary Health, 1*(9), e360–e367. https://www.ncbi.nlm.nih.gov/pubmed/29276803.)*

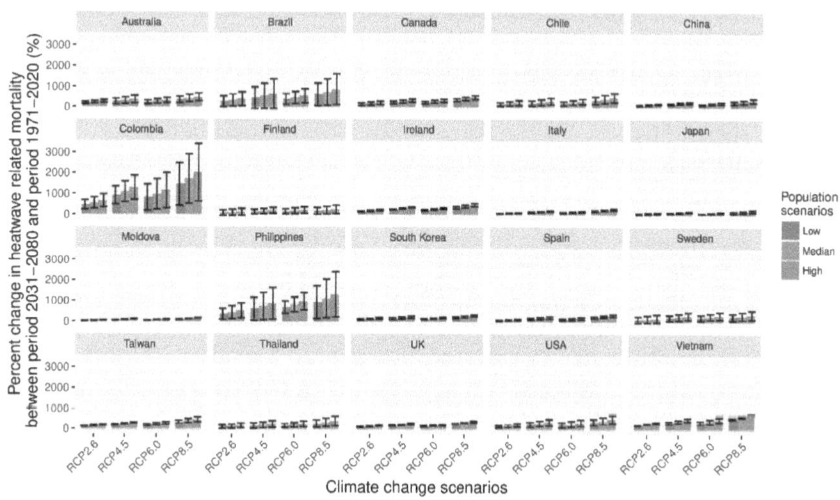

Figure 10.7 Mean percent change heatwave-related excess deaths in 2031–80 in comparison with 1971–2020, in 20 countries/regions under RCP2.6, RCP4.5, RCP6.0, and RCP8.5 scenarios and high-variant, median-variant, and low-variant population scenarios, with assumption of nonadaptation. *(Reproduced with permission from Guo, Y., Gasparrini, A., Li, S., Sera, F., Vicedo-Cabrera, A. M., Coelho, M., et al. (2018). Quantifying excess deaths related to heatwaves under climate change scenarios: A multicountry time series modelling study.* PLOS Medicine, 15*(7), e1002629. https://www.ncbi.nlm.nih.gov/pubmed/30063714.)*

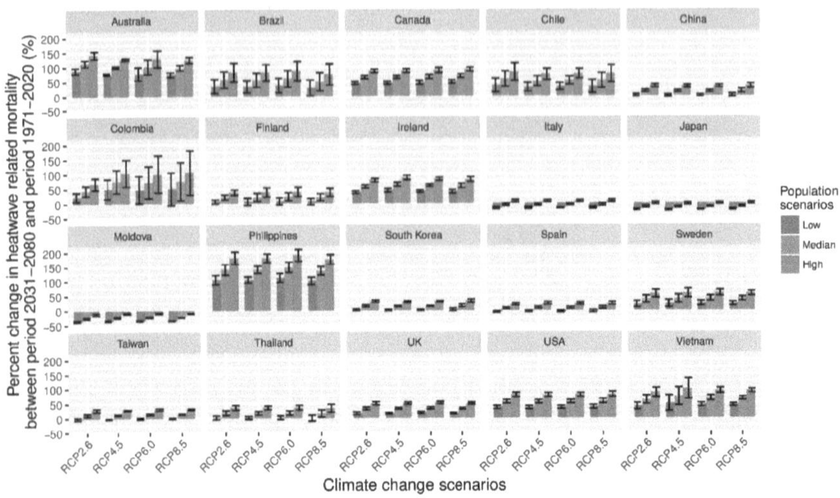

Figure 10.8 Mean percent change heatwave-related excess deaths in 2031–80 in comparison with 1971–2020, in 20 countries/regions under RCP2.6, RCP4.5, RCP6.0, and RCP8.5 scenarios and high-variant, median-variant, and low-variant population scenarios, with assumption of full adaptation. *(Reproduced with permission from Guo, Y., Gasparrini, A., Li, S., Sera, F., Vicedo-Cabrera, A. M., Coelho, M., et al. (2018). Quantifying excess deaths related to heatwaves under climate change scenarios: A multicountry time series modelling study.* PLOS Medicine, *15(7), e1002629. https://www.ncbi.nlm.nih.gov/pubmed/30063714.)*

6. Summary

The strength of heat effect does not equally distribute worldwide. Although individuals living in cold areas may be more vulnerable to high temperature than those in the hot areas. However, some climatic, geographic, sociodemographic, and infrastructural factors may work synergistically, resulting in complicated geographic variation of heat effects across the world. In addition, the change of further temperature will also vary by countries/regions in the context of global warming. Further studies are warranted to clarify the crucial factors that may modify the geographic variation in heat effect for different countries and regions to develop better protective strategies and health promotion policies.

References

Anderson, B. G., & Bell, M. L. (2009). Weather-related mortality: How heat, cold, and heat waves affect mortality in the United States. *Epidemiology, 20*(2), 205.

Anderson, G. B., & Bell, M. L. (2011). Heat waves in the United States: Mortality risk during heat waves and effect modification by heat wave characteristics in 43 US communities. *Environmental Health Perspectives, 119*(2), 210.

Anderson, G. B., Dominici, F., Wang, Y., McCormack, M. C., Bell, M. L., & Peng, R. D. (2013). Heat-related emergency hospitalizations for respiratory diseases in the Medicare population. *American Journal of Respiratory and Critical Care Medicine, 187*(10), 1098–1103.

Armstrong, B. G., Chalabi, Z., Fenn, B., Hajat, S., Kovats, S., Milojevic, A., & Wilkinson, P. (2011). Association of mortality with high temperatures in a temperate climate: England and Wales. *Journal of Epidemiology & Community Health, 65*(4), 340–345.

Åström, D. O., Bertil, F., & Joacim, R. (2011). Heat wave impact on morbidity and mortality in the elderly population: A review of recent studies. *Maturitas, 69*(2), 99–105.

Barnett, A., Hajat, S., Gasparrini, A., & Rocklöv, J. (2012). Cold and heat waves in the United States. *Environmental Research, 112*, 218–224.

Basu, R. (2009). High ambient temperature and mortality: A review of epidemiologic studies from 2001 to 2008. *Environmental Health, 8*(1), 40.

Cheng, J., Xu, Z., Bambrick, H., Su, H., Tong, S., & Hu, W. (2019). Impacts of heat, cold, and temperature variability on mortality in Australia, 2000–2009. *Science of the Total Environment, 651*, 2558–2565.

Cheng, J., Zhang, Y., Zhang, W., Xu, Z., Bambrick, H., Hu, W., & Tong, S. (2018). Assessment of heat-and cold-related emergency department visits in cities of China and Australia: Population vulnerability and attributable burden. *Environmental Research, 166*, 610–619.

Chung, Y., Yang, D., Gasparrini, A., Vicedo-Cabrera, A. M., Fook Sheng Ng, C., Kim, Y., Honda, Y., & Hashizume, M. (2018). Changing susceptibility to non-optimum temperatures in Japan, 1972–2012: The role of climate, demographic, and socioeconomic factors. *Environmental Health Perspectives, 126*(5), 057002.

Deschenes, O. (2014). Temperature, human health, and adaptation: A review of the empirical literature. *Energy Economics, 46*, 606–619.

Ellis, F., Prince, H., Lovatt, G., & Whittington, R. (1980). Mortality and morbidity in Birmingham during the 1976 heatwave. *QJM: An International Journal of Medicine, 49*(1), 1–8.

IPCC (Intergovernmental Panel on Climate Change). (2014). Summary for policymakers. In C. B. Field, V. R. Barros, D. J. Dokken, et al. (Eds.), *Climate change 2014: Impacts, adaptation, and vulnerability. Part A: Global and sectoral aspects* (pp. 1–32). Cambridge, UK, and New York, NY, USA: Cambridge University Press.

Fletcher, B. A., Lin, S., Fitzgerald, E. F., & Hwang, S.-A. (2012). Association of summer temperatures with hospital admissions for renal diseases in New York state: A case-crossover study. *American Journal of Epidemiology, 175*(9), 907–916.

Gasparrini, A., & Armstrong, B. (2011). The impact of heat waves on mortality. *Epidemiology, 22*(1), 68.

Gasparrini, A., Armstrong, B., Kovats, S., & Wilkinson, P. (2012). The effect of high temperatures on cause-specific mortality in England and Wales. *Occupational and Environmental Medicine, 69*(1), 56–61.

Gasparrini, A., Guo, Y., Hashizume, M., Lavigne, E., Zanobetti, A., Schwartz, J., Tobias, A., Tong, S., Rocklöv, J., Forsberg, B., Leone, M., De Sario, M., Bell, M. L., Guo, Y. L., Wu, C. F., Kan, H., Yi, M., de Sousa Zanotti Stagliorio Coelho, M., Saldiva, P. H., & Armstrong, B. (2015). Mortality risk attributable to high and low ambient temperature: A multicountry observational study. *The Lancet, 386*(9991), 369–375.

Gasparrini, A., Guo, Y., Sera, F., Vicedo-Cabrera, A. M., Huber, V., Tong, S., Coelho, M., Saldiva, P., Lavigne, E., Matus Correa, P., Valdes Ortega, N., Kan, H., Osorio, S., Kyselý, J., Urban, A., Jaakkola, J. J. K., Ryti, N. R. I., Pascal, M., Goodman, P. G., &

Armstrong, B. (2017). Projections of temperature-related excess mortality under climate change scenarios. *The Lancet Planetary Health, 1*(9), e360–e367.

Guo, Y. (2017). Hourly associations between heat and ambulance calls. *Environmental Pollution, 220*, 1424–1428.

Guo, Y., Gasparrini, A., Armstrong, B., Li, S., Tawatsupa, B., Tobias, A., Lavigne, E., Coelho, M., Leone, M., Pan, X., Tong, S., Tian, L., Kim, H., Hashizume, M., Honda, Y., Guo, Y. L., Wu, C. F., Punnasiri, K., Yi, S. M., & Williams, G. (2014). Global variation in the effects of ambient temperature on mortality: A systematic evaluation. *Epidemiology, 25*(6), 781.

Guo, Y., Gasparrini, A., Armstrong, B., Tawatsupa, B., Tobias, A., Lavigne, E., Coelho, M., Pan, X., Kim, H., Hashizume, M., Honda, Y., Guo, Y. L., Wu, C. F., Zanobetti, A., Schwartz, J., Bell, M. L., Scortichini, M., Michelozzi, P., Punnasiri, K., & Tong, S. (2017). Heat wave and mortality: A multicountry, multicommunity study. *Environmental Health Perspectives, 125*(8), 087006.

Guo, Y., Gasparrini, A., Li, S., Sera, F., Vicedo-Cabrera, A. M., Coelho, M., Saldiva, P., Lavigne, E., Tawatsupa, B., Punnasiri, K., Overcenco, A., Correa, P. M., Ortega, N. V., Kan, H., Osorio, S., Jaakkola, J. J. K., Ryti, N. R. I., Goodman, P. G., Zeka, A., Michelozzi, P., & Tong, S. (2018). Quantifying excess deaths related to heatwaves under climate change scenarios: A multicountry time series modelling study. *PLOS Medicine, 15*(7), e1002629.

Huang, Z., Lin, H., Liu, Y., Zhou, M., Liu, T., Xiao, J., Zeng, W., Li, X., Zhang, Y., Ebi, K. L., Tong, S., Ma, W., & Wang, L. (2015). Individual-level and community-level effect modifiers of the temperature–mortality relationship in 66 Chinese communities. *BMJ Open, 5*(9), e009172.

Intergovernmental Panel on Climate Change. (2015). *Climate change 2014: Mitigation of climate change*. Cambridge University Press.

Johnson, H., Kovats, R. S., McGregor, G., Stedman, J., Gibbs, M., Walton, H., Cook, L., & Black, E. (2005). The impact of the 2003 heat wave on mortality and hospital admissions in England. *Health Statistics Quarterly*, (25), 6–11.

Kovats, R. S., & Hajat, S. (2008). Heat stress and public health: A critical review. *Annual Review of Public Health, 29*, 41–55.

Lim, Y.-H., Hong, Y.-C., & Kim, H. (2012). Effects of diurnal temperature range on cardiovascular and respiratory hospital admissions in Korea. *Science of the Total Environment, 417*, 55–60.

Ma, W., Chen, R., & Kan, H. (2014). Temperature-related mortality in 17 large Chinese cities: How heat and cold affect mortality in China. *Environmental Research, 134*, 127–133.

Ma, W., Wang, L., Lin, H., Liu, T., Zhang, Y., Rutherford, S., Luo, Y., Zeng, W., Zhang, Y., Wang, X., Gu, X., Chu, C., Xiao, J., & Zhou, M. (2015). The temperature–mortality relationship in China: An analysis from 66 Chinese communities. *Environmental Research, 137*, 72–77.

McMichael, A. J., Wilkinson, P., Kovats, R. S., Pattenden, S., Hajat, S., Armstrong, B., Vajanapoom, N., Niciu, E. M., Mahomed, H., Kingkeow, C., Kosnik, M., O'Neill, M. S., Romieu, I., Ramirez-Aguilar, M., Barreto, M. L., Gouveia, N., & Nikiforov, B. (2008). International study of temperature, heat and urban mortality: The 'ISOTHURM' project. *International Journal of Epidemiology, 37*(5), 1121–1131.

Michelozzi, P., Accetta, G., De Sario, M., D'ippoliti, D., Marino, C., Baccini, M., Biggeri, A., Anderson, H. R., Katsouyanni, K., Ballester, F., Bisanti, L., Cadum, E., Forsberg, B., Forastiere, F., Goodman, P. G., Hojs, A., Kirchmayer, U., Medina, S., Paldy, A., Schindler, C., Perucci, C. A., & PHEWE Collaborative Group. (2009). High temperature and hospitalizations for cardiovascular and respiratory causes in 12 European cities. *American Journal of Respiratory and Critical Care Medicine, 179*(5), 383–389.

NASA. Global temperature. (2019). Available from: https://climate.nasa.gov/vital-signs/global-temperature/.

Pachauri, R. K., Allen, M. R., Barros, V. R., Broome, J., Cramer, W., Christ, R., Church, J., Clarke, L., Dahe, Q., Dasgupta, P., Dubash, N. K., Edenhofer, O., Elgizouli, L., Field, C. B., Forster, P., Friedlingstein, P., Fuglestvedt, J., Gomez-Echeverri, L., Hallegatte, S., Hegerl, G., & van Ypserle, J. P. (2014). *Climate change 2014: Synthesis report. Contribution of working groups I, II and III to the fifth assessment report of the intergovernmental Panel on climate change*. IPCC.

Phung, D., Chu, C., Rutherford, S., Nguyen, H. L. T., Do, C. M., & Huang, C. (2017). Heatwave and risk of hospitalization: A multi-province study in Vietnam. *Environmental Pollution, 220*, 597–607.

Phung, D., Guo, Y., Nguyen, H. T., Rutherford, S., Baum, S., & Chu, C. (2016). High temperature and risk of hospitalizations, and effect modifying potential of socio-economic conditions: A multi-province study in the tropical Mekong Delta Region. *Environment International, 92*, 77–86.

Phung, D., Thai, P. K., Guo, Y., Morawska, L., Rutherford, S., & Chu, C. (2016). Ambient temperature and risk of cardiovascular hospitalization: An updated systematic review and meta-analysis. *Science of the Total Environment, 550*, 1084–1102.

Rogelj, J., Meinshausen, M., & Knutti, R. (2012). Global warming under old and new scenarios using IPCC climate sensitivity range estimates. *Nature Climate Change, 2*(4), 248.

Rydman, R. J., Rumoro, D. P., Silva, J. C., Hogan, T. M., & Kampe, L. M. (1999). The rate and risk of heat-related illness in hospital emergency departments during the 1995 Chicago heat disaster. *Journal of Medicine Systems, 23*(1), 41–56.

Schwartz, J., Samet, J. M., & Patz, J. A. (2004). Hospital admissions for heart disease: The effects of temperature and humidity. *Epidemiology, 15*(6), 755–761.

Tobías, A., Armstrong, B., Gasparrini, A., & Diaz, J. (2014). Effects of high summer temperatures on mortality in 50 Spanish cities. *Environmental Health, 13*(1), 48.

Vaidyanathan, A., Saha, S., Vicedo-Cabrera, A. M., Gasparrini, A., Abdurehman, N., Jordan, R., et al. (2019). Assessment of extreme heat and hospitalizations to inform early warning systems. *Proceedings of the National Academy of Sciences, 116*(12), 5420–5427.

Wang, C., Zhang, Z., Zhou, M., Zhang, L., Yin, P., Ye, W., & Chen, Y. (2017). Nonlinear relationship between extreme temperature and mortality in different temperature zones: A systematic study of 122 communities across the mainland of China. *Science of the Total Environment, 586*, 96–106.

Wichmann, J., Andersen, Z., Ketzel, M., Ellermann, T., & Loft, S. (2011). Apparent temperature and cause-specific emergency hospital admissions in Greater Copenhagen, Denmark. *PLoS One, 6*(7), e22904.

Zhao, Q., Li, S., Coelho, M. S., Saldiva, P. H., Hu, K., Arblaster, J. M., Nicholls, N., Huxley, R. R., Abramson, M. J., & Guo, Y. (2019). Geographic, demographic, and temporal variations in the association between heat exposure and hospitalization in Brazil: A nationwide study between 2000 and 2015. *Environmental Health Perspectives, 127*(01), 017001.

Zhao, Q., Zhang, Y., Zhang, W., Li, S., Chen, G., Wu, Y., Qiu, C., Ying, K., Tang, H., Huang, J. A., Williams, G., Huxley, R., & Guo, Y. (2017). Ambient temperature and emergency department visits: Time-series analysis in 12 Chinese cities. *Environmental Pollution, 224*, 310–316.

CHAPTER 11

Temporal variation in impacts of heat exposure on human health

Katherine Arbuthnott[1], Ana Maria Vicedo-Cabrera[2] and Antonio Gasparrini[3]
[1]The Oxford School of Public Health (OxSPH), United Kingdom; [2]Institute of Social and Preventive Medicine & Oeschger Center for Climate Change Research, University of Bern, Switzerland; [3]Department of Public Health Environments and Society, London School of Hygiene & Tropical Medicine, London, United Kingdom

1. Introduction

1.1 Background/context: why understanding temporal variation in heat-related health risks is important

Under future climate change, we are extremely likely to experience both increases in mean global temperatures and increases in the number and duration of large-scale heatwave events (Pachauri et al., 2014) An understanding of populations' vulnerability to heat-related health impacts is essential to inform public health policies. Appreciation of the distribution and potential magnitude of heat-related impacts is vital for mitigation and adaptation planning.

Heat-related health impacts, however, cannot be viewed as static: for any public health risk assessment and planning, it is critical that we uderstand whether populations susceptibility to heat-related health impacts has changed over time, i.e., to understand how populations can acclimatize or adapt to increasing temperatures, and whether there are limitations to this in terms of magnitude or pace of adaptation. Understanding temporal variation in impacts of heat exposure is an important part of this picture, because of the following:

- It can give valuable insight into whether and to what extent populations have become less susceptible (and possibly adapted) to the effects of heat.
- Understanding of the factors associated with changes in susceptibility to heat effects could further inform future health policies and protect vulnerable populations.
- It informs the epidemiological baselines used for future climate risk/impact assessments, which can underpin management strategies.
- If susceptibly to heat-related health impacts has not reduced in a given location or conversely has increased certain vulnerable populations, then

Heat Exposure and Human Health in the Context of Climate Change
ISBN 978-0-12-819080-7
https://doi.org/10.1016/B978-0-12-819080-7.00007-0
© 2023 Elsevier Inc.
All rights reserved.

this can highlight the need for specific and planned adaptive strategies to reduce future heat-related risk.

1.2 Chapter outline: scope and definitions

Until more recent years, the majority of heat—health research focused on quantifying the risk of heat—health impacts in a given location across a set time period, with less attention focusing on whether this risk has changed over time (Sheridan & Allen, 2018). Many studies have demonstrated the "U-," "V-," or "J-"shaped relationship between the daily risk of mortality and daily ambient temperature, with mortality increasing above given heat thresholds or minimum mortality temperatures (MMTs) (Basu, 2009).

However, with increasing attention now on adaptation to climate change and understanding how this could be achieved, a large number of studies in recent years have looked at temporal changes in heat-related risk and have been the subject of recent reviews (Arbuthnott et al., 2016; Sheridan & Allen, 2018). Most studies have focused on temporal changes in the direct effects of ambient heat or heatwaves on mortality (rather than on say morbidity) and so mortality is the focus of this chapter.

We provide an overview of both temporal variation in general heat-related risk and in heatwave-related risk, as both mean global temperatures and frequency and intensity of heatwave events are expected to increase under climate change (Pachauri et al., 2014).

In this chapter, we give an overview of these studies and discuss the following:

- The types of studies and methods (including some of the limitations) used to examine temporal changes in heat-related mortality
- Studies that have examined changes in both general heat-related mortality and heatwave-related mortality over time, including an overview of how results vary by geography and by time periods studied
- Factors that may account for any demonstrated changes in susceptibility over time, and evidence that distinguishes between specific heat-related adaptation compared with other factors that may attenuate heat-related mortality risk
- The context of climate change and how an understanding of temporal variation in mortality risk can inform adaptation planning, also including an overview of gaps in knowledge and areas for further work in the future

Specifically, this chapter does not cover intraseasonal variation or temperature variability and heat-related risk. Secondly, we do not cover the body of literature which examines deaths coded as being due to heat stress—for example, physiological responses to temperature disturbance (e.g., heat stroke, hyperthermia). These deaths are often associated with occupational or working conditions and potentially underreported as heat exposure is not often confirmed a cause of death (for reviews of occupational-related risks of increased temperatures, see Vanos et al. (2019), Xiang et al. (2014, 2015)).

1.2.1 Definitions

Before considering temporal variation in heat risk, it is worth defining a few key terms that are used throughout this chapter.

Adaptation and acclimatization have been defined as per Fig. 11.1. It is important to draw a distinction between epidemiological evidence of any decrease in heat-related mortality risk and evidence that specific heat-related adaptation has occurred (Arbuthnott et al., 2016; Vicedo-Cabrera et al., 2018). For example, a reduction in heat-related mortality risk could be due to either specific heat adaptation to increasingly warm temperatures or due to other mechanisms that are unrelated to climate change but that reduce heat-related mortality risk (e.g., infrastructural changes and

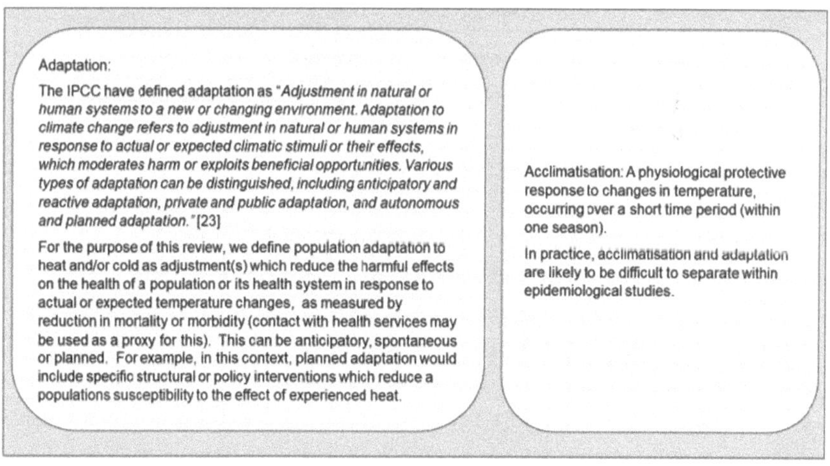

Figure 11.1 Definitions of adaptation and acclimatization. *(From Arbuthnott, K., Hajat, S., Heaviside, C., & Vardoulakis, S. (2016). Changes in population susceptibility to heat and cold over time: Assessing adaptation to climate change.* Environmental Health, 15*(Suppl 1), 33. https://doi.org/10.1186/s12940-016-0102-7. PMID: 26961541; PMCID: PMC4895245.)*

improved healthcare). Specific differentiation between these has been the subject of some studies (e.g., Vicedo-Cabrera et al., 2018) and is detailed further in Section 4 of this chapter.

2. Overview of methods used to study temporal variation impacts of heat exposure on human health

2.1 Study designs: heat exposure

Broadly speaking, there are two types of heat exposure examined when assessing temporal changes in susceptibility to heat—daily measures of ambient heat and specific historical heatwave events. These are typically assessed using the following:

- Observational studies that quantify mortality risk or burden associated with ambient temperatures in a given location(s) at different time points or within different time bands over a specified time period. These are usually time-series studies, where heat-related mortality is quantified above a given heat threshold (e.g., Carson et al., 2006) or MMT (e.g., Gasparrini et al., 2015) or case-crossover studies (e.g., Barnett, 2007). Section 2.3.1 of this chapter gives further details of these studies.
- Observational studies that compare outcomes between one or more discrete extreme temperature events (i.e., heatwaves: extreme events lasting for longer than 1 day, often characterized by the context-specific definition of a heatwave) in a given location. These are usually period analysis studies. Section 2.3.2 of this chapter gives further details of these studies.

In reality, there is some degree of overlap in the interpretation of what is assessed, as many time-series studies examine more extreme ends of the temperature distribution (e.g., heat-related risk at the 98th or 99th percentile compared with the MMT) and many time-series studies that quantify the change in heat-related risk per 1°C above a threshold temperature also include a heatwave term. Beyond the scope of this chapter, a subject of current interest is the specific metric used for ambient heat exposure (e.g., mean temperature, maximum temperature or composite indicators of ambient temperature) and definitions of heatwave events (e.g., see Xu et al., 2016).

2.2 Measures of heat impact and health outcomes

Typically, observational studies that measure changes in heat-related mortality have used all-cause mortality as an outcome, though in some cases changes in mortality from a specific cause (e.g., respiratory or cardiovascular mortality) are quantified.

As a measure of health impact, studies report either or both the relative risk (RR) of heat-related mortality above a given threshold temperature or the MMT across different time periods or the heat impact in terms of excess heat-related deaths or mortality burdens attributable to heat. There are pros and cons to each of these measures of effect: the RR gives an estimate, which is not affected by the number of days above a given heat threshold in each time period or the baseline mortality, and therefore allows a comparison uninfluenced by temperature differences between the time points that are being compared. Attributable burdens give a measure of actual impact and therefore are considered by many to be important for informing policy planning and implementation (Gasparrini & Leone, 2014; Vicedo-Cabrera et al., 2018). They will, however, vary with a given time periods baseline mortality and the number of days above a given threshold, and this should be taken into account in either the analysis or interpretation of results.

As mentioned in Section 1, there is little literature that examines impact in terms of temporal changes in morbidity (e.g., non-fatal cardiovascular events, or indicators of morbidity such as health service use—e.g., admissions to emergency departments) over time, and this chapter therefore only covers temporal variation in measures of heat-related mortality.

2.3 Evaluating changes over time in epidemiological studies

2.3.1 Studies of temporal changes in susceptibility to ambient heat

The evidence for temporal changes in susceptibility to ambient heat is based on studies examining the overall statistical relationship between increased temperatures and health outcomes. Time periods are compared either by analyzing annual changes in temperature-related health risk (e.g., Vicedo-Cabrera et al., 2018); subdividing the overall data into time bands and evaluating changes in outcomes over time across multiple time periods (e.g., Carson et al., 2006); subdividing data according to a singular event (e.g., to test whether heat-related mortality had changed in Japan following decreased electricity use after the Fukushima disaster (Kim et al., 2017); or comparing excess heat-related mortality or the RR of/or excess heat-related mortality at two times representing the beginning and end points of a given time period (e.g., Bobb et al., 2014; Donaldson et al., 2003; Ha & Kim, 2013)).

Evaluating whether there have been temporal shifts in heat-related mortality requires some technical knowledge of the epidemiological methods used to quantify heat-related risk/impacts.

Where papers model relative risk above a heat threshold or MMT, studies either fix the temperature thresholds or MMT (above which heat effects are modeled) across the whole time period or allow them to vary within the time periods analyzed (or undertake a sensitivity analysis using both approaches—e.g., Arbuthnott et al., 2020). These approaches matter because the slope of the regression line is linked to the heat threshold (both in terms of how well the model fits the data and also in that the higher the heat threshold the steeper the regression line). Understanding how changes have been modeled will allow a better interpretation of the results. For example, it is possible that if the threshold is allowed to vary and increases over time but the RR remains constant, this could in fact represent a form of reduced susceptibility over time even though no change in risk is demonstrated. Similar issues arise in studies examining changes in risk at a given temperature compared with the MMT (e.g., deciding whether the MMT should be fixed across the time period or be allowed to vary), and some papers have considered a shift in MMT or heat threshold as evidence of temporal changes in susceptibility to the effects of heat (e.g., Folkerts et al., 2020). Additionally, where attributable burdens of heat-related mortality are quantified, these are affected by changes in baseline mortality and temperature over time. Discussion of the advantages and disadvantages of different methods used is included in recent reviews on this topic (e.g., Arbuthnott et al., 2016).

2.3.2 Studies of heatwave impacts

Most evidence quantifying the impacts of heatwaves on health is based on studies using an episode analysis approach to compare expected and actual deaths during heatwaves. When comparing the impact of different heatwaves within a given location, challenges arise, as no two heatwaves will have the same characteristics (intensity or duration and when in the warm season they occur), which have all been shown to affect impact (Anderson & Bell, 2011; Xu et al., 2016; D'Ippoliti et al., 2010). To allow for this and for comparisons over time, a variety of approaches have been taken, for example, using model parameters from previous years with weather data from heatwaves in later years or vice versa or including qualitative information on the characteristic of the heatwaves alongside the excess mortality figures (for further discussion, see Arbuthnott et al., 2016). The literature examining differential impacts of heatwaves includes studies that have examined the impact of interventions such as heat–health warning systems (HHWSs) (reviewed also in Toloo et al., 2013).

3. Epidemiological evidence of temporal variation in heat–health impacts

In this section, temporal variation in susceptibilty to both heatwaves and ambient heat is described. The majority of evidence consists of studies examining changes in the risk of heat-related mortality or burdens attributable to increased temperatures (rather than specific heatwave events) over time.

Many studies have demonstrated a decrease in population-level susceptibility to the (risk or burden of) heat-related mortality. However, observed temporal changes do vary with location, time period analyzed, and in some instances by specific cause of mortality. Even in the many locations where attenuation in heat risks has been demonstrated, significant risks remain emphasizing the importance of heat-related risks to health both currently and in the context of climate change.

In some instances where decreases are reported but with no indication of whether trends are statistically significant, evidence of change is hard to interpret (see Arbuthnott et al., 2016; Sheridan & Allen, 2018).

3.1 Evidence of changes in susceptibly to ambient heat

3.1.1 Temporal variation by time period studied

The majority of studies have analyzed changes in heat related risk over recent decades (e.g., from the 1960 to 1970s onward). Some studies, however, have assessed declines in heat-related mortality from periods at the start of the previous century or before. Of note, studies that examine risk over such a long time period have mostly used data aggregated to weekly or monthly levels and as such can be difficult to compare to more recent studies. These studies have shown a decrease in population-level heat-related mortality in the United Kingdom (Carson et al., 2006), Sweden (Astrom et al., 2013), the Netherlands (Ekamper et al., 2009), and the United States (Barreca et al., 2016; Petkova et al., 2014). In Europe, most change appears to occur between the first (where risks are higher) and the latter part of the last century (e.g., Astrom et al., 2013; Carson et al., 2006). It has been hypothesized that this reflects much broader changes in housing, lifestyle, sanitation, health systems, and economic prosperity.

While many studies of more recent time periods have indicated a decrease in susceptibility to ambient heat and hotter days, in some locations, there is some evidence that the decline in vulnerability has slowed over more recent years.

3.1.2 Temporal variation by study location

The majority of evidence analyzing temporal variation in heat-related risk comes from studies that have analyzed data from a single country (by region or city in most cases). Although results are generally grouped here by location, different methods and definitions of heat used in studies make it difficult to compare the extent of any temporal changes across locations unless data have been analyzed using a common analytical framework. This has been undertaken in some instances (e.g., see Gasparrini et al., 2015; Vicedo-Cabrera et al., 2018; De'Donato et al., 2015; Chung et al., 2017). Fig. 11.2 shows the changes in risk–response curves between two time periods—1993 (black lines on the graph) and 2006 (blue lines on the graph)—from one such study by Gasparrini et al. (2015). The analysis in the study suggested a decrease in heat-related mortality risk in Japan, Spain, and the United States and some evidence of a decrease in Canada. For example, in the United States, the RR of heat-related mortality at the 99th percentile compared with the MMT dropped from 1.115 (95% CI: 1.095, 1.136) to 1.022 (95% CI: 0.998, 1.048) between the two time periods. Results were less clear for Australia and South Korea, and there was no evidence of a change in mortality risk associated with high temperatures in the United Kingdom (see Table 11.1 for further details).

Results from the many studies examining temporal changes in vulnerability over time are summarized by geographical area in the following and also in Table 11.1. Notably the majority of studies are based in North America, Europe, or Northeast Asia, leaving a large gap in understanding across many locations.

North America: Across the United States, evidence points to a decrease in heat-related mortality risk over time (Barreca et al., 2016; Bobb et al., 2014; Davis et al., 2003a, 2003b; Donaldson et al., 2003; Kalkstein et al., 2011; Nordio et al., 2015; Petkova et al., 2014; Sheridan et al., 2009; Sheridan & Lin, 2014). Spatial variation in vulnerability to heat risk across

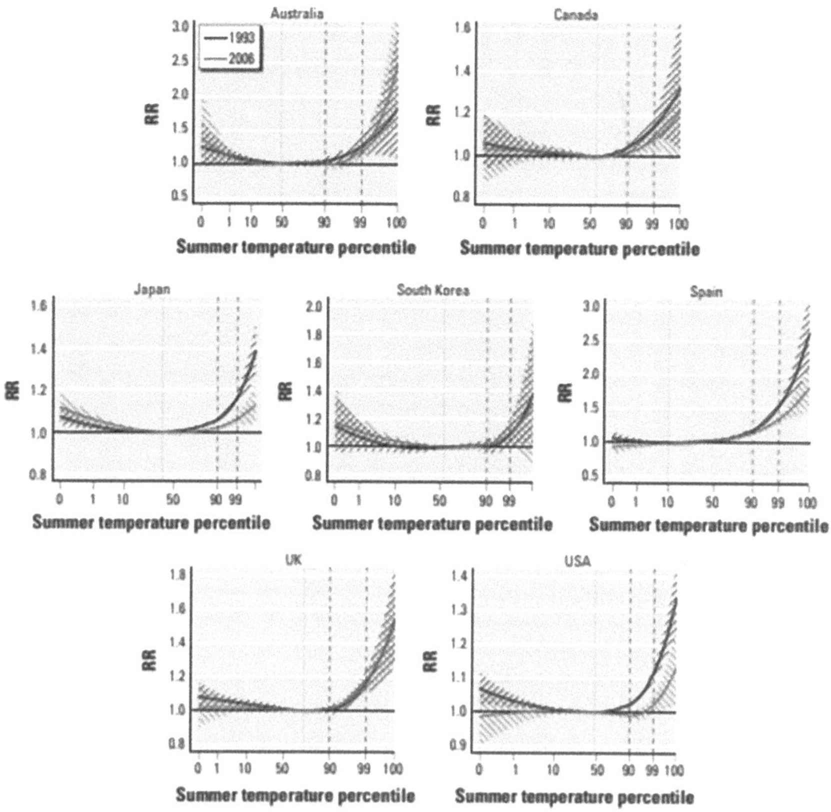

Figure 11.2 Variation in heat-related mortality exposure–response relationships by time and location. This figure shows overall cumulative exposure–response relationships between heat and mortality for 1993 (black) and 2006 (blue) across seven different countries. *(From Gasparrini, A., Guo, Y., Hashizume, M., et al. (2015). Temporal variation in heat-mortality associations: A multicountry study.* Environmental Health Perspectives, 123*(11), 1200–1207. https://doi.org/10.1289/ehp.1409070.)*

the United States has been well documented (e.g., see Braga et al., 2001; Nordio et al., 2015). Some US-based studies have indicated that locations that originally had higher risks (such as cities with cooler climates in the northeastern, midwestern, and northwestern US) have seen the largest reductions in vulnerability to heat (Barnett, 2007; Bobb et al., 2014; Davis, Novicoff, et al., 2003; Sheridan & Dixon, 2017; Sheridan et al., 2009), though others have not found this (e.g., see Kalkstein et al., 2011). This has led to the hypothesis that cities within hotter regions in the United States have already to some extent adapted to the effects of heat. This pattern,

Table 11.1 Summary of studies examining temporal relationship between ambient heat exposure and heat-related mortality. Note, this table does not include studies specifically on heatwaves or evaluations of heatwave plans.

Region	Location and reference	City/Region/ part of multicountry study	Population studied	Time period evaluated	Change in susceptibility: Decrease, no change, increase, or mixed (i.e., different patterns across either regions or by outcome studied)	
North America	United States (Bobb et al., 2014)	105 cities	All ages	1987–2005	**Decrease**: Significant for all ages (greatest in >75 years) and for heat-related respiratory and CVD mortality. Declines larger in northern regions and cities with cooler climates. Overall heat-related deaths per 1000 deaths (all cities): 51 (95% PI: 42,61) in 1987 compared with 19 (95% PI: 12,27) in 2005.	●
	United States (Petkova et al., 2014)	New York	>65 years	1900–48 and 1973–2006	**Decrease:** Heat-related mortality seen in the first part of the century substantially reduced in second half. Overall decrease in RR at 29°C versus 22°C of 4.6% (CI: 2.4,6.7) per decade (all ages). >65 years: highest initial risk and most decline in RR over time.	●
	United States (Vicedo-Cabrera et al., 2018)	135 cities. Part of a multicountry study	All ages	1985–2006	**Decrease** in heat-related mortality burdens. Mortality fraction (%) attributable to heat: 1985–89 = 0.45 [0.40; 0.49], 1990–94 = 0.40 [0.36; 0.43], 1995–99 = 0.36 [0.32; 0.40], 2000–04 = 0.26 [0.22; 0.29].	●

	United States (Gasparrini et al., 2015)	Part of a multicountry study	All ages	1985–2006	**Decrease:** RR at 90th percentile of summer temperature distribution, compared with MMT (58th percentile): in 1993 = 1.024 (1.016, 1.033), in 2006 = 0.992 (0.980, 1.004). RR at 99th percentile of summer temperature distribution, compared with MMT in 1993 = 1.115 (1.095, 1.136), in 2006 = 1.022 (0.998, 1.048).	●
	United States (Barnett, 2007)		Elderly	1987–2000	**Decrease**: CVD mortality in the elderly. In summer 1987, CVD mortality increased by 4.7% per 10°F increase in temperature (95% posterior interval (PI) = 3.0%–6.5%). In 2000, the risk was −0.4% (95% PI: −3.2% to 2.5%).	●
	United States (Barreca et al., 2016)	United States and by US climate regions	All ages	1900–2004	**Decrease**: Mortality impact of days with mean temperature >80°F declined by 75%. Majority of decline occurred after 1960. Air-conditioning analyzed as an explanatory factor.	●
	United States (Davis, Knappenberger, et al., 2003)	28 major US cities	Age-standardized population	1964–98	**Mixed, but decrease in most locations:** Heat-related mortality (HRM) rates declined significantly over time in 19 of the 28 cities. Most	○

Continued

Table 11.1 Summary of studies examining temporal relationship between ambient heat exposure and heat-related mortality. Note, this table does not include studies specifically on heatwaves or evaluations of heatwave plans.—cont'd

Region	Location and reference	City/Region/ part of multicountry study	Population studied	Time period evaluated	Change in susceptibility: Decrease, no change, increase, or mixed (i.e., different patterns across either regions or by outcome studied)
					decline in 1980s in the South and NE cities. Seattle and Washington showed increased HRM in later decades compared with the 1960s.
	United States (Donaldson et al., 2003)	North Carolina. Part of a multicountry study	Age >55 years	1971–96	**Decrease**: All-cause heat-related mortality, significant (adjusted for age/sex).
	United States (Kalkstein et al., 2011)	40 metropolitan areas		1975–2004	**Decrease** in 35/40 areas.
	United States (Nordio et al., 2015)	211 US cities, grouped by temperature and humidity			**Decrease**: Across the United States, pooled risk of mortality decreased over time, from 10.61% (95% CI: 9.10, 12.13) during the 1962–66 period to 0.87% (95% CI: −0.21, 1.97) in 1994–99. Patterns of decrease varied by city-cluster. Additionally, a 50% increase in air-conditioning use was associated 1.37% decrease in heat-related mortality.

	United States (Sheridan et al., 2021)		All ages	1975–2018	**Results varied by age group**: In the past decade analyzed, decrease in heat-related mortality continued over the past decade in over 65s but not in middle-aged males: In many locations, there was an increase in mortality among men 45–64 (+1.3 deaths per year), especially across southern and southwestern United States.	○
	Canada (Gasparrini et al., 2015)	25 cities. Part of a multicountry study		1986–2011	**Nonsignificant decrease**: RR at 90th percentile of summer temperature distribution, compared with MMT (58th percentile): in 1993 = 1.043 (1.015, 1.072) in 2006 = 1.018 (0.993, 1.044). RR for 99th percentile compared with MMT: in 2006 = 1.143 (1.084, 1.204) in 1993 = 1.083 (1.030, 1.138).	◍
	Canada (Vicedo-Cabrera et al., 2018)	25 cities. Part of a multicountry study		1986–2012	**Decrease** in heat-related mortality burdens. Mortality fraction (%) attributable to heat: 1985–89 = 0.69 [0.55; 0.83], 1990–94 = 0.42 [0.33; 0.51], 1995–99 = 0.49 [0.38; 0.58], 2000–04 = 0.43 [0.33; 0.53], 2005–09 = 0.38 [0.20; 0.53].	●

Continued

Table 11.1 Summary of studies examining temporal relationship between ambient heat exposure and heat-related mortality. Note, this table does not include studies specifically on heatwaves or evaluations of heatwave plans.—cont'd

Region	Location and reference	City/Region/part of multicountry study	Population studied	Time period evaluated	Change in susceptibility: Decrease, no change, increase, or mixed (i.e., different patterns across either regions or by outcome studied)	
South America	Brazil (Vicedo-Cabrera et al., 2018)	18 cities. Part of a multicountry study		1997–2011	**Stable** heat-related mortality burdens. Mortality fraction (%) attributable to heat: 1995–99 = 0.78 [0.57; 0.97], 2000–04 = 0.53 [0.41; 0.63], 2005–09 = 0.68 [0.50; 0.85].	○
Europe	Austria (Muthers et al., 2010)	Vienna	All ages	1970–2007	**Decrease** in heat-related mortality, reported as % change in heat related mortality per decade (1970–2007) for given ranges of (physiological equivalent temperature [PET]). Significance varied according to the PET thresholds analyzed.	●
	England (Carson et al., 2006)	London	All ages	1900–96	**Decrease**: RR at 90th percentile of summer temperature distribution, compared to MMT (40th percentile): For analysis of all ages, and overall significant reduction in RR of heat-related mortality and % deaths to heat.	●
		10 regions. Part of a	All ages	1993–2006	**No significant change:** RR at 90th percentile of summer temperature	○

England (Gasparrini et al., 2015)	multicountry study			distribution, compared to MMT (40th percentile): For 1993 = 1.005 (0.983, 1.027), 2006 = 1.014 (0.995, 1.032). RR at 99th percentile of summer temperature distribution, compared with MMT for 1993 = 1.158 (1.093, 1.227), for 2006 = 1.168 (1.111, 1.229).	
England (Vicedo-Cabrera et al., 2018)	10 regions. Part of a multicountry study	All ages		**No change** in heat-related mortality burdens. Mortality fraction (%) attributable to heat: 1990–94 = 0.23 [0.18; 0.27], 1995–99 = 0.42 [0.36; 0.48], 2000–04 = 0.32 [0.27; 0.37], 2005–09 = 0.28 [0.23; 0.32].	○
England (Donaldson et al., 2003)	SouthEast England (part of multilocation study).	Age >55 years	1971–96	**Decrease:** Significant decrease in heat-related mortality between beginning and end time points, adjusted for age and sex.	●
England (De'Donato et al., 2015)	London	All ages	1996–2011	**No significant** change between the two time periods in RR of mortality at 99th compared with 75th percentile of the summer distribution. All-cause mortality RR in 1996–2002 = 1.20 [1.16–1.25]. RR in 2004–10 = 1.18 [1.12–1.23].	○

Continued

Table 11.1 Summary of studies examining temporal relationship between ambient heat exposure and heat-related mortality. Note, this table does not include studies specifically on heatwaves or evaluations of heatwave plans.—cont'd

Region	Location and reference	City/Region/ part of multicountry study	Population studied	Time period evaluated	Change in susceptibility: Decrease, no change, increase, or mixed (i.e., different patterns across either regions or by outcome studied)	
	England (Arbuthnott et al., 2020)	London, West Midlands, Greater Manchester	All ages	1996–2013	**Stable**: No obvious increase or decrease seen for all-cause mortality and YLL attributable to heat. Similar results seen whether thresholds were allowed to vary by time band analyzed or were fixed over the whole time period.	○
	England and Wales (Christidis et al., 2010)		All ages	1976–2005	**Mixed:** Decrease in RR, but Increase in HRM (increased temperatures over time period analyzed). Paper compared change in HRM obtained using regression slopes from different time periods (1976 compared with 2005) to demonstrate no adaptation or early adaptation.	○
	France (De'Donato et al., 2015)	Paris. Part of a multicountry study	All ages	1996–2011	**Decrease** between the two time periods in RR of mortality at 99th compared with 75th percentile of the summer distribution. All-cause mortality RR in 1996–2002 = 1.31 [1.24–1.37]. RR in 2004–10 = 1.11 [1.06–1.17]. *P* value < .001. Similar patterns for cardiovascular and respiratory mortality.	●

	Greece (De'Donato et al., 2015)	Athens. Part of a multicountry study	All ages	1996–2011	**Decrease** between the two time periods in RR of mortality at 99th compared to 75th percentile of the summer distribution. Similar patterns of change for total, respiratory, and cardiovascular mortality. All-cause mortality RR in 1996–2002 1.63 [1.53–1.75], RR in 2004–10 = 1.35 [1.29–1.42] *P* value < .001.	●
	Hungary (De'Donato et al., 2015)	Budapest. Part of a multicountry study	All ages	1996–2011	**Mixed:** Some increases seen between the two time periods in the RR of mortality at 99th compared with 75th percentile of the summer distribution, only significant for cardiovascular mortality. All-cause mortality RR in 1996–2002 = 1.29 [1.22–1.37], RR in 2004–10 = 1.33 [1.27–1.40] *P* value = .451.	○
	Italy (De'Donato et al., 2015)	Rome. Part of a multicountry study	All ages	1996–2011	**Decrease** between the two time periods in RR of mortality at 99th compared with 75th percentile of the summer distribution. All-cause mortality RR in 1996–2002 = 1.53 [1.45–1.61], RR in 2004–10 = 1.27 [1.19–1.35] *P* value < .001.	●

Continued

Table 11.1 Summary of studies examining temporal relationship between ambient heat exposure and heat-related mortality. Note, this table does not include studies specifically on heatwaves or evaluations of heatwave plans.—cont'd

Region	Location and reference	City/Region/ part of multicountry study	Population studied	Time period evaluated	Change in susceptibility: Decrease, no change, increase, or mixed (i.e., different patterns across either regions or by outcome studied)	
	Island of Ireland (Vicedo-Cabrera et al., 2018)	6 region (across republic of Ireland and Northern Ireland). Part of a multicountry study	All ages	1985–2007)	**No significantchange** for heat-related mortality burdens. Mortality fraction (%) attributable to heat: 1985–89 = 0.04 [−0.03; 0.10], 1990–94 = 0.02 [−0.02; 0.06], 1995–99 = 0.03 [−0.06; 0.11], 2000–04 = 0.00 [−0.02; 0.01], 2005–09 = −0.02 [−0.06; 0.01].	○
	Spain (Vicedo-Cabrera et al., 2018)	50 cities. Part of a multicountry study	All ages	1990–2010	**Decrease** in heat-related mortality burdens. Mortality fraction (%) attributable to heat: 1990–94 = 1.39 [1.1; 1.65], 1995–99 = 0.90 [0.81; 0.99], 2000–04 = 1.12 [1.05; 1.19], 2005–09 = 0.93 [0.84; 1.01].	●
	Spain (Gasparrini et al., 2015)	50 cities. Part of a multicountry study	All ages	1990–2010	**Decrease:** RR of heat-related mortality at 90th percentile of summer temperature distribution, compared with MMT (40th percentile): For 1993 = 1.191 (1.151, 1.232), for 2006 = 1.157 (1.128, 1.187). RR at 99th percentile of	●

				summer temperature distribution, compared with MMT in 1993 = 1.559 (1.478, 1.643), in 2006 = 1.367 (1.310, 1.426).	
Spain (De'Donato et al., 2015)	Barcelona. Part of a multicountry study	All ages	1996–2011	**No change** between the two time periods in RR of mortality at 99th compared to 75th percentile of the summer distribution. All-cause mortality RR in 1997–2002 = 1.27 [1.18–1.36]. RR in 2004–10 = 1.26 [1.17–1.36].	○
Spain (De'Donato et al., 2015)	Valencia. Part of a multicountry study	All ages	1996–2011	**Nonsignificant increase**: RR of all-cause mortality at 99th compared with 75th percentile of the summer distribution. All-cause mortality RR in 1996–2002 = 1.11 [1.04–1.17], RR in 2004–10 = 1.18 [1.03–1.36] *P* value = .216.	●
Spain (Miron et al., 2015)	Central Spain	All ages	1975–2008	**Decrease:** On days where the maximum temperature exceeded 37°C, CVD-related mortality decreased, respiratory mortality did not. The percentage increase in mortality due to CVD was 19.3% (17.3–21.3) in 1975–85, 30.3% (28.3	●

Continued

Table 11.1 Summary of studies examining temporal relationship between ambient heat exposure and heat-related mortality. Note, this table does not include studies specifically on heatwaves or evaluations of heatwave plans.—cont'd

Region	Location and reference	City/Region/ part of multicountry study	Population studied	Time period evaluated	Change in susceptibility: Decrease, no change, increase, or mixed (i.e., different patterns across either regions or by outcome studied)	
					−32.3) in 1986−96, and 7.3% (6.2 −8.4) in 1997−2008.	
	Sweden (Astrom et al., 2013)	Stockholm	All ages	1901−2009	**Decrease**: Significant decline in mortality risk for elderly and combined age categories.	●
	Sweden (De'Donato et al., 2015)	Stockholm. Part of a multicountry study	All ages	1996−2011	**Mixed:** Little change between the two time periods in the RR of all-cause mortality at 99th compared with 75th percentile of the summer distribution. All-cause mortality RR in 1996−2002 = 1.10 [1.04−1.17], RR in 2004−10 = 1.12 [1.06−1.19] *P* value = .628. Increase more prominent for cardiovascular deaths (though not significant).	○
	Finland (Donaldson et al., 2003)	Southern Finland. Part of a multicountry study	Age >55 years	1972−	**Decrease**: Significant decrease in heat-related mortality between beginning and end time points, adjusted for age and sex.	●

	Finland (De'Donato et al., 2015)	Helsinki. Part of a multicountry study		1996–2002 and 2004–10	**Increase** in RR of mortality for increases in mean temperature from the 75th to 99th percentiles of the summer distribution between the two time periods. All-cause mortality RR in 1996–2002 1.02 [0.93–1.12], RR in 2004–10 = 1.24 [1.14–1.35] *P* value = .003	◍
	Switzerland (Ragettli et al., 2017)		Age >65 years	1995 and 2013	**Decrease: Nonsignificant** reduction in effect of high temperatures on mortality in the more recent time period. Overall the RR of heat-related mortality on days on days where temperatures exceeded the 98th percentile of summer temperatures compared with the 50th percentile was 1.16 (1.09–1.23) for the mean temperature.	○
	Switzerland (Vicedo-Cabrera et al., 2018)	8 cities. Part of a multicountry study		1995–2012	**Decrease** in heat-related mortality burdens. Mortality fraction (%) attributable to heat: 1995–99 = 1.66 [0.72; 2.56], 2000–04 = 1.32 [0.67; 1.94], 2005–09 = 0.54 [0.27; 0.78].	●
	Netherlands (Folkerts et al., 2020)		Age >65 years	1995–2017	**Decrease**: Authors used a variety of methods to examine whether vulnerability to heat decreased over	●

Continued

Table 11.1 Summary of studies examining temporal relationship between ambient heat exposure and heat-related mortality. Note, this table does not include studies specifically on heatwaves or evaluations of heatwave plans.—cont'd

Region	Location and reference	City/Region/ part of multicountry study	Population studied	Time period evaluated	Change in susceptibility: Decrease, no change, increase, or mixed (i.e., different patterns across either regions or by outcome studied)	
					the time period, including examining shifts in the MMT and using different models to examine changes in RR over different time periods. A decrease in vulnerability was seen across all methods used.	
	Netherlands (Ekamper et al., 2009)	Zeeland		1855–2006	**Decrease:** Regression coefficients for HRM decreased over time (no test for significance). Demonstrated shift in MMT to higher temperatures in later time periods: MMT slightly below 15°C for 1855–97 and around 17°C for 1905–29 and 1930–54.	○
Asia	South Korea (Ha & Kim, 2013)	Seoul	All ages	1993–2009 (1994 excluded: extreme HW)	**Decrease:** Decrease largest for deaths of all circulatory causes, and for late summers for all causes (compared with early summers across the two time periods). An increase for all -cause nonaccidental mortality seen for early and all summer (compared	○

				with late summers). No changes significant on testing (decreases or increases).	
South Korea (Heo et al., 2016)	City-level analysis	All ages	1996–2012	**Mixed**: For all regions combined, heat-related mortality risk decreased in 2008–12 compared with 1996–2000. However, increasing mortality risk in the later time period was observed hotter cities (RR of 1.07 (95% CI 1.02 to 1.12) in 1996–2000 and RR od 1.10 (95% CI 1.06 to 1.14) in 2008–12 for all-cause mortality). The increases in mortality in the hotter cities were larger for younger adults, those with less education and blue-collar workers.	○
South Korea (Vicedo-Cabrera et al., 2018)	7 cities. Part of a multicountry study	All ages	1992–2010	**Decrease** in heat-related mortality burdens. Mortality fraction (%) attributable to heat: 1990–94 = 0.53 [0.28; 0.75], 1995–99 = 0.33 [0.18; 0.48], 2000–04 = 0.22 [0.11; 0.33], 2005–09 = 0.17 [0.03; 0.29].	●
		All ages	1992–2010	No significant change. RR at 90th percentile of summer temperature	○

Continued

Table 11.1 Summary of studies examining temporal relationship between ambient heat exposure and heat-related mortality. Note, this table does not include studies specifically on heatwaves or evaluations of heatwave plans.—cont'd

Region	Location and reference	City/Region/part of multicountry study	Population studied	Time period evaluated	Change in susceptibility: Decrease, no change, increase, or mixed (i.e., different patterns across either regions or by outcome studied)
	South Korea (Gasparrini et al., 2015)	6 cities. Part of a multicountry study			distribution, compared with MMT (40th percentile): for 1993 = 1.010 (0.966, 1.056), for 2006 = 1.021 (0.989, 1.053). RR at 90th percentile of summer temperature distribution, compared with MMT for 1993 = 1.094 (1.003, 1.193) for 2006 = 1.106 (1.010, 1.211). Low statistical power.
	South Korea (Kim et al., 2015)	Seoul	All ages	1995–2011	**Mixed:** Variability in temporal trend by cause and by lag period (concurrent day vs. cumulative risks over 21 days) examined. The RR of mortality at the 99th percentile compared to the 90th percentile for short lag periods decreased over time for all-cause mortality but increased for cardiovascular-related mortality. RR of heat-related mortality at longer lags increased over time, except for respiratory outcomes.

South Korea (Chung et al., 2017)	6 cities. Part of a multicountry study		1992–2010	**Decrease:** Significant decrease in heat-related risk. This was largest for CVD and respiratory deaths. There was a rapid decrease in heat-related risk in the 1990s and small fluctuations in the mid-2000s, followed by a strong decrease again in the late 2000s.	●
Japan (Chung et al., 2017)	6 cities. Part of a multicountry study		1972–2009	**Decrease:** Overall significant decrease. **H**eat-related risk dramatically decreased in the 1970 and 1980s and slowed in the mid-1990s through the 2000s. Decrease more dramatic for respiratory deaths than for CVD or noncardiorespiratory deaths.	●
Japan (Gasparrini et al., 2015)	47 prefectures. Part of a multicountry study		1985–2012	**Decrease:** RR at 90th percentile of summer temperature distribution, compared with MMT (40th percentile): for 1993 = 1.059 (1.047, 1.072), for 200 6 = 1.022 (1.013, 1.031). For the 99th percentile compared to MMT: 1993 = 1.161 (1.133, 1.190) for 2006 = 1.062 (1.038, 1.086).	●

Continued

Table 11.1 Summary of studies examining temporal relationship between ambient heat exposure and heat-related mortality. Note, this table does not include studies specifically on heatwaves or evaluations of heatwave plans.—cont'd

Region	Location and reference	City/Region/ part of multicountry study	Population studied	Time period evaluated	Change in susceptibility: Decrease, no change, increase, or mixed (i.e., different patterns across either regions or by outcome studied)	
	Japan (Vicedo-Cabrera et al., 2018)	47 prefectures. Part of a multicountry study		1985–2012	**Decrease** in heat-related mortality burdens. Mortality fraction (%) attributable to heat: 1985–89 = 0.45 [0.40; 0.49], 1990–94 = 0.63 [0.57; 0.69], 1995–99 = 0.45 [0.40; 0.49], 2000–04 = 0.34 [0.30; 0.39], 2005–09 = 0.15 [0.11; 0.18].	●
	Japan (Ng et al., 2016)	47 prefectures	All ages	1972–2010	**Decrease** in attributable mortality: At national level, the paper found a reduction of 20 (95% CI: 17, 22), 21 (95% CI: 18, 25), and 46 (95% CI: 36, 55) cases of total, cardiovascular, and respiratory heat-related deaths (per 1000 deaths). Increased air-conditioning prevalence was not associated with a reduction of excess mortality over time. Larger decline of excess mortality was seen in populations with improved economic status.	●

Japan (Onozuka & Hagihara, 2015)	Fukuoka	All ages	1973–2012	**Decrease**: For whole time period: RRs of mortality during the heat extremes were 1.06 (95% CI = 1.00 −1.13). Changes over time were analyzed by either fixing thresholds over the entire time period (model 1) or allowing them to vary by decade (model 2). For 1973–82, the RR was 1.05 (0.99, 1.12) for model 1 and 1.07 (1.03, 1.10) for model 2. For 2003–12, RR was 1.03(1.01, 1.05) for model 1 and 1.03 (1.00, 1.05) for model 2.	○
Japan (Kim et al., 2017)	Prefectures, analyzed by whether most or least affected by earthquake			**Mixed:** In prefectures most affected by earthquake and greatest reduction in electricity consumption, the risk of mortality due to heat decreased, but there was little change in heat-related mortality in the less affected prefecture. In 13 of 15 of the affected prefectures, RR of mortality associated with the 95th versus 50th percentile of daily mean temperature decreased: summary RRs across these prefectures decreased, from 1.08 [95% (CI): 1.05, 1.11] to 1.02 (95% CI:	●

Continued

Table 11.1 Summary of studies examining temporal relationship between ambient heat exposure and heat-related mortality. Note, this table does not include studies specifically on heatwaves or evaluations of heatwave plans.—cont'd

Region	Location and reference	City/Region/ part of multicountry study	Population studied	Time period evaluated	Change in susceptibility: Decrease, no change, increase, or mixed (i.e., different patterns across either regions or by outcome studied)	
					1.00, 1.05), resulting in a 5% decrease in the estimated risk of heat-related all-cause mortality (RRR = 0.95; 95% CI: 0.92, 0.99). However, for less affected prefectures, summary RRRs were close to the null (e.g., RRR for the 95th percentile = 1.01; 95% CI: 0.98, 1.04).	
	Taiwan (Chung et al., 2017)	3 cities. Part of a multicountry study		1994–2007	**Decrease overall**: Pattern varied by decade: heat-related risk increased in the 1990s and decreased in the 2000s for CVD and respiratory deaths, while noncardiorespiratory risk decreased throughout the study period. Decrease not significant.	●
	China (Yang et al., 2015)	Urban district of Shanghai		1981–2012	**No decrease:** In moderate or extreme risk of cardiovascular mortality. For overall time period, RR 90th and 99th percentile with	○

					the reference temperature (26°C) over the cumulative lags of 0–14 day was 1.03 (95% CI: 1.01, 1.05), and 1.14 (95% CI: 1.07, 1.21).	
	China (Li et al., 2017)	Ningbo	All ages, YLL		**Increase:** In general, increase in RR of heat-related YLL in the later time period for all-cause mortality and whole population. When results were stratified by modifying factors, significant increase over the time periods demonstrated for YLL due to heat-related respiratory mortality in married individuals.	●
Australasia	Australia (Gasparrini et al., 2015)	3 cities. Part of a multicountry study		1988–2009	**Unclear**: Risk of all-cause heat-related mortality at 90th percentile of temperature distribution compared with MMT (48th percentile): 1993 = 1.038 (0.986, 1.092), 2006 = 0.997 (0.943, 1.055). At 99th percentile of temperature distribution compared with MMT: 1993 = 1.256 (1.117, 1.412), 2006 = 1.205 (1.072, 1.355). Nonsignificance of results attributed to low statistical power.	○

Continued

Table 11.1 Summary of studies examining temporal relationship between ambient heat exposure and heat-related mortality. Note, this table does not include studies specifically on heatwaves or evaluations of heatwave plans.—cont'd

Region	Location and reference	City/Region/ part of multicountry study	Population studied	Time period evaluated	Change in susceptibility: Decrease, no change, increase, or mixed (i.e., different patterns across either regions or by outcome studied)	
	Australia (Vicedo-Cabrera et al., 2018)	3 cities. Part of a multicountry study		1988–2008	**Increase** in heat-related mortality burdens. Mortality fraction (%) attributable to heat: 1990–94 = 0.05 [−0.11; 0.18], 1995–99 = 0.26 [0.07; 0.43], 2000–04 = 0.52 [0.32; 0.71], 2005–09 = 0.67 [0.42; 0.90].	●

Key: Significant decrease in vulnerability found ●; nonsignificant decrease found (or significance not reported) ○; Mixed results (direction trend depends on outcome, region, or method) ○: nonsignificant increase in vulnerability found ●; significant increase in vulnerability found ●.

however, has not been replicated in European or Asian studies (see following sections).

While most studies have consistently shown decreases in risk for the United States in recent years, there are some indications this may have slowed over the most recent decade (Sheridan et al., 2021).

Reductions in the RR and AF of heat-related mortality over recent years have also been demonstrated for Canada (Vicedo-Cabrera et al., 2018), though decreases in risk for Canada have not been statistically significant in some studies (Gasparrini et al., 2015).

South America: There are few studies that have included data from South America to examine changes in heat-related mortality risk over time, representing a gap in knowledge. One study found little change in heat-related mortality burdens across 18 cities in Brazil (Vicedo-Cabrera et al., 2018).

Europe: The picture across Europe indicates a general reduction in heat-related mortality risk for those countries studied, but with considerable heterogeneity between studies and locations (see Table 11.1). For example, de'Donato et al. demonstrated that in 9 European cities, deaths attributable to heat between 1996–2002 and 2004–10 only decreased significantly in Rome, Paris, and Athens and marginally in London, but that in other locations the risk remained stable or increased—for example, in Helsinki, Finland (21) (though by contrast, others have found decreases in heat related mortality risk in Finland, albeit over a different time period (41)). In another multilocation study, heat-related mortality burdens decreased over time in Spain and Switzerland but not in the United Kingdom and Ireland (Vicedo-Cabrera et al., 2018). The decreasing trend has been supported for Switzerland (though decreases seen were nonsignificant; Ragettli et al., 2017). One country-level study in Spain found changes in mortality risk on hot days varied according to outcome (Miron et al., 2015).

Asia: Studies of temporal changes in heat-related risk have used data from Japan, China, Taiwan, and South Korea. Results have again varied across location and outcome studied. Chung et al. found that across 15 cities in northeastern Asia, heat-related mortality decreased between 1972 and 2009 in Northeast Asia, with larger decreases across earlier decades and for older populations (Chung et al., 2017). For Japan, most evidence points to a decrease in risk (Gasparrini et al., 2015; Ng et al., 2016; Onozuka & Hagihara, 2015; Vicedo-Cabrera et al., 2018), though one study found that trends over time varied by cause of death (Kim et al., 2015) (and see Section 3.1.3). In South Korea, results have varied according to location, outcome, and lag period studied. Some have pointed to nonsignificant decreases or stable trends (Ha & Kim, 2013; Vicedo-Cabrera et al., 2018). One study found evidence of increasing heat-related mortality in hotter cities and decreasing patterns elsewhere (Heo et al., 2016). In China, there is less evidence of a decrease in susceptibility. Yang et al. found no evidence for a decrease in heat-related cardiovascular mortality in Shanghai between 1981 and 2012 (Yang et al., 2015), and Li et al. found an increase in YLL attributable to heat in recent years in Ningbo (Li et al., 2017).

Australasia: Temporal changes in heat-related mortality risk and burdens in Australia have been examined as part of larger data sets. However, no statistically significant decrease in heat-related mortality risk was observed in one study (Gasparrini et al., 2015), whereas another found an increase in heat-related mortality burdens (AFs) (Vicedo-Cabrera et al., 2018). Though not specifically an outcome detailed in this chapter, one study has specifically looked at heat-related emergency cardiovascular admissions Queensland, Australia, from 1995 to 2016 and found that these have increased in recent years (Lu et al., 2020).

3.1.3 Temporal variation by outcome (RR, AF, type of mortality)

Across studies, temporal variation can vary according to the outcome or mortality measure studied.

Firstly, whether susceptibility to heat is estimated in terms of relative risks or heat-related mortality burdens is important, as burdens reflect the frequency of heat events, days above a given threshold temperature and baseline mortality (see Section 2). This can be important, for example, in some cases where the heat risk function has decreased over time but is offset by increasing temperatures (e.g., Christidis et al., 2010).

Secondly, the cause of mortality studied (typically all-cause, cardiovascular or respiratory) can affect the observed trends. Some studies have observed similar results across both cardiovascular and respiratory mortality outcomes (Bobb et al., 2014; Ng et al., 2016). However, many have found varying patterns. For example, some studies have found that heat-related cardiovascular mortality has declined but not respiratory or all-cause mortality (Ha & Kim, 2013; Miron et al., 2015), and some have found that heat-related cardiorespiratory mortality has declined faster than all-cause mortality (Chung et al., 2017). Others have found no decrease in cardiovascular heat effects (Yang et al., 2015) or in some cases an increase in heat related cardiovascular mortality (Kim et al., 2015), and some have demonstrated variable patterns in cardio and respiratory causes according to location (De'Donato et al., 2015).

3.1.4 Temporal variation by population characteristics

Given that populations are projected to age over coming years, understanding temporal trends by demographic factors is important. A number of studies have analyzed outcomes by age group, and some have only analyzed trends in older ages (e.g., Barnett, 2007; Donaldson et al., 2003). Many have shown decreases in heat-related risks across the age groups, with some finding greater declines in older ages (e.g., Astrom et al., 2013; Bobb et al., 2014; Petkova et al., 2014), and some have shown variation in declines across age bands and different locations (De'Donato et al., 2015). Of note, while Sheridan et al. found evidence of a continuing decrease in heat-related risk for over 65s in recent years, the risk for men aged 45–64 had increased, especially in southern and southwestern United States (Sheridan et al., 2021).

Where differences in temporal trend have been examined by gender, then trends have either been comparable or showed a greater decrease for women (starting from a higher vulnerability) (see Sheridan & Allen, 2018 for further details).

3.2 Evidence of temporal variation in the health impact of heatwaves

Across most locations that have been studied, the impact of heatwaves on excess mortality has decreased over the time periods analyzed. The time periods over which these heatwave studies have been conducted vary from the late 1990s to more recent years.

In the United States, for example, studies have mostly shown reduced impacts in the 1999 heatwaves compared with the 1995 Chicago heatwave (Palecki et al., 2001; Weisskopf et al., 2002). One study has shown that in St Louis, United States, the lower impact of the 1995 heatwave event in the over 65s compared to a 1980 event could have been due to differences in intensity, duration, and timing in the summer (Smoyer, 1998). When models used to compare events accounted for these factors, then a reduction in population vulnerability was not clear, despite increased air-conditioning prevalence and public health measures.

Within Europe, the 2003 heatwave is the focus of much health impact literature. In France and England, where the impacts of the 2003 were severe, there is evidence of a reduction in heatwave-related mortality since this time (Fouillet et al., 2008; Green et al., 2016). Elsewhere in Europe (where the 2003 heatwave impacts were less severe), some studies have demonstrated reduced excess mortality during the 2003 heatwave compared with heatwaves in preceding years (e.g., in Spain,(DeCastro et al., 2011) and the Czech republic (Kyselý and Kříž, 2008; Kyselý & Plavcová, 2012). However, in the Czech republic, excess mortality during the 2015 heatwave was similar to that in a previous 1994 event, showing that at least in this location the decline in heatwave-related mortality in the years up to 2003 cannot be assumed to continue (Urban et al., 2017).

In Asia, reduced mortality during heatwaves between 1998 and 2003 has been demonstrated in Shanghai, China (Tan et al., 2007). More recent studies examining extreme heat (heat effects at the 99th compared with 90th percentile of the temperature distribution, but not necessarily as part of a sustained heatwave) have indicated that, for 15 cities across Japan, Taiwan, and Korea, mortality related to extreme heat has decreased over recent years (Chung et al., 2017).

Differences over time in susceptibility to heat waves and to ambient heat: an example from the United Kingdom

Many specific measures, such as heat health warning systems (HHWS), are only fully implemented during an extreme event (Toloo et al., 2013, Toloo et al., 2014), although many do include adaptive and planning measures to be taken year round (e.g. (Public Health England., 2015) . In addition, political will to react to extreme events, such as the 2003 heatwave (often referred to as the trigger for many European countries' HHWS) may be greater (Tompkins et al., 2010): they have an immediate and high impact compared to slowly changing environmental risk. A recent evaluation of England's heat wave plan (HWP) found that despite recent heat wave events in England not being associated with large excesses in mortality, there little evidence for a reduction in the risk of annual heat related mortality outside heatwave events since the HWP's introduction (Williams L., 2019). This is important, as in general across a given year the biggest heat related health burdens in the UK lie across the larger number of moderately hot days. It is consistent with other work from the UK, where analysis has shown cumulative heat-related mortality during extreme events has decreased, (for example in the 2013 heatwave compared to those in 2003 and 2006, despite similar periods of prolonged high temperatures (Green et al., 2016)), but where over more recent years, studies have not consistently shown a decrease in the risk of heat related mortality in response to increases in general ambient temperatures (Vicedo-Cabrera et al., 2018, Arbuthnott et al., 2020, Gasparrini et al., 2015).

By contrast, in other locations, such as in the US, both heat related mortality associated with extreme events and general ambient temperatures has decreased (e.g.(Bobb et al., 2014, Petkova et al., 2014) and in some locations such as in Adelaide, Australia, heat wave deaths have increased in later years (Nitschke et al., 2011).

Figure 11.3 Differences over time in susceptibility to heatwaves and to ambient heat.

In Australia, however, one study examined heatwave impacts over time in Adelaide and demonstrated greater heat–health impacts in the 2009 heatwave compared with previous events (Nitschke et al., 2011).

In any given location, temporal changes in vulnerability may not be consistent between extreme events and general increased ambient temperatures—see Fig. 11.3.

4. Factors accounting for changing impacts over time

Of relevance to public health and adaptive planning is not just whether populations have become less susceptible to the effects of heat over time, but understanding the factors that have driven this change (where it has occurred).

A reduction in heat-related mortality risk can be driven by either specific acclimatization or adaptation—i.e., to temperatures that have increased, or by general decreasing susceptibility to heat arising through mechanisms related to general improvements to health or social care, which

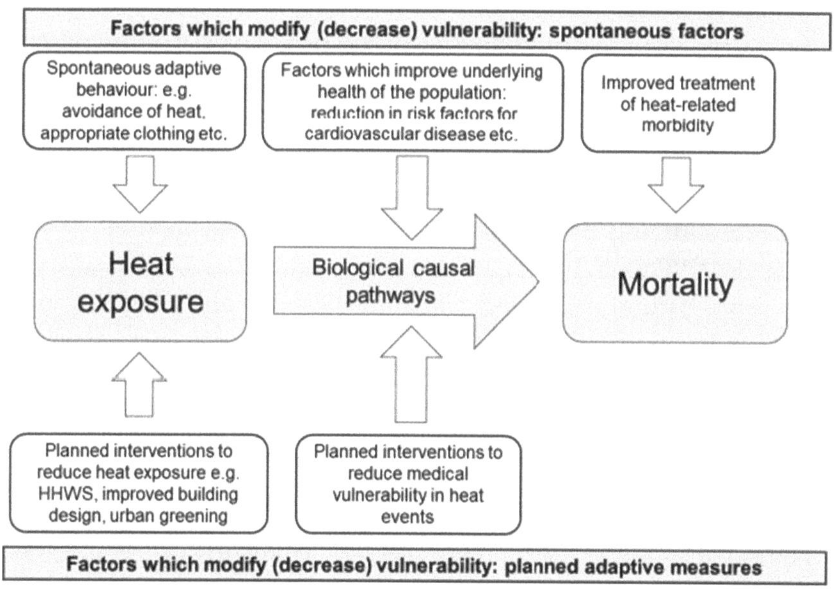

Figure 11.4 Factors accounting for changes in heat-related susceptibility over time. *(From Arbuthnott, K., Hajat, S., Heaviside, C., & Vardoulakis, S. (2016). Changes in population susceptibility to heat and cold over time: Assessing adaptation to climate change.* Environmental Health, 15*(Suppl 1), 33. https://doi.org/10.1186/s12940-016-0102-7. PMID: 26961541; PMCID: PMC4895245.)*

can happen simultaneously but not specifically in response to changing temperatures (Arbuthnott et al., 2016). Understanding the evolution of temporal changes that have occurred and evaluating the impact of specific adaptive plans (for example, HHWSs) can give valuable insight into potential future adaptive options and adaptive capacity. Potential pathways for reductions in heat susceptibility are illustrated in Fig. 11.4.

The specific factors contributing to any attenuation in observed heat—health impacts are likely to vary according to historical and geographical context. Attributing declines in vulnerability to specific spontaneous or planned factors depends on available, reliable, and complete data on hypothesized modifying factors. These data are often limited or capture only broad changes at the ecological level, whereas differences in mortality may be due to a number of factors interacting across a complex systems.

In this section, we bring together explanations that have been offered and in some cases quantified, for observed changes in susceptibility to heat. This is done bearing the aforementioned limitations in mind, and acknowledging that the explanatory factors discussed will not represent the full range of possible and potential adaptive strategies. Some of these have reviewed elsewhere (e.g., in Boeckmann & Rohn, 2014), and the impact of some potential adaptive interventions has been demonstrated through modeling studies (e.g., the impact of cool roofs in reducing heat-related mortality (Macintyre & Heaviside, 2019))

4.1 Differentiating between spontaneous and heat-related adaptive mechanisms

Over earlier parts of the century and in the Global North especially, many societies have undergone broad changes and improvements in general living standards (following industrial revolutions and accompanying epidemiological transitions). Research highlighting the large decreases in heat-related risk over earlier time periods has offered these broad societal changes, as part of the explanation of the decreases in risk seen (Astrom et al., 2013; Carson et al., 2006).

One study made use of two different counterfactual scenarios to determine the specific contributions of changes in temperature distribution and evolution of temperature mortality curves to changes in heat-related mortality across different locations. By comparing temporal variation uniquely driven by changes in response—exposure function and temporal variation uniquely driven by changes in temperature distribution, the research demonstrated that where heat mortality had decreased, attenuation in most of the countries was beyond that simply expected from the pure adaptation to the changing climate: the pace of decrease over the time period studies was faster than observed warming (Vicedo-Cabrera et al., 2018). This pattern suggests that non—climate-related mechanisms (e.g., infrastructure changes, improved healthcare) have made a large contribution to reduced heat susceptibility and as such that there is still much scope for the continued development of specific heat-adaptive strategies to reduce current and future heat-related risk.

4.2 Attributing declines in heat-related susceptibility to specific modifying factors and interventions

The study describedabove (section 4.1) by Vicedo-Cabrera et all. helped deconstruct contributions of spontaneous and heat-related adaptive changes to heat related risk. However, within these two broad categories it is helpful to understand more about specific factors that have led to decreases in vulnerability.

Factors at the city or urban level: In studies covering recent time periods, a number of explanations for temporal reductions in risk have been proposed (but not quantified). These range from the individual to the built environment to local- and national-level policies. Specifically, these include improved urban planning (including urban greening) and building design (e.g., as in Petkova et al., 2014; Davis, Knappenberger, et al., 2003; Gasparrini et al., 2015; De'Donato et al., 2015), the use of air-conditioning (see the following) and medical factors such as improved cardiovascular or respiratory morbidity (e.g., Petkova et al., 2014), and improved healthcare (Vicedo-Cabrera et al., 2018).

Air-conditioning: Air-conditioning receives much attention in adaptive literature. Nordio et al. (2015), examining temporal declines in vulnerability in the United States, found that a 50% increase in air-conditioning prevalence was associated with a 1.37% decrease in heat-related mortality. However, other studies both in North America (e.g., Bobb et al., 2014) and in Asia (Heo et al., 2016; Ng et al., 2016) have not demonstrated a statistically significant link between decreasing heat vulnerability and increased air-conditioning prevalence. In a recent multicountry, multicity longitudinal study, Sera et al. quantified the role of air-conditioning in observed attenuation of heat-related mortality in Canada, Japan, Spain, and the United States (Sera et al., 2020). The increases

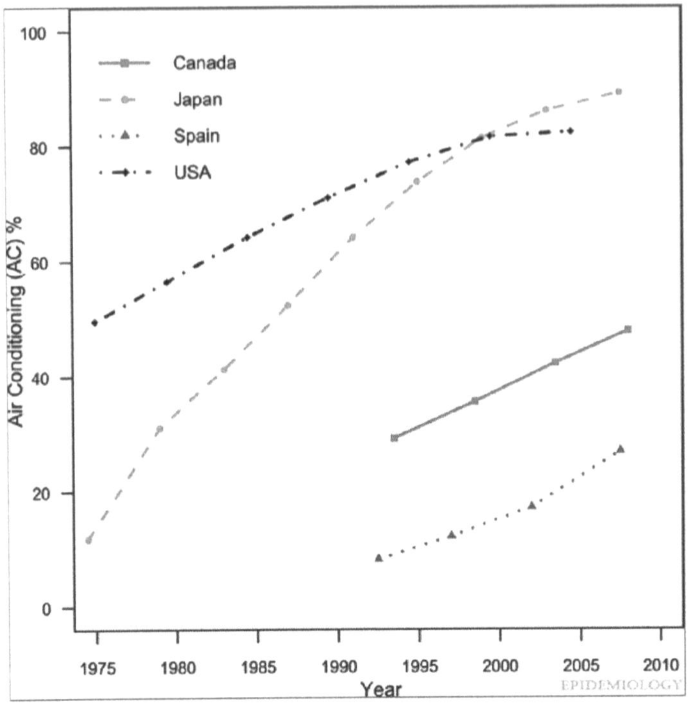

Figure 11.5A Air-conditioning prevalence (%) by year in Canada, Japan, Spain, and the United States.

in air-conditioning prevalence across each of the four locations are illustrated in Fig. 11.5A. Fig. 11.5B shows the average exposure–response curves (in RR) across the four countries, predicted at the beginning and end of the study periods. By comparing excess mortality under scenarios of observed increases in air conditioning and counterfactuals of no change in air-conditioning prevalence (see Fig. 11.6), they estimated that observed air-conditioning prevalence is responsible for part of the observed attenuation in heat attributable mortality. Across the study periods, excess deaths due to heat decreased from 1.40% to 0.80% in Canada, 3.57%–1.10% in Japan, 3.54%–2.78% in Spain, and 1.70%–0.53% in the United States. The increased prevalence of air-conditioning observed across the locations was estimated to account for 16.7% of the attenuation in heat-related excess deaths in Canada, 20.0% in Japan, 14.0% in Spain, and 16.7% in the United States, suggesting that there are also additional factors that have played an important role in reducing heat-related vulnerability in these populations.

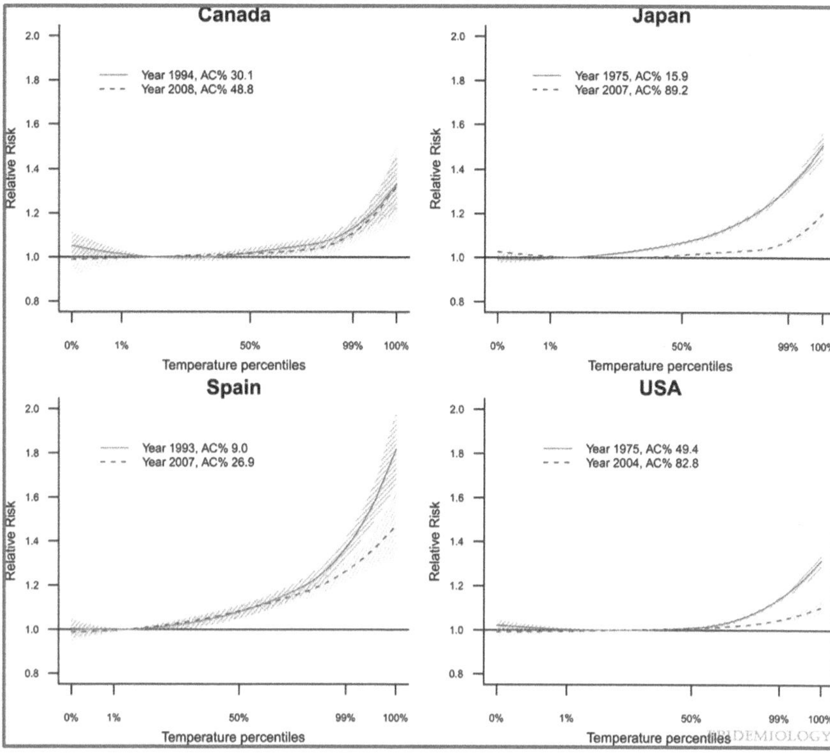

Figure 11.5B Country-average exposure–response curves (in RR) predicted at the beginning and end of the study periods in Canada, Japan, Spain, and the United States. The x-axis represents relative temperatures in percentiles, but rescaled using the average distribution of absolute temperature across cities in each country. *(From Sera, F., Hashizume, M., Honda, Y., Lavigne, E., Schwartz, J., Zanobetti, A., Tobias, A., Iñiguez, C., Vicedo-Cabrera, A.M., Blangiardo, M., & Armstrong, B. (2020). Air conditioning and heat-related mortality: A multi-country longitudinal study.* Epidemiology, 31*(6), 779–87. https://doi.org/10.1097/EDE.0000000000001241.)*

There are, however, potential disadvantages to promoting air-conditioning as an adaptive strategy, including potential equity issues in terms of affordability and use of air-conditioning units (e.g., see Smoyer, 1998; Toloo et al., 2013), though access to shared air-conditioned space may help. Extensive use of air-conditioning may also increase energy demands and itself contribute to increased temperatures through exacerbation of the urban heat island effect (Salamanca et al., 2014).

Specific socioeconomic and demographic factors: There are many and socioeconomic characteristics that modify the heat–mortality relationship over time (e.g., population age structures). Susceptible subgroups may

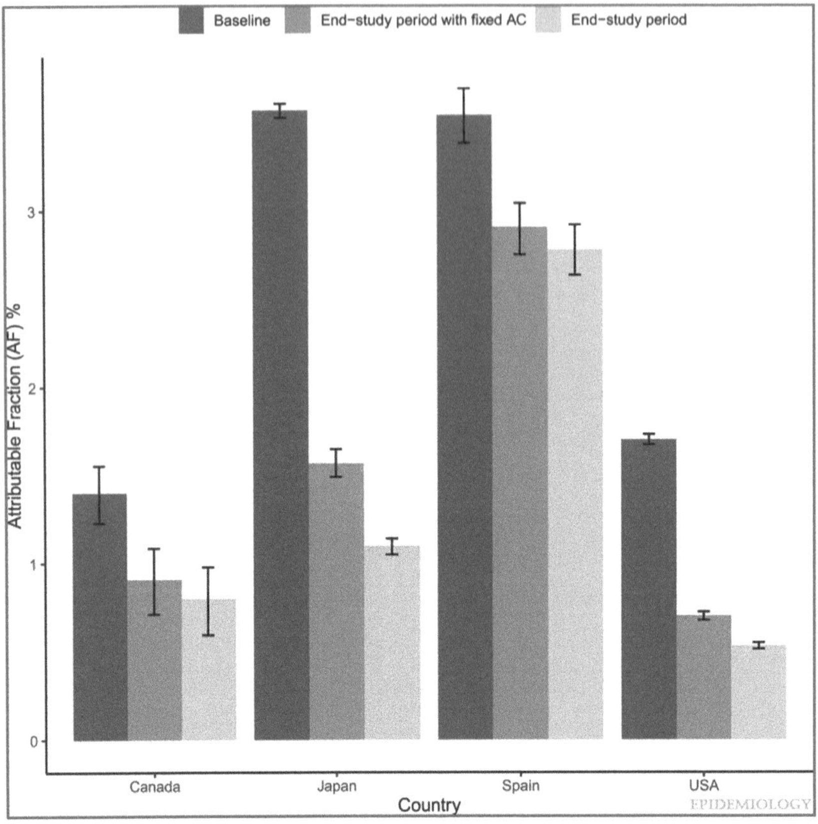

Figure 11.6 Excess mortality associated to heat reported as attributable fraction (AF%) estimated at the beginning (baseline, dark) and end of the study period assuming no change (end-study period with fixed air-conditioning, medium) or with the observed change (end-study period, light) in air-conditioning (AC) prevalence. *(From Sera, F., Hashizume, M., Honda, Y., Lavigne, E., Schwartz, J., Zanobetti, A., Tobias, A., Iñiguez, C., Vicedo-Cabrera, A.M., Blangiardo, M., & Armstrong, B. (2020). Air conditioning and heat-related mortality: A multi-country longitudinal study.* Epidemiology, 31*(6), 779–87. https://doi.org/10.1097/EDE.0000000000001241.)*

change over the longer time periods analyzed, and this needs to be considered when analyzing or interpreting temporal variations of heat–health outcomes. However, in many cases, populations have aged, which would be expected to increase heat-related health risks (therefore where decreases in heat-related risk are observed over long time periods, these are often despite an ageing population). Some studies have specifically analyzed changes in heat-related risk in older aged populations (see Section 3).

Heat-Heatlh Warning Systems (HHWSs): Many of the studies examining specific heatwave events over time have found reductions in mortality impacts. Some caution needs to be exercised in interpreting results, as important and varying features of any given heatwave (intensity, duration, when in the season it occurs, etc.) may not be adequately captured by heatwave definitions or models used. Therefore, while studies have examined heatwave effects pre-and post-HHWS implementation, establishing a causal link between the intervention and reduced mortality in later heatwaves can be difficult. That said, a number have directly or indirectly cited HHWS as a factor in reducing heatwave-related mortality (e.g., see Ebi et al., 2004; Fouillet et al., 2008; Weisskopf et al., 2002). The overall effectiveness of HHWS has been reviewed in Toloo et al. (2013), which concluded that HHWS can be effective in reducing mortality and potentially morbidity, but that improved research and into the cost-effectiveness of such systems is needed. A more recent study used a difference-in-differences approach to examine the effect of a heat action plan on heat-related mortality on hot days in Quebec, Canada, and found some evidence that the heat action plan had contributed to a decrease in heat-related mortality, especially in older people and in neighborhoods with lower socioeconomic status (Benmarhnia et al., 2016).

Reductions in susceptibility to heat which have occurred are likely to be due to a combination of many of the factors described. An understanding of these specific factors, together with evidence of factors modifying of heat–health impacts across geographical locations (e.g., Hajat & Kosatky, 2010; Sera et al., 2019) and modeling studies of specific heat-adaptive measures (e.g., Macintyre & Heaviside, 2019), can help identify potential future adaptive actions. It can also help in identifying locations where more adaptive efforts could be focused to improve health outcomes.

5. Understanding temporal variation in heat–health impacts in the context of climate change: implications for impact assessments and future adaptation

Many of the health challenges that future populations face will be determined by our social and environmental contexts (Van Den Bosch & Nieuwenhuijsen, 2017). Understanding the role that climate change will play in this is critical, especially considering the health impacts of heat. Potential climate change impacts on health are assessed through projections of climate-related impacts: where epidemiological coefficients of heat-related mortality are applied to different climate projections (for a methodological overview or review, see Huang et al., 2011; Vicedo-Cabrera et al., 2019). Understanding how susceptibility has changed in recent years can improve both our projections of heat—health-related impacts, and our appreciation of the role adaptation could play in the face of increasing heat-related risks.

5.1 The use of historical changes to inform climate risk assessments and limits in understanding

The uncertainties which exist when projecting future heat-related mortality, are variably considered in analyses. These include uncertainties in climate projections; demographic projections; and the extent to which populations will adapt (reflected in the epidemiological coefficients or thresholds used in models). The effect of age and potential adaptation on exposure response curves is illustrated in Fig. 11.7.

A number of previous health impact assessments have made assumptions about population-level adaptation to heat (Huang et al., 2011). However, justification of a given approach is rarely given, despite results being

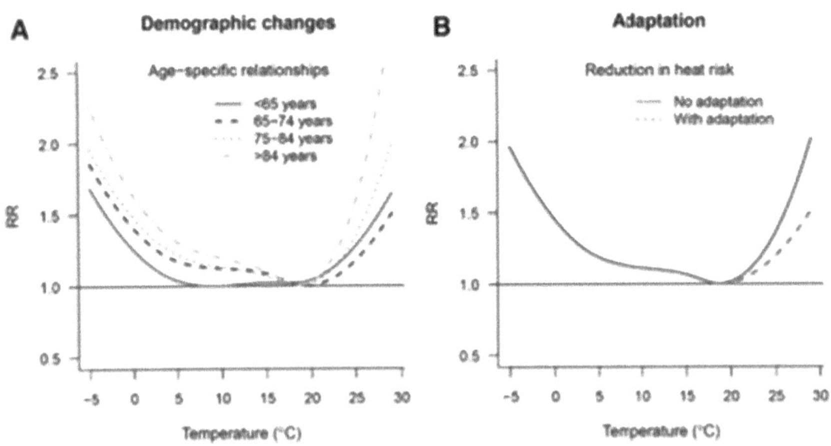

Figure 11.7 Accounting for complex scenarios arising in projecting heat-related mortality under climate change. This figure illustrates the potential impact of demography and hypothesized adaptation on exposure—response curves. *(From Vicedo-Cabrera, A.M., Sera, F., & Gasparrini, A. (2019). Hands-on tutorial on a modeling framework for projections of climate change impacts on health.* Epidemiology, 30*(3), 321—329. https://doi.org/10.1097/EDE.0000000000000982. PMID: 30829832; PMCID: PMC6533172.)*

sensitive to the approach used (Gosling et al., 2017). Approaches have ranged from assuming populations adapt to a certain number of degrees that populations in one location will behave like those in another hotter locations exhibiting less heat sensitivity in years to come (analogue cities) or that minimum mortality temperatures will shift to the right as the climate warms (Kinney et al., 2008). In general, these have not been informed by current or past trends of changes in susceptibility (with the exception of some, e.g., Petkova et al., 2017).

Quantifying historical changes in vulnerability to heat impacts can improve future projections of heat-related mortality. Understanding whether the baseline risk has changed over recent years highlights the importance of the use of contemporary epidemiological data to inform heat-related mortality projections. If susceptibility to heat has reduced, but baseline epidemiological data used to inform heat-related mortality projections are not current, then future climate impacts may be overestimated. Conversely, if a given population has not adapted to heat in recent years, then the validity of many "adaptation assumptions" used in climate heat—health impact assessments should be carefully considered. One step further is to use evidence of past adaptation to inform the magnitude of heat exposure response used in the climate impact assessment (e.g. as in Petkova et al., 2017). However, even when past trends are used to inform future assessment, some degree of caution is necessary. The climate is projected to warm faster over the next century than it has in the past : whether populations that experienced some temporal reduction in risk will be able to continue to adapt at the same rate is not known i.e.here is no way of knowing whether recent trends in changes in risk (if any) will continue into the future. Limits to the extent and the pace of adaptation are largely unknown and are likely to depend on a wide range of societal and context dependent factors—for example, the pace of adaptation may slow as the relative risk of heat-related mortality approaches 1, or for areas with higher initial risks, the pace of adaptation may be faster if local planned adaptive measures put in place. An improved understanding of historical drivers of adaptation would help in understanding this. If mechanisms and pathways that lead to a reduction in vulnerability are better understood, then projected future trends of these or scenarios that include these could inform heat—health impact assessments under climate change.

5.2 Implications for current and future adaptation to heat

The continued heat-related risk of mortality in the majority of locations (even where there has been some temporal attenuation of risk), highlights the need for continuing commitment to climate mitigation and adaptation. The persistence of ambient heat risk to population health emphasizes the need both to better understand and further explore specific adaptive actions that can reduce general heat-related risk. Many policies relating to urban adaptation needs are likely to be system-wide and transdisciplinary in nature (crossing public health, urban planning, design and architecture, etc.). Ideally adaptive policies would be designed to improve health and reduce inequities (for example, through healthy place shaping).

5.3 Current gaps in understanding

While the evidence on temporal changes in heat-related health risks has increased in recent years, many gaps in our knowledge remain.

The majority of studies are based in the United States, Europe, or Northeast Asia, leaving a large gap in understanding across many locations. Notable by their absence from this body of evidence are published studies that have examined temporal shifts in heat-related risks in the Global South. This represents a major and important limitation in our understanding , especially given concurrent urbanization, demographic shifts, and climate change.

There are also gaps in knowledge at more local levels (though these studies may be more limited in statistical power to detect change) and of changes in particularly vulnerable subgroups. In addition, further work using morbidity outcomes would be welcome to gain a more comprehensive understanding of heat—health impacts on populations and on our health systems.

An increased understanding of specific factors that have contributed to declines in heat-related mortality would be beneficial for public health planning.

6. Conclusion

The number of studies examining changes in heat-related mortality risk and burdens has increased over recent years, and there is a growing body of evidence demonstrating a decrease in susceptibility over time. However, this is not consistent across all locations, time periods, exposures (heatwaves compared with less extreme heat), and outcomes studied. There remain significant gaps in knowledge for many geographical locations and morbidity-related outcomes.

Characterizing the decrease in mortality and pace of change is important to improve projections of future climate-related heat–health impacts. Further work to understand factors that have contributed to attenuation in risk, where this has been seen, could help with better understanding potential adaptive options.

References

Anderson, G. B., & Bell, M. L. (2011). Heat waves in the United States: Mortality risk during heat waves and effect modification by heat wave characteristics in 43 U.S. Communities. *Environmental Health Perspectives, 119*, 210–218.

Arbuthnott, K., Hajat, S., Heaviside, C., & Vardoulakis, S. (2016). Changes in population susceptibility to heat and cold over time: Assessing adaptation to climate change. *Environmental Health, 15*, S33.

Arbuthnott, K., Hajat, S., Heaviside, C., & Vardoulakis, S. (2020). Years of life lost and mortality due to heat and cold in the three largest English cities. *Environmental International, 144*, 105966.

Astrom, D. O., Forsberg, B., Edvinsson, S., & Rocklov, J. (2013). Acute fatal effects of short-lasting extreme temperatures in Stockholm, Sweden: Evidence across a century of change. *Epidemiology, 24*, 820–829.

Barnett, A. G. (2007). Temperature and cardiovascular deaths in the US elderly: Changes over time. *Epidemiology, 18*, 369–372.

Barreca, A., Clay, K., Deschenes, O., Greenstone, M., & Shapiro, J. S. (2016). Adapting to climate change: The remarkable decline in the US temperature-mortality relationship over the twentieth century. *Journal of Political Economy, 124*, 105–159.

Basu, R. (2009). High ambient temperature and mortality: A review of epidemiologic studies from 2001 to 2008. *Environmental Health, 8*, 40.

Benmarhnia, T., Bailey, Z., Kaiser, D., Auger, N., King, N., & Kaufman, J. S. (2016). A difference-in-differences approach to assess the effect of a heat action plan on heat-related mortality, and differences in effectiveness according to sex, age, and socioeconomic status (Montreal, Quebec). *Environmental Health Perspectives, 124*, 1694–1699.

Bobb, J. F., Peng, R. D., Bell, M. L., & Dominici, F. (2014). Heat-related mortality and adaptation to heat in the United States. *Environmental Health Perspectives, 122*, 811–816.

Boeckmann, M., & Rohn, I. (2014). Is planned adaptation to heat reducing heat-related mortality and illness? A systematic review. *BMC Public Health, 14*, 1–13.

Braga, A. L., Zanobetti, A., & Schwartz, J. (2001). The time course of weather-related deaths. *Epidemiology, 12*.

Carson, C., Hajat, S., Armstrong, B., & Wilkinson, P. (2006). Declining vulnerability to temperature-related mortality in London over the 20th century. *American Journal of Epidemiology, 164*, 77–84.

Christidis, N., Donaldson, G. C., & Stott, P. A. (2010). Causes for the recent changes in cold- and heat-related mortality in England and Wales. *Climatic Change, 102*, 539–553.

Chung, Y., Noh, H., Honda, Y., Hashizume, M., Bell, M. L., Guo, Y. L., & Kim, H. (2017). Temporal changes in mortality related to extreme temperatures for 15 cities in Northeast Asia: Adaptation to heat and maladaptation to cold. *American Journal of Epidemiology, 185*, 907–913.

D'ippoliti, D., Michelozzi, P., Marino, C., DE'donato, F., Menne, B., Katsouyanni, K., Kirchmayer, U., Analitis, A., MEDINA-Ramon, M., Paldy, A., Atkinson, R., Kovats, S., Bisanti, L., Schneider, A., Lefranc, A., Iniguez, C., & Perucci, C. A. (2010). The impact of heat waves on mortality in 9 European cities: Results from the Euro-HEAT project. *Environmental Health: A Global Access Science Source, 9*, 37.

Davis, R. E., Knappenberger, P. C., Michaels, P. J., & Novicoff, W. M. (2003). Changing heat-related mortality in the United States. *Environmental Health Perspectives, 111*, 1712–1718.

Davis, R. E., Knappenberger, P. C., Novicoff, W. M., & Michaels, P. J. (2003). Decadal changes in summer mortality in U.S. cities. *International Journal of Biometeorology, 47*, 166–175.

Decastro, M., Gomez-Gesteira, M., Ramos, A., Alvarez, I., & Decastro, P. (2011). Effects of heat waves on human mortality, Galicia, Spain. *Climate Research, 48*, 333–341.

DE'donato, F. K., Leone, M., Scortichini, M., DE Sario, M., Katsouyanni, K., Lanki, T., Basagaña, X., Ballester, F., Åström, C., & Paldy, A. (2015). Changes in the effect of heat on mortality in the last 20 years in nine European cities. Results from the PHASE project. *International Journal of Environmental Research and Public Health, 12*, 15567–15583.

Donaldson, G. C., Keatinge, W. R., & Nayha, S. (2003). Changes in summer temperature and heat-related mortality since 1971 in North Carolina, South Finland, and Southeast England. *Environmental Research, 91*, 1–7.

Ebi, K. L., Teisberg, T. J., Kalkstein, L. S., Robinson, L., & Weiher, R. F. (2004). Heat watch/warning systems save lives: Estimated costs and benefits for Philadelphia 1995?98. *Bulletin of the American Meteorological Society, 85*, 1067–1074.

Ekamper, P., Duin, C., & Garssen, J. (2009). 150 Years of temperature-related excess mortality in The Netherlands. *Demographic Research, 21*, 385–426.

Folkerts, M. A., Bröde, P., Botzen, W. J. W., Martinius, M. L., Gerrett, N., Harmsen, C. N., & Daanen, H. A. M. (2020). Long term adaptation to heat stress: Shifts in the minimum mortality temperature in The Netherlands. *Frontiers in Physiology, 11*, 225.

Fouillet, A., Rey, G., Wagner, V., Laaidi, K., Empereur-Bissonnet, P., LE Tertre, A., Frayssinet, P., Bessemoulin, P., Laurent, F., De Crouy-Chanel, P., Jougla, E., & Hemon, D. (2008). Has the impact of heat waves on mortality changed in France since the European heat wave of summer 2003? A study of the 2006 heat wave. *International Journal of Epidemiology, 37*, 309–317.

Gasparrini, A., Guo, Y., Hashizume, M., Kinney, P. L., Petkova, E. P., Lavigne, E., Zanobetti, A., Schwartz, J. D., Tobias, A., & Leone, M. (2015). Temporal variation in heat–mortality associations: A multicountry study. *Environmental Health Perspectives, 123*, 1200–1207.

Gasparrini, A., & Leone, M. (2014). Attributable risk from distributed lag models. *BMC Medical Research Methodology, 14*, 55.

Gosling, S. N., Hondula, D. M., Bunker, A., Ibarreta, D., Liu, J., Zhang, X., & Sauerborn, R. (2017). Adaptation to climate change: A comparative analysis of modeling methods for heat-related mortality. *Environmental Health Perspectives, 125*, 087008.

Green, H. K., Andrews, N., Armstrong, B., Bickler, G., & Pebody, R. (2016). Mortality during the 2013 heatwave in England – how did it compare to previous heatwaves? A retrospective observational study. *Environmental Research, 147*, 343–349.

Hajat, S., & Kosatky, T. (2010). Heat-related mortality: A review and exploration of heterogeneity. *Journal of Epidemiology and Community Health, 64*, 753–760.

Ha, J., & Kim, H. (2013). Changes in the association between summer temperature and mortality in Seoul, South Korea. *International Journal of Biometeorology, 57*, 535–544.

Heo, S., Lee, E., Kwon, B. Y., Lee, S., Jo, K. H., & Kim, J. (2016). Long-term changes in the heat–mortality relationship according to heterogeneous regional climate: A time-series study in South Korea. *BMJ Open, 6*, e011786.

Huang, C., Barnett, A. G., Wang, X., Vaneckova, P., Fitzgerald, G., & Tong, S. (2011). Projecting future heat-related mortality under climate change scenarios: A systematic review. *Environmental Health Perspectives, 119*, 1681–1690.

Kalkstein, L. S., Greene, S., Mills, D. M., & Samenow, J. (2011). An evaluation of the progress in reducing heat-related human mortality in major U.S. cities. *Natural Hazards, 56*, 113–129.

Kim, Y., Gasparrini, A., Hashizume, M., Honda, Y., Ng, C. F. S., & Armstrong, B. (2017). Heat-related mortality in Japan after the 2011 Fukushima disaster: An analysis of potential influence of reduced electricity consumption. *Environmental Health Perspectives, 125*, 077005.

Kim, H., Heo, J., Kim, H., & Lee, J.-T. (2015). Has the impact of temperature on mortality really decreased over time? *Science of The Total Environment, 512–513*, 74–81.

Kinney, P. L., O'neill, M. S., Bell, M. L., & Schwartz, J. (2008). Approaches for estimating effects of climate change on heat-related deaths: Challenges and opportunities. *Environmental Science & Policy, 11*, 87–96.

Kyselý, J., & KŘÍŽ, B. (2008). Decreased impacts of the 2003 heat waves on mortality in the Czech republic: An improved response? *International Journal of* Biometeorology, *52*, 733–745.

Kyselý, J., & Plavcová, E. (2012). Declining impacts of hot spells on mortality in the Czech republic, 1986–2009: Adaptation to climate change? *Climatic Change, 113*, 437–453.

Li, G., Huang, J., Xu, G., Pan, X., Qian, X., Xu, J., Zhao, Y., Zhang, T., Liu, Q., Guo, X., & He, T. (2017). Temporal variation in associations between temperature and years of life lost in a southern China city with typical subtropical climate. *Scientific Reports, 7*, 4650.

Lu, P., Xia, G., Zhao, Q., Xu, R., Li, S., & Guo, Y. (2020). Temporal trends of the association between ambient temperature and hospitalisations for cardiovascular diseases in Queensland, Australia from 1995 to 2016: A time-stratified case-crossover study. *PLoS Medicine, 17*, e1003176.

Macintyre, H. L., & Heaviside, C. (2019). Potential benefits of cool roofs in reducing heat-related mortality during heatwaves in a European city. *Environment International, 127*, 430–441.

Miron, I. J., Linares, C., Montero, J. C., CRIADO-Alvarez, J. J., & Díaz, J. (2015). Changes in cause-specific mortality during heat waves in central Spain, 1975–2008. *International Journal of Biometeorology, 59*, 1213–1222.

Muthers, S., Matzarakis, A., & Koch, E. (2010). Climate change and mortality in Vienna–a human biometeorological analysis based on regional climate modeling. *International Journal of Environmental Research and Public Health, 7*, 2965–2977.

Ng, C. F. S., Boeckmann, M., Ueda, K., Zeeb, H., Nitta, H., Watanabe, C., & Honda, Y. (2016). Heat-related mortality: Effect modification and adaptation in Japan from 1972 to 2010. *Global Environmental Change, 39*, 234–243.

Nitschke, M., Tucker, G. R., Hansen, A. L., Williams, S., Zhang, Y., & Bi, P. (2011). Impact of two recent extreme heat episodes on morbidity and mortality in Adelaide, South Australia: A case-series analysis. *Environmental Health, 10*, 42.

Nordio, F., Zanobetti, A., Colicino, E., Kloog, I., & Schwartz, J. (2015). Changing patterns of the temperature–mortality association by time and location in the US, and implications for climate change. *Environment International, 81*, 80–86.

Onozuka, D., & Hagihara, A. (2015). Variation in vulnerability to extreme-temperature-related mortality in Japan: A 40-year time-series analysis. *Environmental Research, 140*, 177–184.

Pachauri, R. K., Allen, M. R., Barros, V. R., Broome, J., Cramer, W., Christ, R., Church, J. A., Clarke, L., Dahe, Q., & Dasgupta, P. (2014). *Climate change 2014: Synthesis report. Contribution of working groups I, II and III to the fifth assessment report of the intergovernmental panel on climate change*. IPCC.

Palecki, M. A., Changnon, S. A., & Kunkel, K. E. (2001). The nature and impacts of the July 1999 heat wave in the Midwestern United States: Learning from the lessons of 1995. *Bulletin of the American Meteorological Society, 82*, 1353–1368.

Petkova, E. P., Gasparrini, A., & Kinney, P. L. (2014). Heat and mortality in New York City since the beginning of the 20th century. *Epidemiology, 25*, 554–560.

Petkova, E. P., Vink, J. K., Horton, R. M., Gasparrini, A., Bader, D. A., Francis, J. D., & Kinney, P. L. (2017). Towards more comprehensive projections of urban heat-related mortality: Estimates for New York City under multiple population, adaptation, and climate scenarios. *Environmental Health Perspectives, 125*, 47–55.

Public Health England. (2015). *Heatwave plan for England* [Online]. Available: https://www.gov.uk/government/uploads/system/uploads/attachment_data/file/429384/Heatwave_Main_Plan_2015.pdf.

Ragettli, M. S., Vicedo-Cabrera, A. M., Schindler, C., & Röösli, M. (2017). Exploring the association between heat and mortality in Switzerland between 1995 and 2013. *Environmental Research, 158*, 703–709.

Salamanca, F., Georgescu, M., Mahalov, A., Moustaoui, M., & Wang, M. (2014). Anthropogenic heating of the urban environment due to air conditioning. *Journal of Geophysical Research: Atmospheres, 119*, 5949–5965.

Sera, F., Armstrong, B., Tobias, A., Vicedo-Cabrera, A. M., Åström, C., Bell, M. L., Chen, B. Y., De Sousa Zanotti Stagliorio Coelho, M., Matus Correa, P., Cruz, J. C., Dang, T. N., Hurtado-Diaz, M., DO Van, D., Forsberg, B., Guo, Y. L., Guo, Y., Hashizume, M., Honda, Y., Iñiguez, C., … Gasparrini, A. (2019). How urban characteristics affect vulnerability to heat and cold: A multi-country analysis. *International Journal of Epidemiology, 48*, 1101–1112.

Sera, F., Hashizume, M., Honda, Y., Lavigne, E., Schwartz, J., Zanobetti, A., Tobias, A., Iñiguez, C., VICEDO-Cabrera, A. M., Blangiardo, M., Armstrong, B., & Gasparrini, A. (2020). Air conditioning and heat-related mortality: A multi-country longitudinal study. *Epidemiology, 31*, 779–787.

Sheridan, S. C., & Allen, M. J. (2018). Temporal trends in human vulnerability to excessive heat. *Environmental Research Letters, 13*, 043001.

Sheridan, S. C., & Dixon, P. G. (2017). Spatiotemporal trends in human vulnerability and adaptation to heat across the United States. *Anthropocene, 20*, 61–73.

Sheridan, S. C., Dixon, P. G., Kalkstein, A. J., & Allen, M. J. (2021). Recent trends in heat-related mortality in the United States: An update through 2018. *Weather, Climate, and Society, 13*, 95–106.

Sheridan, S. C., Kalkstein, A. J., & Kalkstein, L. S. (2009). Trends in heat-related mortality in the United States, 1975–2004. *Natural Hazards, 50*, 145–160.

Sheridan, S. C., & Lin, S. (2014). Assessing variability in the impacts of heat on health outcomes in New York city over time, season, and heat-wave duration. *EcoHealth, 11*, 512–525.

Smoyer, K. E. (1998). A comparative analysis of heat waves and associated mortality in St. Louis, Missouri–1980 and 1995. *International Journal of Biometeorology, 42*, 44–50.

Tan, J., Zheng, Y., Song, G., Kalkstein, L. S., Kalkstein, A. J., & Tang, X. (2007). Heat wave impacts on mortality in Shanghai, 1998 and 2003. *International Journal of Biometeorology, 51*, 193–200.

Toloo, G., Fitzgerald, G., Aitken, P., Verrall, K., & Tong, S. (2013). Are heat warning systems effective? *Environmental Health, 12*, 27.

Toloo, G. S., Fitzgerald, G., & Tong, S. (2014). Managing health risks of extreme weather events: Need for a systematic approach. *Australian and New Zealand Journal of Public Health, 38*, 90–91.

Tompkins, E. L., Adger, W. N., Boyd, E., Nicholson-Cole, S., Weatherhead, K., & Arnell, N. (2010). Observed adaptation to climate change: UK evidence of transition to a well-adapting society. *Global Environmental Change, 20*, 627–635.

Urban, A., Hanzlíková, H., Kyselý, J., & Plavcová, E. (2017). Impacts of the 2015 heat waves on mortality in the Czech Republic—a comparison with previous heat waves. *International Journal of Environmental Research and Public Health, 14*, 1562.

Van Den Bosch, M., & Nieuwenhuijsen, M. (2017). No time to lose-Green the cities now. *Environment International, 99*, 343.

Vanos, J., Vecellio, D. J., & Kjellstrom, T. (2019). Workplace heat exposure, health protection, and economic impacts: A case study in Canada. *American Journal of Industrial Medicine, 62*, 1024–1037.

Vicedo-Cabrera, A. M., Sera, F., & Gasparrini, A. (2019). Hands-on tutorial on a modeling framework for projections of climate change impacts on health. *Epidemiology, 30*, 321–329.

Vicedo-Cabrera, A. M., Sera, F., Guo, Y., Chung, Y., Arbuthnott, K., Tong, S., Tobias, A., Lavigne, E., Coelho, M. D. S. Z. S., & Saldiva, P. H. N. (2018). A multi-country analysis on potential adaptive mechanisms to cold and heat in a changing climate. *Environment International, 111*, 239–246.

Weisskopf, M. G., Anderson, H. A., Foldy, S., Hanrahan, L. P., Blair, K., Török, T. J., & Rumm, P. D. (2002). Heat wave morbidity and mortality, Milwaukee, Wis, 1999 vs 1995: An improved response? *American Journal of Public Health, 92*, 830–833.

Williams, L.,E. B., Ettelt, S., Hajat, S., Manacorda, T., & Mays, N. (2019). *Evaluation of the heatwave plan for England.*

Xiang, J., Bi, P., Pisaniello, D., & Hansen, A. (2014). Health impacts of workplace heat exposure: An epidemiological review. *Industrial Health, 52*, 91–101.

Xiang, J., Hansen, A., Pisaniello, D., & Bi, P. (2015). Extreme heat and occupational heat illnesses in South Australia, 2001–2010. *Occupational and Environmental Medicine, 72*, 580–586.

Xu, Z., Fitzgerald, G., Guo, Y., Jalaludin, B., & Tong, S. (2016). Impact of heatwave on mortality under different heatwave definitions: A systematic review and meta-analysis. *Environment International, 89–90*, 193–203.

Yang, C., Meng, X., Chen, R., Cai, J., Zhao, Z., Wan, Y., & Kan, H. (2015). Long-term variations in the association between ambient temperature and daily cardiovascular mortality in Shanghai, China. *Science of The Total Environment, 538*, 524–530.

CHAPTER 12

Strategies to reduce the health impacts of heat exposure

Yuming Guo, Yiwen Zhang, Pei Yu, Yao Wu, Bo Wen, Zhengyu Yang, Rahini Mahendran and Shanshan Li
Climate, Air Quality Research Unit, School of Public Health and Preventive Medicine, Monash University, Melbourne, Australia

1. Introduction

Exposure to heat can be hazardous to human health when the average temperature of the surrounding air is high during a period (e.g., firefighting) and when a heat extreme lasts for several days, which is known as heatwaves (Gasparrini et al., 2015). A heatwave is usually measured relative to the usual weather in the area and relative to normal temperatures for the season, which is often related to atmospheric heat stress (PJ, 2001).

Although definitions of heatwaves vary regarding temperature thresholds, the associations between heat exposure and adverse health consequences (morbidity and mortality) have been widely reported on a global scale (Guo et al., 2017; Xu et al., 2016). It was estimated that 0.91% of all deaths could be attributed to heat exposure from 2000 to 2019 around the globe, corresponding to 489,075 excess deaths (Zhao et al., 2021). For young children and the elderly who are more susceptible to heat exposure, it was reported that they are suffering 3.1 billion additional person-days of heatwave exposure and 626 million additional person-days of heatwave exposure in 2020 compared with the 1986–2005 baseline average, respectively (Romanello et al., 2021). Heatwaves or hot weather are related to heatstroke and kidney disease, especially urolithiasis (Bouchama & Knochel, 2002; Liu et al., 2021a, 2021b). Hot ambient temperatures can contribute to premature deaths from cardiovascular diseases and respiratory morbidities (hospital admissions and emergency department visits) in diverse regions (Lv et al., 2022; Cheng et al., 2019). It is found in New York city that high temperatures can increase hospital admissions for cardiovascular and respiratory disorders, especially for the elderly residents, which is also found in the Assessment and Prevention of Acute Health Effects of Weather Conditions in Europe (PHEWE), conducted among 12 European cities (Lin

Heat Exposure and Human Health in the Context of Climate Change
ISBN 978-0-12-819080-7
https://doi.org/10.1016/B978-0-12-819080-7.00009-4
© 2023 Elsevier Inc.
All rights reserved.

et al., 2009; Michelozzi et al., 2009). Not the only case, a case study of Shanghai found heightened heat-related mortality in urban regions, resulting in the heat island effect (Tan et al., 2010). The association between heatwaves and health effects has also been proved to have spatial—demographical heterogeneity (Campbell et al., 2018). For example, the older in the United States is proved to be the most vulnerable to severe heat conditions due to their lower physical status, isolation, and possibly reduced incomes (Gamble et al., 2013). With the growing breathing system and the rapid breathing rates, young children can be exposed to a higher risk of asthma and other respiratory disorders exacerbated by ozone air pollution, which can be enhanced by heatwaves (D'Amato et al., 2014). Individuals with low economic status can be more vulnerable to heatwaves owing to limited access to air conditioning and a poor living environment with inadequate ventilation (Gronlund, 2014). Population with chronic diseases, disabilities, and obesity are also more vulnerable to heatwave-related damage (Kenny et al., 2010; Meade et al., 2020). In addition to physiological health, the prevalence of mental-related diseases including mood disorders, schizophrenia, neurotic, and anxiety disorders tend to grow significantly when ambient temperatures rise (Liu et al., 2021a, 2021b).

Rising ambient temperature can also have indirect impacts on population health by accelerating the transmission of some communicable diseases via effects on pathogens distribution and abundance (Wu et al., 2016). Rising temperatures can also increase interpersonal violent crimes such as sex offences and assaults (Mahendran et al., 2021). From a more macroscopic view, hot pavement and rooftop surfaces can impair the water quality of streams, ponds, and rivers as well, by heating stormwater runoff, which can bring stress to aquatic ecosystem (Somers et al., 2016).

With the ongoing climate change trajectory, heatwaves are expected to increase in frequency and intensity, which could compound the effects of heatwaves, accompanied by longer durations, and occurrences in more countries or regions (Jay et al., 2021; Yang et al., 2019). As a result, heatwave-related deaths and diseases are predicted to forge, considering the progressive population globally aging (Peng et al., 2011). Hence, heat exposure is a critical public health problem in both the short-term and long run. The population can adapt to the local climate within their physiological limits through thermoregulation, but this will put much burden on physiological systems (e.g., cardiovascular health) (Guo et al., 2014; Liu et al., 2015). Hence, adaptation interventions can better release the health burden brought by the increasingly hot temperatures, especially for occasions in heat-vulnerable settings such as aged care homes, slums, workplaces, schools, refugee camps, and sports (Jay et al., 2021). This is also supported by research demonstrating that countries near the equator experience a significant increase in heatwave-related deaths in the absence of intervention compared with the intervention scenario (Guo et al., 2018). Hence, this chapter aims to elaborate on the six-level adaptation intervention strategies further summarized by a previous research (Guo et al., 2018) based on the six essential Steps for Quality Intervention Development (6SQulD) (Wight et al., 2016), to reduce the health burden of heat exposure for the public.

2. Adaptation interventions

Luckily, mortality and illnesses associated with heat exposure are preventable from six levels: individual, interpersonal, community, institutional, environmental, and public policies (Fig. 12.1).

2.1 Individual

Many measures have been taken to reduce the health impacts of heat exposure, including heat response planning, heat warning systems, and

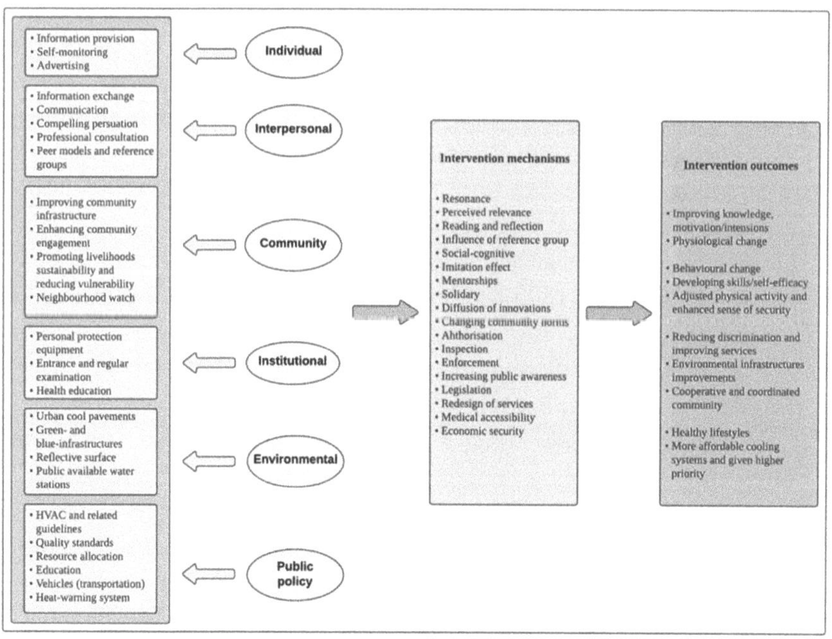

Figure 12.1 Six-level adaptation intervention strategy framework.

improving infrastructures (cooling centers, green roofs, or cool pavements). Moreover, it is vital to ensure that individuals have access to information to cope with the adverse impacts of heat exposure. Thus, the first adaption strategy at the individual level is to provide information about potential risk factors, symptoms of heat stress, and recommendations on responses.

2.1.1 Information provision and resonance

Extreme heat can pose a threat to everyone's health. For the general public, the basic strategy to cope with the hot weather is to stay cool, stay hydrated, and stay informed. Many governments and authorities have published guidelines on recommendations in the hot weather for their citizens. For example, the Centers for Disease Control and Prevention (CDC) in the United States recommended that individuals should wear appropriate clothing, stay in an air-conditioned place, and limit their outdoor activity to the coolest hours to stay cool in the hot weather. To stay hydrated, individuals are recommended to drink more fluids before feeling thirsty. Besides, it is recommended to check local news for alerts and tips during the hot weather to stay informed. Similarly, the Department of Health of Victoria in Australia prepared a suite of communication resources to educate individuals about the effects of extreme heat and how they can prepare for heat exposure. One of the most important issues for individuals in response to heat exposure is to receive accurate information. For example, it is straightforward to use electric fans or air conditioning during hot weather. It is reasonable because this strategy could substantially reduce the physiological heat strain induced by heat exposure, such as increased core temperature, cardiovascular strain, and dehydration (Jay et al., 2021).

Individuals should also be aware of the potential drawbacks of these measures. Electric fans could accelerate convective heat loss by forcing air across the skin, which is more accessible and requires less electricity compared with air conditioning (Morris et al., 2021). However, fans would be less effective and even harmful when air temperatures exceed a threshold (e.g., 35°C). In this case, using air conditioning or moving to an air-conditioned place (cooling centers or swimming pools) would be more practical. It should also be noted that extreme heat may have impacts on community infrastructures, for example, power supply (Bartos et al., 2016), which means that individuals should prepare for the potential power failure. A battery-operated or hand-held fan would be helpful when there is a power failure due to extreme heat.

2.1.2 Potential symptoms and self-monitoring

In addition to these preventive measures, individuals are also recommended to monitor the potential symptoms and try to seek medical help when needed. Heat exposure will affect human body temperature regulation and may cause dehydration by sweating heavily (Miescher & Fortney, 1989). Excessive heat exposure may lead to some mild symptoms such as heat rash or heat cramps and other severe illnesses such as heat exhaustion and heatstroke (Khan, 2019). Besides, the underlying conditions could be exacerbated by heat exposure due to physiological heat stress responses. The primary risk factors of heat-related deaths include cardiovascular diseases, respiratory diseases, and cerebrovascular diseases (Ebi et al., 2021); many other conditions are potential risk factors for mortality due to heat exposure, including genitourinary disease, diabetes, and mental disorder (Ebi et al., 2021). As a result, individuals with preexisting conditions are more vulnerable to heat exposure. During the hot weather, they should be taken better care of, in addition to conventional preventive measures by having someone (relatives, friends, or neighbors) check on them.

2.1.3 Vulnerable groups

There are many other vulnerable groups, including infants or children, outdoor workers, and those with low socioeconomic status or homeless. Infants and children are more susceptible to heat exposure mainly due to their immature thermoregulatory system and greater body surface area-to-mass ratio (Ebi et al., 2021). In addition, infants or children are often left alone in parked vehicles, which made them at higher risk for heatstroke and even death. It should be ensured that everyone is out of the car. During the hot weather, outdoor activities should be limited while it may be infeasible for outdoor workers. Compared with the general public, outdoor workers have to work with heat exposure and would be prone to being dehydrated. More importantly, heat exposure will affect their work performance and may further increase the risk of occupational injuries (Fatima et al., 2021). It was estimated that each 1°C increase in temperature above reference values is associated with a 1% higher risk of occupational injuries (Fatima et al., 2021). Thus, outdoor workers should drink more fluids, wear suitable clothes (e.g., sunscreen, hat, and light-colored clothing), and take more breaks in air-conditioned places. Individuals with a low socioeconomic status tend to be at higher risk of heat exposure because they may have limited access to air conditioning and other related information about extreme hot weather (Hsu et al., 2021). Hence, it is recommended to find air-conditioned cooling centers and stay in a shopping mall or public library to reduce the adverse impacts of heat exposure if air-conditioning is not available in one's home.

2.1.4 Behavioral change and challenges

Governments and academic institutions have made great efforts to disseminate prevention measures for heat exposure to the public. However, it should be noted that the heat warning information and knowledge would not lead to effective actions at the individual level (Jay et al., 2021). Similarly, numerous public health interventions and education have been conducted in different countries and regions while the increasing trend of a global burden on noncommunicable diseases is still not effectively controlled. In other words, it is easy for individuals to receive health information (e.g., strategies to reduce heat impacts), but difficult to change their behavior. There are many reasons for the failure of behavior change (Whitehead & Russell, 2004). People sometimes have to overcome potential barriers to change their behavior, or the process might conflict with their routine behavior. As a result, the first step to reducing the health impacts of heat exposure is to improve the knowledge at the individual level. After that, it is important to understand how to change an individual's behavior. Individuals need support and motivation to change their behavior (Kelly & Barker, 2016). For example, individuals could be encouraged to use air-conditioning in hot weather by financial measures or subsidies.

2.2 Interpersonal

In addition, to ensure the adaptation interventions individually, populations should also pay attention to the adaptation ability of their family, colleagues, and neighbors. That refers to the interpersonal level interventions, which can depend heavily on people's networks and relationships. The strength of the tendency and motivation for applying interventions individually can be influenced by education communication, peer models, reference groups, and self-efficacy. Self-efficacy plays a crucial role in supporting people to adapt long-term interventions proactively and consistently, facing the potentially long-term sustained global warming and continued aging population.

2.2.1 Information exchange and trust building

The effectiveness of individual interventions depends mainly on the information messengers and their relationship with the recipients (Wight et al., 2016; O'Neill et al., 2009). Some intimate connections, such as friends, family, neighbors, and coworkers, can exchange information and encourage others to promote healthy lifestyles such as wearing sunscreen and sunglasses, choosing suitable clothes, and consuming more water daily. It is straightforward for individuals to receive the information through spontaneous information exchange, enhancing motivation/intention building. It can be explained psychosocially that populations have a higher probability of trusting those around them, like friends (Sera et al., 2020), which can indirectly foster or change behavior on a psychological level over time. One limitation of this strategy is that individuals are more likely to keep crucial private information (Guardaro et al., 2020). Therefore, the efficiency and effectiveness of information exchange can be largely influenced by the relationships between information sharers and their willingness to share helpful information. Hence, it is critical to develop interpersonal trust building, which can be achieved from a relational signal perspective via exchanging more positive signals and avoiding negative signals (e.g., dealing with troubles) (Lioubimtseva & Da Cunha, 2022, pp. 257–276; Moser & Pike, 2015).

2.2.2 Communication and education

Communication involving education can also enhance conservation behaviors and thus reduce vulnerability to climate change, including rising ambient temperatures (Weir, 2002). That can be achieved from two aspects. On the one hand, relevant skills can be developed to resist heatwaves through peer education. For example, outdoor workers can promote sun protection products among their colleagues, while people living in hot zones can encourage air conditioning in the neighborhood. On the other hand, experts (e.g., researchers) can broadly deliver education on heat exposure's health hazards and the related adaptation strategies, through relevant training, workshops, and mentorships. This approach is more psychologically compelling, as individuals are more inclined to be persuaded by persuasive arguments brought by those with a professional background. From a more specific point, educating students about heatwaves and adaptation interventions on campuses is possible, via warning them to reduce the time spent outdoors in hot weather or consider indoor activities (e.g., indoor sports like table tennis), to avoid potential heat and UV-related injuries. Such educational communication is possible through mentorships in both enterprises and academies.

2.2.3 Reference group and peer models

Meanwhile, social interaction has a general imitation effect, originating from domain-general learning mechanisms (O'Neill et al., 2009; Sera et al., 2020). People tend to unconsciously imitate the behaviors of those around them, like partners, also known as the "chameleon effect" (Guardaro et al., 2020). Besides, people are also likely to be influenced by reference groups they wish to be a part of (Moser & Pike, 2015). From a more specific example, we have known that individual decision-making and consumer behavior are influenced by familial or peer-based reference groups (Lioubimtseva & Da Cunha, 2022, pp. 257–276). Meanwhile, in medicine, the term "KOL" has become synonymous with a respected physician or scientist (Weir, 2002). It has been discovered that "KOLs" (key opinion leaders) influence people's opinions and preferences more than other pressures, such as advertising (Ma et al., 2015). Not the only case, autistic children are proven to learn by observing peer models better than traditional procedures (Kim & Kim, 2017). Hence, it is reasonable to speculate that people can be influenced by reference groups and peer models when adopting adaptation interventions routinely, probably resulting from the herd mentality or unconscious imitation of authority (Kabir & Serrao-Neumann, 2020, pp. 167–179).

Consequently, adaptation interventions are supposed to spread and circulate through mentorships, peer education, or parent–child relationships. One limitation is that inappropriate imitation can occur, especially among children who learn about the social world by observing others' acts (Abeje et al., 2019). Unlike purchasing behavior, inappropriate adaptation interventions to heatwaves can cause harm to human health. Therefore, scientific guidance is required.

2.2.4 Compelling persuasion

Exhortations can also generate effects. A general physician (GP), a dietitian, or a beautician can counsel their patients (or customers) to promote the adaptation interventions on a one-to-one basis. For example, urge individuals to develop healthy lifestyles such as controlling the time spent outdoors, choosing comfortable and breathable or sun-protective clothing, eating a light diet, trying to guarantee access to cooling systems, and so on. Through regular follow-up visits and repeated persuasion, people are expected to gradually accept these suggestions from their hearts and thus change their behaviors. Those first successful recipients, in turn, can serve as role models to influence and inspire others around them to further popularize the intervention concept and promote adaptation measures.

2.2.5 Self-efficacy

For individuals, especially those who live alone, practicing interventions in daily lives is more about the initiative besides the influence of the environment, which refers to the self-efficacy from social—cognitive theories. Self-efficacy shows an individual's confidence in their capacity to exercise control over their motivation, behavior, and social environment to achieve performance outcomes (O'Neill et al., 2009; Sera et al., 2020; Guardaro et al., 2020). The construct of self-efficacy is widely applied in research, education, and clinical practice such as self-management of chronic disease, alcohol use, eating, exercise, and the prevention of HIV (Lioubimtseva & Da Cunha, 2022, pp. 257—276; Moser & Pike, 2015; Weir, 2002). Gradually, the self-efficacy theory (SET) is also used in investigating people's attitudes toward climate change (Ma et al., 2015). Self-efficacy has been mentioned as a possible determinant in deciding whether a person would participate in adaptation therapies and to what extent (Kim & Kim, 2017). Interpersonal information sharing and education communication, discussed previously, can contribute to the better construction of self-efficacy. At the same time, peer models have the potential to encourage and arouse individuals to engage in adaptation actions favorably.

2.3 Community

There are many measures that the community could take to reduce the adverse impacts of heat exposure, such as strengthening community infrastructure, encouraging community engagement, and developing vulnerable people groups.

2.3.1 Infrastructures

Increasing the prevalence of air-conditioners and cooling centers is the possible way for the community to reduce heat exposure (O'Neill et al., 2009). As one of the most straightforward strategies, it has been proved that increased air-conditioning prevalence was associated with lower heat-related mortality risk (Sera et al., 2020). For those unable to afford air-conditioning, cooling centers (e.g., public libraries, community centers, or police stations) could effectively provide air conditioning for a large group of people. For example, some cities have been setting up cooling centers and urging people to drink more water as a record-breaking heatwave swept across the Pacific Northwest region in the United States. By using air-conditioning and cooling centers, indoor temperatures could be regulated directly, and thus the burden on the human thermoregulatory function in hot weather could be reduced (Sera et al., 2020). Besides, air-conditioning could also improve indoor air quality by filtering air, which may bring more benefits to human health. Except for cooling centers, communities can also provide other infrastructures and facilities to residents. According to the US Environmental Protection Agency (EPA), heat mitigation strategies focusing on the urban areas could include installing green roofs, planting trees and vegetation, and building cool pavements. These community infrastructures will not only help to reduce the heat exposure and urban heat island effects but also help to reduce the consumption of fuels and improve air quality.

2.3.2 Community engagement

Community engagement is the process for residents to be involved in the actions of planning and implementing strategies to reduce the impacts of heat exposure. Poor and marginalized groups are vulnerable to heat exposure, and they were often overlooked in the traditional climate agenda. For example, some people may not have available transportation to the cooling centers during the hot weather. To address this issue, community engagement is important to consider the interests of various segments of the population. Based on the community organization, residents living in the community and their neighbors were recruited to participate in a set of activities aiming to provide critical knowledge, communicate with policy-makers, refine the proposed solutions, and form the final action plan (Guardaro et al., 2020). In this process, external experts and researchers could be included to provide professional knowledge and necessary skills. More importantly, the core of community engagement is to advance adaptation actions by involving individuals emotionally, behaviorally, and politically (Moser & Pike, 2015). It is thus vital to ensure the diversity of the stakeholders, including residents and residents' groups, environmental and social justice advocacy groups, local governments, local services, and other agencies and groups, such as academia, private experts contractors, and local businesses (Lioubimtseva & Da Cunha, 2022, pp. 257–276). Through this process, the adaptation actions could be easily accepted by residents, and the community may create more effective and practical strategies by drawing on the local knowledge and experience.

2.3.3 Identify vulnerable groups

Another strategy for the community to reduce the health impacts of heat exposure is to identify vulnerable groups of people and develop targeted policies. Generally, everyone will be affected by heat exposure, while some people are more vulnerable, including infants and children, elderly people, homeless or poor people, people with underlying conditions or illnesses, and outdoor workers (Weir, 2002). For example, some studies found that heat-related mortality risk was higher for elderly people and those living in districts with a high deprivation index (Ma et al., 2015; Kim & Kim, 2017). Thus, it is necessary to give priority to the vulnerable populations when the hot weather occurs. However, it may be difficult as well as inaccurate to identify these populations. Many efforts have been made up to now. For example, the IDPoor in Cambodia is a program proposed by the government to identify poor households and to further reduce poverty. At the community level, a Village Representative Group (VRG) will be elected by villagers in the first step. Then, the first draft list of poor households will be reviewed and commented on in a Village Consultation Meeting. After the meeting, the final list of poor households will be identified. It is clear to see that community engagement plays an important role in the identification of vulnerable groups.

2.3.4 Livelihood initiatives

In addition to the aforementioned measures, livelihood initiatives are also important for adaptions at the community level. Extreme hot weather has posed a great threat to human health as well as livelihoods. Different from other extreme weather events, heat exposure would not cause obvious threats to livelihood, such as damage to houses and roads due to flooding. Heat exposure could affect the work performance and health conditions of workers and will further affect their financial conditions (Kabir & Serrao-Neumann, 2020, pp. 167−179). For example, it was estimated that 295 billion hours of potential work were lost around the world due to heat exposure in 2020 (Romanello et al., 2021). The impacts of heat exposure on livelihood suggested that extreme weather would cause more harm than imagined, especially for those living in less developed areas. As a result, policies should be developed to support livelihood by increasing resilience or the ability to restore livelihood quickly after the hot weather. Potential initiatives may include identifying susceptible populations, increasing the accessibility of insurance, and building diverse livelihoods for residents. The main aim of the initiatives is to strengthen the adaptive capacity, reduce livelihood vulnerability, and help residents to maintain sustainable livelihoods (Abeje et al., 2019). Another solution is the neighborhood watch, which aims to educate residents within the community on knowledge to cope with the threats and encourage them to look out for potential health outcomes in the community due to heat exposure. The community organization based on the neighborhood watch could be efficient in its role because of its knowledge of the needs and resource allocation of the community.

2.3.5 Solidarity

Strategies as a response to heat exposure at the community level are not limited to these measures. Overall, the primary purpose and mechanism of community adaptation solutions are to increase community solidarity. By connecting people living in the same community, these measures could ensure that residents will not only give and receive help from the network referring to physical damage but also receive emotional and psychological support (Harrington & Cole, 2021). In the process of constructing community solidarity, the diffusion of adaption strategies is completed as well. For example, some people may not be aware of the adverse health consequences that can result from heat exposure, and they will not take any measures even after the alert of a heat warning system. Thus, another purpose for the strategies at the community level is to improve the motivation or intentions of the residents and finally achieve behavioral change. With these measures, the vulnerability of the community to heat exposure will be reduced, and the future risks will be also lessened through long-term planning.

2.4 Institutional

The health impact of occupational heat exposure has attracted attention due to its long length, strong strength, and potential inequality. The institution is responsible to manage the safety of the workforce and abide the law as a compounding role and thus is of great importance in the adaptation of heat exposure. Institutional heat exposure adaptation could reduce the health impact from four aspects including (1) introducing new technology; (2) improving working flow; (3) strengthening occupational health surveillance; (4) establishing individual protection checklist; and (5) providing an emergency plan for heat extremes (Fig. 12.2).

2.4.1 Introducing new technology

Introducing new technologies, which would reduce heat exposure in working places, is the fundamental measure for a healthy operating environment. Automated production could significantly reduce heat exposure

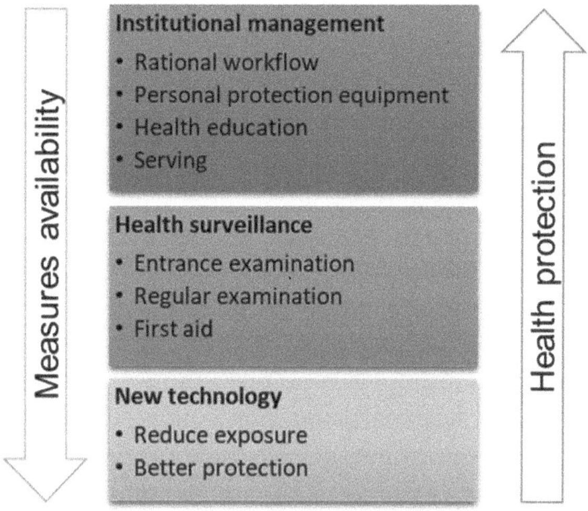

Figure 12.2 Institutional adaptation measures.

compared with open or semiopen production instruments. Though significant progress has been made in the past century, there are still several jobs that have to be done by individuals, such as firefighters, city constructors, and cleaners. Since the unavoidable heat exposure for these populations, new technologies and innovative prevention methods should be applied to individual protective equipment. Taking the fire-protective clothing for firefighters who are exposed to extreme high-temperature exposure in the burn as an example, the development and application of fire-protective clothing have a history of many years (Morel et al., 2014). Increasing innovative technologies are still under investigation for better protection, such as triple-network (TN) hydrogel laminated on cotton fabric instead of the traditional flame retardant (Yu et al., 2021). A cooling vest is also recommended for construction workers which is suggested significantly to alleviate heat strain and improve thermal comfort (Zhao et al., 2017). It is the responsibility of institutes to track the industry's latest technologies to reduce the health impact of heat exposure.

Though office works are more moderate in thermal comfortable, new technologies also make sense in reducing heat-related health impact on working days. Air-conditioners are widely used in buildings. However, the health impact of air-conditioner use is still controversial. The air-conditioner is also a major source of greenhouse gases emission (Lundgren-Kownacki et al., 2018). Besides the long-term effect, the use of an air-conditioner would directly influence the outdoor temperature via the exhaust fan. New building materials could be considered to reduce heat stress in a sustainable way (Wi et al., 2021).

However, the update of the equipment may lead to an increase in costs. The balance between occupational health and operating costs needs to be carefully assessed and maximizing protection against occupational exposure, including heat exposure where possible is highly recommended.

2.4.2 Strengthening occupational health surveillance

Due to the impaired heat tolerance of people with underlying medical conditions, the entry examination for heat operation should be cautious. Some people with predisposing conditions should be detected to avoid adverse outcomes after heat exposure. The working regulation is various in countries, but according to the previous studies, people with uncontrolled diseases are not recommended to work in high temperatures (Xiang et al., 2014). In the evaluation of outdoor occupational heat-related fatal illnesses events, 9 of 14 workers (64.3%) have known present predisposing personal risk factors including obesity, diabetes, hypertension, cardiac disease, and use of certain medications or illicit drugs in the United States from 2011 to 2016 (Tustin et al., 2018). The regulation for different working positions, duration, and intensity should be considered separately. One study summarized the occupational heat stress in Australia, which may help to find the differences among regions and working categories (Jay & Brotherhood, 2016). The health conditions of people may be affected after accumulative heat exposure. Thus, regular examinations are needed to protect the safety of workers. One study introduced remote advanced electrocardiography for improved cardiovascular risk assessment in career firefighters (Dolezal et al., 2014).

The monitoring is essential when working, for both environmental conditions and workers. Environmental conditions monitoring would help to assess and identify the occupational risks, especially in summer, hot seasons, and heatwave weather (Peterson et al., 2013). The institutes should comply with applicable workplace heat regulations or make more strict regulations to meet the institutional requirements. Other factors including radiation and humidity should also be considered (Chen et al., 2019; Wehrli et al., 2019). Cooperation is recommended to avoid long time exposure for each person. Multiple supervision could also help in identifying risks and giving immediately action on risks. For people working alone, the institute should provide a monitor to give first aid to minimize the risks.

2.4.3 Institutional management

The institute is responsible for management via providing rational working flow, flexible rotation, and health education. As described before, the cost of inviting new technologies is high, which may be beyond the budget of the institute. Rational working flow is another key strategy in reducing heat exposure in working places. Limited work on heat extreme days to only cooling time in the day is recommended especially for outdoor tasks. During high-temperature weather, the employer shall, according to the production characteristics and specific conditions, take measures such as arranging working hours reasonably, rotating operations, appropriately increasing the rest time of workers in a high-temperature working environment, reducing working intensity, and reducing outdoor operations during high-temperature periods. Employers should conduct regular occupational hygiene training for workers before and during their employment and popularize occupational hygiene knowledge such as high-temperature protection and heatstroke first aid. An individual protection checklist is recommended.

Personal protection equipment, such as a thermal radiation cap, thermal radiation mask, goggles, high temperature—resistant flame-retardant gloves, and high-temperature protective shoes, should be provided by the institute. Some equipment may lose protection after years of use or extreme heat exposure (Rossi et al., 2008). In consequence, a regular check is needed to ensure the protection effectiveness of the equipment.

The training content should be detailed including but not limited to the following aspects. Awareness should be delivered to workers that risk identification is closely related to their safety of themselves. The attitude toward personal protection could affect the daily routine of risk detection and prevention (Ferrer & Klein, 2015). The introduction of the heat management plan should be introduced in plain language and make sure that people from different cultural backgrounds understood. Exercises on how to identify the hazard, work safely, and recognize and avoid heat-related illnesses should be provided regularly to ensure the effectiveness of the occupational safety education. How to use, maintain, and check the personal protective equipment correctly is an essential task as well. Furthermore, workers should know how to report the heat-related illness to the facility and seek consultation from doctors. First aid should also be included in occupational health education since immediate action could significantly increase the survival possibility.

Services provided by the institute could prevent heat-related illness. People who work in heat or low humidity may be prone to dehydration (Bi et al., 2011; Popkin et al., 2010). Providing cool drinking water and salt-containing beverages with easy access could help workers replenish in time after the loss of body water and salt. Meanwhile, adequate sleep and rest should be ensured, and dietary health should be strengthened. Employers should set up rest places in high-temperature working environments; rest places should be equipped with seats, keep good ventilation, or be equipped with air-conditioners and other heatstroke prevention and cooling facilities. Considering equity and convenience issues, adequate male and female toilets should be appropriately arranged for us. This can avoid the problem of unwillingness to drink the water due to the inconvenience of going to the toilet.

2.5 Environmental

Urban areas generally have a higher average temperature than adjacent rural and suburban areas because urban environments tend to trap and store more heat than natural environments, which is known as the "urban heat island (UHI) effect." Compared with natural environments, urban environments have surfaces massively consisting of buildings or materials that are highly efficient at absorbing and storing heat (e.g., bricks, concrete, pitch, and metal), as the replacement of vegetation, which moderates microclimates of natural environments through transpiration with leaves and evaporation from soil moisture (Chen & You, 2020). The heat stored in the urban surfaces will be released after the air has cooled, leading to higher temperatures during the night in addition during the day. What is more, urban areas have more and denser anthropogenic heat sources (e.g., transport systems, lights, air conditioners) apart from natural sources (mostly solar). As a result, the average ambient temperatures in the center of some large cities can be 1–6°C higher than in the adjacent suburban and rural areas (Dousset et al., 2019). Targeting the aforementioned urban characteristics that lead to the UHI effect, various strategies and technologies have been proposed and applied to minimize the heat impacts on human health, especially in urban areas.

2.5.1 Landscape

2.5.1.1 Green infrastructures

Green infrastructures covered by vegetation, including parks, gardens, and urban forests, have been generally regarded as one of the most effective techniques for counteracting the UHI effect (Jay et al., 2021; Tong et al.,

2021). Through transpiration from plants, evaporation from soil moisture, and casting shade, landscapes with adequate tree canopies can perform the so-called urban cool island effect, providing thermal relief to nearby residential and commercial precincts (Jay et al., 2021). This is the most fundamental strategy for adapting to increasing heat exposure.

An observative study reported that the average air temperatures in parks can be up to 2°C lower than the background air temperature (Kurn et al., 1994). Another study suggested that the temperatures of shaded surfaces can be 12–20°C lower than the temperatures of unshaded surfaces (Akbari et al., 1997). Vegetation is the most effective intervention to the health impacts of heat exposure when they are implemented strategically around buildings or to shade pavement, outdoor car parks, and streets. For example, planting deciduous trees or vines on the west side of a building is typically the most effective for cooling the indoor temperature, especially if they shade windows and part of the building's roof (US EPA O). A simulation study reported that surface temperatures were reduced by 1–2°C when 1% of new green infrastructures were optimally added (Zhang et al., 2017).

A major issue of this strategy is water consumption. Implementing more spaces covered by vegetation may significantly increase water demand. Therefore, the application will be impractical in areas with insufficient water resources. This issue may be moderated by incorporating mixed types of vegetation. For example, a study conducted in the Negev Desert reported that grass shaded by shade trees or shade mesh yielded a greater cooling effect with 50% less water use compared with unshaded grass (Shashua-Bar et al., 2009). Another issue is land use. The cost of converting land-use types in urban areas, especially in built-up precincts (e.g., commercial precincts), may be unaffordable. Therefore, the implementation of landscapes covered by vegetation (e.g., gardens and parks) may be limited in urban areas. Some innovations have been applied. For example, green roofs and green walls (Section 2.5.2 Building) do not require converting land-use types and perform a direct cooling effect on buildings.

Besides the health impacts of heat exposure, green infrastructure is also a technique for other challenges. First, it can reduce the energy consumption used for cooling by decreasing the demand for air-conditioning. Therefore, energy-related greenhouse gases emission and air pollution, which lead to global warming and impact human health respectively, will decrease. Green infrastructures also remove carbon dioxide and some air pollutants. Third, some green infrastructures (e.g., parks and roof gardens) have been found beneficial to mental and physical health by promoting social connections and physical activities and reducing noise (Fong et al., 2018). Fourth, some types of green infrastructures (e.g., urban forests and rain gardens) are regarded as strategies to increase the resilience to floods and to cope with intense rainfall because they can absorb, store, and filter rainwater.

2.5.1.2 Blue infrastructures

Due to its high heat capacity, evaporative, and convection cooling effect, blue (i.e., water) infrastructures can moderate urban microclimate. A simulation study suggested that the UHI effect would be alleviated from 1.24 to 1.16°C when the proportion of water infrastructures grow from 0% to 10% (Chen & You, 2020). Besides, blue infrastructures are found to have a cooling effect at pedestrian level (1.5 m above ground), contributing to pedestrian thermal comfort in residential districts (Chen & You, 2020).

The cooling effect is stronger with a higher wind speed, drier air, a larger contact surface, and when the water is moving (Oke, 1987, p. 464). Therefore, for cities with abundant wind and water resources, large blue infrastructures are highly recommended because the cooling effects may be enhanced with a high wind speed and a large lake surface. For residential precincts or cities with limited water resources, where the population is dense, fountains may be a better option since they accelerate the evaporative cooling effect with spray and moving water. In areas with heavy rainfalls, facilities for rainwater, stormwater, and flood management (e.g., rainwater gardens and water storage networks) can also be adapted to counteract the health impacts of heat exposure.

2.5.2 Building

2.5.2.1 Green roofs

Green roofs are a layer of vegetation planted on a rooftop. They provide cooling effects similar to landscapes (e.g., parks and gardens) covered by vegetation, through shading, evaporation, and transpiration. An advantage is that they mitigate building surfaces or indoor temperature directly rather than through an urban cool island effect, through which green landscapes affect adjacent indoor environments. Therefore, this strategy may be more effective. Since green roofs do not require converting land-use types and can be installed on a wide range of buildings, from industrial facilities to private residences, they are good complements to the conventional vegetation applications (e.g., green landscapes) used to counteract urban heat exposure. Another strength is that the water cost may be reduced. Unlike parks and gardens, which necessitate establishing a new irrigation system, green roofs can use the water supply system of the buildings directly. In addition, this also provides an opportunity to make use of wastewater regularly produced by the building. The burden of sewage purification may be greatly reduced if green roofs are widely installed.

Green roof surface temperatures can be 17–22°C lower than those of conventional roofs, and a large-scale application can reduce city-wide ambient temperatures by up to 2.7°C (US EPA O). In addition, green roofs can reduce building energy use by 0.7% compared with conventional roofs and lead to an annual savings of $0.23 per square foot of the roof surface (Administration, 2011; Sailor et al., 2012). These temperature reduction and energy efficiency benefits are the main reasons for the growing prevalence of green roofs in the United States. In North America, the green roof industry is estimated to have risen by more than 10% in 2016 compared with that in 2015, maintaining the growth trend over the past decade (Cities, 2016).

2.5.2.2 Reflective surface

Besides through roofs, heat enters indoor environments also via conductions through walls. This process varies considerably with the color and material properties (e.g., conductivity and heat capacity) of walls. Therefore, a normal way to reduce indoor heat is by increasing the solar reflectivity of building coating. By doing this, building surfaces absorb and transfer less heat to the buildings, so the indoor environments stay cooler, resulting in less energy used for air-conditioning. High emittance is another material property that improves the performance of the coating, which allows indoor heat emitted to cold outdoor spaces efficiently. Combined with high solar reflectivity and near-perfect thermal radiation emittances, a kind of hierarchically porous polymer building coatings was reported in 2018, achieving net radiative losses as high as 96 W/m^2 under a solar intensity of 750 W/m^2 and resulting in passive daytime radiative cooling effects as much as −9 K (Mandal et al., 2018). The limitation of surface coatings is that it may exacerbate thermal pollution by reflecting solar radiation on surrounding buildings and pavements. To minimize the impacts of this, the location and orientation of the building coatings should be designed strategically.

2.5.3 Pavement level

2.5.3.1 Cool pavements

Pavements, including pedestrians and roads, comprise as much as 30%–45% of urban land areas. Conventional paving materials can reach

48–78°C during summer, impacting human body temperatures or drastically increasing energy used by vehicles (for air-conditioning). Therefore, creating cool pavements is important to reduce the impact of heat exposure. Options for cool pavements include reflective pavements and wall coatings. However, they may be less effective because the reflected heat can be absorbed by adjacent surfaces (e.g., walls and buildings). On this occasion, strategic locating and orienting buildings, trees, and pedestrians to maximize shading (e.g., shading the south side of pavements in the northern hemisphere) may be a better choice.

2.5.3.2 Water stations

Water stations may be a strategy to reduce the health impacts of heat exposure. Staying cool and staying hydrated are the two main preventions as well as treatments for heat-related illnesses (e.g., heat stroke) (Peiris et al., 2017). During extremely hot weather, intermittently applying cool tap water (around 20°C) to skin surfaces where large blood vessels are located alleviates cardiovascular strain and thermal discomfort, and dehydration can be reduced by up to 65% (Jay et al., 2021). With water stations strategically placed, or even accompanied with heat stroke first-aid guidelines, people will be able to prevent themselves from heat-related illnesses as well as to provide help when they find someone with signs of heatstroke (Peiris et al., 2017).

2.6 Public policy

These policies are mostly evidence based, properly communicated, and frequently evaluated by real-time surveillance data to be effective (Jay et al., 2021). Policy-level interventions to overcome the health effects of hot weather developed through the collaboration of interdisciplinary teams could maximize the people's acceptability and practicality (Wight et al., 2016), leading to behavioral change, and may also indirectly influence physiological change (Guo et al., 2018). These large-scale interventions to reduce adverse health outcomes from extreme heat discussed in this chapter involve short- to long-term such as distribution and installation of heating, ventilation, and air-conditioning (HVAC) systems in healthcare facilities and nursing homes for the elderly and the general population; uninterrupted power and water supply; long-term vehicle and transport regulations; health education and communication; and heat warning systems.

2.6.1 Heating, ventilation, and air-conditioning and related guidelines

Installation or presence of heating systems in hospitals could help reduce mortality risk during extreme heat episodes. Evidence suggests that the installation of HVAC or fans in general wards and intensive care units has shown improvement in patients with respiratory disease, heat and cardiac stress, and heat-related illness and made hospital treatment more efficient during heatwaves (Lenzer et al., 2020). For example, the presence of air-conditioning in hospital wards of Portugal during the 2003 heatwave reduced the mortality risk of inpatients by 40% (Nunes et al., 2011). Therefore, distributing heating systems to healthcare facilities could be a promising strategy to mitigate adverse health effects of those who come to hospitals during hot weather.

Nursing homes are also at higher temperature—mortality risk for their composition of the vulnerable elderly population and poor HVAC facilities (Klenk et al., 2010). Previous evidence from the United States suggests, a 2.3-fold increased mortality risk in nursing homes in New York without air-conditioning to those with air-conditioning (Marmor, 1978) and in Florida, air-conditioning failure in nursing homes accounted for deaths due to hyperpyrexia during heatwaves (Sullivan-Bolyai et al., 1979). It clearly shows that there should be universal standards and guidelines for structural aspects of the HVAC system and practicing standards in hospitals (Saran et al., 2020). These universal standards could also be incorporated into existing nursing home guidelines, which could reduce preventable deaths from hot weather and heat extremes.

2.6.2 Air-conditioning and disparities in the society

In many countries, loss of air-conditioning due to power disruption during heatwaves contributed to a rise in heat-related morbidity and mortality (Skarha et al., 2021; Zhang et al., 2017; Son et al., 2019). A single-day power outage in Melbourne could have attributed to 40—79 excess deaths in the 2009 heatwave and reduced to 24—47 additional deaths in the 2014 heatwave (Nairn & Williams, 2019, pp. 1—37), possibly due to the implementation of the Victorian Heatwave Plan involving communities and local government (Department of Human Services, 2009). Furthermore, having an air-conditioning at home was reported to be protective against adverse health effects of hot weather (Zhang et al., 2017; Son et al., 2019) suggesting fans and air-conditioner distribution programs and continuous power supply for an air-conditioner during extreme heat must be guaranteed by relevant local authorities (White-Newsome et al., 2014). However, there are urban–rural disparities in the use of air-conditioning, and it remains unaffordable due to electricity cost and maintenance to the people with low socioeconomic status who live in poor quality houses (Son et al., 2019).

2.6.3 Other sustainable cooling strategies

The use of air-conditioning can benefit in the short term but indirectly contributes to an increase in global greenhouse gas emissions, which can exacerbate the heatwave in the long run (Jay et al., 2021). Hence, broadly accessible and sustainable cooling strategies during hot weather are a pressing need for economically deprived and clinically vulnerable population groups. Applying cool tap water to the skin surface showed a reduction in heart rate, sweating, and thermal discomfort, and cold water foot immersion lowered thermal discomfort in hot and humid conditions (Morris et al., 2019). Unlike air-conditioning, water does not require electricity; hence, the use of cool tap water on external body surfaces can be recommended as an alternative cooling strategy. Furthermore, public health authorities recommend drinking more water during extremely hot weather. Uninterrupted water supply and safe drinking water to all sections of society, in settings without any electricity supply or during unplanned power outages especially in the times of hot weather, is another cost-effective strategy to mitigate adverse health effects of extreme heat (White-Newsome et al., 2014).

2.6.4 Vehicles and transportation

Vehicles and transportation are other contributing sectors to hot weather for which long-term policy-level regulations can be introduced. Replacement of conventional vehicles with electric vehicles has the benefits of much less emission of heat and CO_2. For example, in Beijing, the average heat emission in 2012 by electric vehicles (1.25 million joules) was much less than that of conventional vehicles (6.31 million joules), which could reduce the CO_2 emissions by 10,686 tonnes per day (Li et al., 2015). Other strategies may include changing traffic infrastructure, which can reduce vehicle density on the roads, and improving public transport, cycling, and urban walking infrastructures may lead to a physically active population (Jay et al., 2021).

2.6.5 Health education and effective communication

Health education plays a major role in publicizing heat-related information to the public. At mass gatherings in hot weather, educating people on hydration and preventive practices to increase their compliance with protective behaviors and providing heat prevention information such as signs and symptoms of heat illness, prevention, and treatment are a few acceptable and practical strategies (Jay et al., 2021). In the United States, awareness programs on heat-related health risks to school students targeting youths, homebound elders, isolated populations, homeless, at-risk groups such as children and pregnant mothers, government and nongovernment officials, and healthcare practitioners were used to reduce the harmful effects of heat (White-Newsome et al., 2014). Public warning using printed leaflets and mass media is the most common strategy to increase awareness of heat exposure risks and ways of reducing its impacts (Koppe et al., 2004; Matthies & Menne, 2009). A structured communication strategy is an essential part of conveying clear and specific information to the public before and during a heatwave. It can include alerting public health practitioners with surveillance messages indicating health statistics from hot weather in their local areas, a messaging system containing information about cooling center locations and other relevant resources to the public in advance of heatwaves and a hotline helpline (Department of Human Services, 2009; White-Newsome et al., 2014).

2.6.6 Heat warning systems

Another widely used strategy to reduce the burden of heat extremes is heat health warning systems that use local weather forecasts to predict adverse health outcomes of heat (Jay et al., 2021; White-Newsome et al., 2014). In German, an online tool developed by the country's Weather Service forecasts the probability of heatwaves and supports the health services to plan in advance (Matthies & Menne, 2009). The heat alert system in Victoria, Australia, is a collaboration between the Department of Health, Bureau of Meteorology, and local councils. It forecasts 7-day maximum and minimum temperatures that can predict heatwave between 1 and 7 days before the event (Department of Human Services, 2009). The effectiveness of heat warning systems relies on adverse health effects associated with weather conditions, the accuracy of the weather forecast, and the timeliness of warning concerning the response (Jay et al., 2021; Koppe et al., 2004).

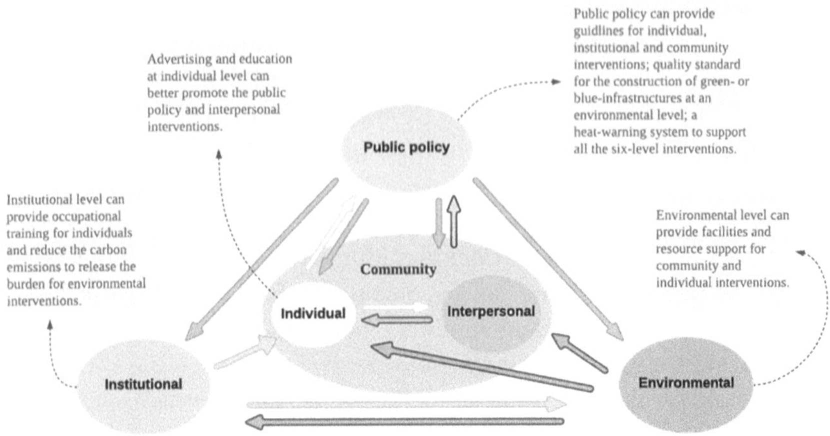

Figure 12.3 Associations among six-level adaptation intervention strategy framework.

3. Conclusion

As discussed in various previous studies or climate reports, extreme heatwaves will occur with higher frequency and intensity in the future (Beckmann et al., 2021; Steffen et al., 2014; Trancoso et al., 2020). Public health is thus put at a higher risk, both short term and long run. People are expected to adapt to that increasing heat exposure through adaptation strategies from physiological, behavioral, infrastructural, and technological aspects (Hondula et al., 2015). This chapter proposed a six-level adaptation framework, consisting of individual, interpersonal, institutional, community, and public policy (Fig. 12.3). Different levels of interventions are associated and can reinforce each other. For example, advertising at the individual level and promotion at the community level can better achieve public policy interventions such as broadly applying and promoting the heatwave-warning systems. Information provision and awareness strengthens of individuals can contribute to higher-efficiency information exchange and broader neighborhood watch at the interpersonal level. Institutional policies and quality standards can provide guidelines for the community and environmental interventions (e.g., community engagement and infrastructure construction). Environmental interventions such as green infrastructures and water stations can provide resource and facility support for individuals (e.g., individuals with vulnerable socioeconomic status can get cool water from water stations since their accessibility to air conditioning can be limited). Successful individual and interpersonal interventions can contribute to the solidary of the community and make better use of the facilities built to combat the heatwaves (e.g., cool pavement), while active community participation can enhance the self-efficacy of individuals in turn. Also, a practical and accurate heatwave-warning system can fine prepare communities and individuals to withstand the incoming heat exposure.

Meanwhile, we should also pay attention to the potential long-term compounds of the adaptation strategies. For example, the use of air-conditioning can release the heat burden. However, energy consumption can be raised simultaneously, especially in countries where most buildings are equipped with air-conditioners such as in America, which can promote the emissions of greenhouse gas, the main trigger of global warming (Morris et al., 2019; Li et al., 2015). Hence, sustainable strategies should be applied at all levels; related institutions should guide and educate the public, providing related facilities and material security; public policies should be formulated considering both short-term and long-term effects.

By enhancing the coordination of six levels in resource allocation, communication, education, advocacy, policy reformation, and improvement of health services, the six-level adaptation intervention framework aims to establish a collaborative, practical, adaptive, and sustainable strategy systems to reduce the health impacts of heat exposure for the public.

References

Abeje, M. T., et al. (2019). Communities' livelihood vulnerability to climate variability in Ethiopia. *Sustainability, 11*(22), 6302.

Administration, G. S. (2011). *The benefits and challenges of green roofs on public and commercial buildings*.

Akbari, H., et al. (1997). Peak power and cooling energy savings of shade trees. *Energy and Buildings, 25*(2), 139–148.

Bartos, M., et al. (2016). Impacts of rising air temperatures on electric transmission ampacity and peak electricity load in the United States. *Environmental Research Letters, 11*(11), 114008.

Beckmann, S. K., et al. (2021). Heat adaptation measures in private households: An application and adaptation of the protective action decision model. *Humanities and Social Sciences Communications, 8*(1), 227.

Bi, P., et al. (2011). The effects of extreme heat on human mortality and morbidity in Australia: Implications for public health. *Asia Pacific Journal of Public Health, 23*(2_Suppl. 1), 27S–36S.

Bouchama, A., & Knochel, J. P. (2002). Heat stroke. *The New England Journal of Medicine, 346*(25), 1978–1988.

Campbell, S., et al. (2018). Heatwave and health impact research: A global review. *Health & Place, 53*, 210–218.

Chen, X., et al. (2019). Global heat wave hazard considering humidity effects during the 21st century. *International Journal of Environmental Research and Public Health, 16*(9), 1513.
Cheng, J., et al. (2019). Cardiorespiratory effects of heatwaves: A systematic review and meta-analysis of global epidemiological evidence. *Environmental Research, 177*, 108610.
Chen, R., & You, X.-y. (2020). Reduction of urban heat island and associated greenhouse gas emissions. *Mitigation and Adaptation Strategies for Global Change, 25*(4), 689–711.
Cities, G. R.f. H. (2016). *Annual green roof industry survey: Executive summary*, 2017.
D'Amato, G., et al. (2014). Climate change and respiratory diseases. *European Respiratory Review, 23*(132), 161–169.
Department of Human Services. (2009). *Heatwave planning guide: Development of heatwave plans in local councils in Victoria*. Victorian Government Department of Human Services Melbourne.
Dolezal, B. A., et al. (2014). Deployment of remote advanced electrocardiography for improved cardiovascular risk assessment in career firefighters. *Telemedicine and e-Health, 20*(7), 660–663.
Dousset, B., Luvall, J. C., & Hulley, G. C. (2019). 7 - surface temperatures in the urban environment. In G. C. Hulley, & D. Ghent (Eds.), *Taking the temperature of the earth* (pp. 203–226). Elsevier.
Ebi, K. L., et al. (2021). Hot weather and heat extremes: Health risks. *The Lancet, 398*(10301), 698–708.
Fatima, S. H., et al. (2021). Extreme heat and occupational injuries in different climate zones: A systematic review and meta-analysis of epidemiological evidence. *Environment International, 148*, 106384.
Ferrer, R., & Klein, W. M. (2015). Risk perceptions and health behavior. *Current Opinion in Psychology, 5*, 85–89.
Fong, K., Hart, J. E., & James, P. (2018). A review of epidemiologic studies on greenness and health: Updated literature through 2017. *Current Environmental Health Reports, 5*(1), 77–87.
Gamble, J. L., et al. (2013). Climate change and older Americans: State of the science. *Environmental Health Perspectives, 121*(1), 15–22.
Gasparrini, A., et al. (2015). Mortality risk attributable to high and low ambient temperature: A multicountry observational study. *Lancet, 386*(9991), 369–375.
Gronlund, C. J. (2014). Racial and socioeconomic disparities in heat-related health effects and their mechanisms: A review. *Current Epidemiology Reports, 1*(3), 165–173.
Guardaro, M., et al. (2020). Building community heat action plans story by story: A three neighborhood case study. *Cities, 107*, 102886.
Guo, Y., et al. (2014). Global variation in the effects of ambient temperature on mortality: A systematic evaluation. *Epidemiology, 25*(6), 781–789.
Guo, Y., et al. (2017). Heat wave and mortality: A multicountry, multicommunity study. *Environmental Health Perspectives, 125*(8), 087006.
Guo, Y., et al. (2018). Quantifying excess deaths related to heatwaves under climate change scenarios: A multicountry time series modelling study. *PLoS Medicine, 15*(7), e1002629.
Harrington, E., & Cole, A. (2021). *Typologies of mutual aid in climate resilience: Variation in reciprocity, solidarity, self-determination, and resistance*. Environmental Justice.
Hondula, D. M., et al. (2015). Rising temperatures, human health, and the role of adaptation. *Current Climate Change Reports, 1*(3), 144–154.
Hsu, A., et al. (2021). Disproportionate exposure to urban heat island intensity across major US cities. *Nature Communications, 12*(1), 2721.
Jay, O., & Brotherhood, J. R. (2016). Occupational heat stress in Australian workplaces. *Temperature (Austin), 3*(3), 394–411.
Jay, O., et al. (2021). Reducing the health effects of hot weather and heat extremes: From personal cooling strategies to green cities. *Lancet, 398*(10301), 709–724.

Kabir, M. E., & Serrao-Neumann, S. (2020). *Climate change effects on people's livelihood*. Springer International Publishing.
Kelly, M. P., & Barker, M. (2016). Why is changing health-related behaviour so difficult? *Public Health, 136*, 109–116.
Kenny, G. P., et al. (2010). Heat stress in older individuals and patients with common chronic diseases. *CMAJ : Canadian Medical Association journal = journal de l'Association medicale canadienne, 182*(10), 1053–1060.
Khan, A. A. (2019). Heat related illnesses. Review of an ongoing challenge. *Saudi Medical Journal, 40*(12), 1195–1201.
Kim, E. J., & Kim, H. (2017). Effect modification of individual- and regional-scale characteristics on heat wave-related mortality rates between 2009 and 2012 in Seoul, South Korea. *Science of The Total Environment, 595*, 141–148.
Klenk, J., Becker, C., & Rapp, K. (2010). Heat-related mortality in residents of nursing homes. *Age and Ageing, 39*(2), 245–252.
Koppe, C., et al. (2004). *Heat-waves: Risks and responses. Health and global environmental change series, No. 2*. Copenhagen: World Health Organization (Regional Office for Europe).
Kurn, D. M., et al. (1994). *The potential for reducing urban air temperatures and energy consumption through vegetative cooling*. , CA (United States): Lawrence Berkeley Lab.
Lenzer, B., et al. (2020). Health effects of heating, ventilation and air conditioning on hospital patients: A scoping review. *BMC Public Health, 20*(1), 1287.
Li, C., et al. (2015). Hidden benefits of electric vehicles for addressing climate change. *Scientific Reports, 5*(1), 9213.
Lin, S., et al. (2009). Extreme high temperatures and hospital admissions for respiratory and cardiovascular diseases. *Epidemiology, 20*(5), 738–746.
Lioubimtseva, E., & Da Cunha, C. (2022). *Community engagement and equity in climate adaptation planning: Experience of small- and mid-sized cities in the United States and in France*. Springer International Publishing.
Liu, J., et al. (2021a). Hot weather as a risk factor for kidney disease outcomes: A systematic review and meta-analysis of epidemiological evidence. *Science of the Total Environment, 801*, 149806.
Liu, J., et al. (2021b). Is there an association between hot weather and poor mental health outcomes? A systematic review and meta-analysis. *Environment International, 153*, 106533.
Liu, C., Yavar, Z., & Sun, Q. (2015). Cardiovascular response to thermoregulatory challenges. *American Journal of Physiology. Heart and Circulatory Physiology, 309*(11), H1793–H1812.
Lundgren-Kownacki, K., et al. (2018). Challenges of using air conditioning in an increasingly hot climate. *International Journal of Biometeorology, 62*(3), 401–412.
Lv, L. S., et al. (2022). Impact of ambient temperature on life loss per death from cardiovascular diseases: A multicenter study in central China. *Environmental Science and Pollution Research International, 29*(11), 15791–15799.
Ma, W., et al. (2015). The short-term effect of heat waves on mortality and its modifiers in China: An analysis from 66 communities. *Environment International, 75*, 103–109.
Mahendran, R., et al. (2021). Interpersonal violence associated with hot weather. *Lancet Planet Health, 5*(9), e571–e572.
Mandal, J., et al. (2018). Hierarchically porous polymer coatings for highly efficient passive daytime radiative cooling. *Science, 362*(6412), 315–319.
Marmor, M. (1978). Heat wave mortality in nursing homes. *Environmental Research, 17*(1), 102–115.
Matthies, F., & Menne, B. (2009). Prevention and management of health hazards related to heatwaves. *International Journal of Circumpolar Health, 68*(1), 8–12.

Meade, R. D., et al. (2020). Physiological factors characterizing heat-vulnerable older adults: A narrative review. *Environment International, 144*, 105909.

Michelozzi, P., et al. (2009). High temperature and hospitalizations for cardiovascular and respiratory causes in 12 European cities. *American Journal of Respiratory and Critical Care Medicine, 179*(5), 383–389.

Miescher, E., & Fortney, S. M. (1989). Responses to dehydration and rehydration during heat exposure in young and older men. *American Journal of Physiology-Regulatory, Integrative and Comparative Physiology, 257*(5), R1050–R1056.

Morel, A., et al. (2014). A review of heat transfer phenomena and the impact of moisture on firefighters' clothing and protection. *Ergonomics, 57*(7), 1078–1089.

Morris, N. B., et al. (2019). A preliminary study of the effect of dousing and foot immersion on cardiovascular and thermal responses to extreme heat. *JAMA, 322*(14), 1411–1413.

Morris, N. B., et al. (2021). Electric fan use for cooling during hot weather: A biophysical modelling study. *The Lancet Planetary Health, 5*(6), e368–e377.

Moser, S. C., & Pike, C. (2015). Community engagement on adaptation: Meeting a growing capacity need. *Urban Climate, 14*, 111–115.

Nairn, J., & Williams, S. (2019). *Power outages during heatwaves: Predicting mortality burden in Australian cities*. Australia: Bureau of Meteorology.

Nunes, B., et al. (2011). Air conditioning and intrahospital mortality during the 2003 heatwave in Portugal: Evidence of a protective effect. *Occupational and Environmental Medicine, 68*(3), 218.

Oke, T. R. (1987). *Boundary layer climates* (2 ed.). London: Routledge.

O'Neill, M. S., et al. (2009). Preventing heat-related morbidity and mortality: New approaches in a changing climate. *Maturitas, 64*(2), 98–103.

Peiris, A. N., Jaroudi, S., & Noor, R. (2017). Heat stroke. *JAMA, 318*(24), 2503.

Peng, R. D., et al. (2011). Toward a quantitative estimate of future heat wave mortality under global climate change. *Environmental Health Perspectives, 119*(5), 701–706.

Peterson, T. C., et al. (2013). Monitoring and understanding changes in heat waves, cold waves, floods, and droughts in the United States: State of knowledge. *Bulletin of the American Meteorological Society, 94*(6), 821–834.

PJ, R. (2001). On the definition of a heat wave. *Journal of Applied Meteorology and Climatology, 40*(4), 762–775.

Popkin, B. M., D'Anci, K. E., & Rosenberg, I. H. (2010). Water, hydration, and health. *Nutrition Reviews, 68*(8), 439–458.

Romanello, M., et al. (2021). The 2021 report of the lancet countdown on health and climate change: Code red for a healthy future. *The Lancet, 398*(10311), 1619–1662.

Rossi, R. M., Bolli, W., & Stampfli, R. (2008). Performance of firefighters' protective clothing after heat exposure. *International Journal of Occupational Safety and Ergonomics, 14*(1), 55–60.

Sailor, D. J., Elley, T. B., & Gibson, M. (2012). Exploring the building energy impacts of green roof design decisions – a modeling study of buildings in four distinct climates. *Journal of Building Physics, 35*(4), 372–391.

Saran, S., et al. (2020). Heating, ventilation and air conditioning (HVAC) in intensive care unit. *Critical Care, 24*(1), 194.

Sera, F., et al. (2020). Air conditioning and heat-related mortality. *Epidemiology, 31*(6), 779–787.

Shashua-Bar, L., Pearlmutter, D., & Erell, E. (2009). The cooling efficiency of urban landscape strategies in a hot dry climate. *Landscape and Urban Planning, 92*(3), 179–186.

Skarha, J., et al. (2021). Association of power outage with mortality and hospitalizations among Florida nursing home residents after Hurricane Irma. *JAMA Health Forum, 2*(11). p. e213900–e213900.

Somers, K. A., et al. (2016). Downstream dissipation of storm flow heat pulses: A case study and its landscape-level implications. *JAWRA Journal of the American Water Resources Association, 52*(2), 281–297.

Son, J.-Y., Liu, J. C., & Bell, M. L. (2019). Temperature-related mortality: A systematic review and investigation of effect modifiers. *Environmental Research Letters, 14*(7), 073004.

Steffen, W., Hughes, L., & Perkins, S. (2014). *Heatwaves: Hotter, longer, more often*. Climate Council.

Sullivan-Bolyai, J. Z., et al. (1979). Hyperpyrexia due to air-conditioning failure in a nursing home. *Public Health Reports (Washington, D.C. : 1974), 94*(5), 466–470.

Tan, J., et al. (2010). The urban heat island and its impact on heat waves and human health in Shanghai. *International Journal of Biometeorology, 54*(1), 75–84.

Tong, S., et al. (2021). Urban heat: An increasing threat to global health. *BMJ*, n2467.

Trancoso, R., et al. (2020). Heatwaves intensification in Australia: A consistent trajectory across past, present and future. *Science of The Total Environment, 742*, 140521.

Tustin, A. W., et al. (2018). Evaluation of occupational exposure limits for heat stress in outdoor workers - United States, 2011–2016. *MMWR Morbidity and Mortality Weekly Report, 67*(26), 733–737.

US EPA, O. Using green roofs to reduce heat islands. [Overviews and factsheets] 2014 2014-06-17T12:16:19-04:00.

US EPA, O. Using trees and vegetation to reduce heat islands. [Overviews and factsheets] 2014 2014-06-17T12:16:11-04:00.

Wehrli, K., et al. (2019). Identifying key driving processes of major recent heat waves. *Journal of Geophysical Research: Atmospheres, 124*(22), 11746–11765.

Weir, E. (2002). Heat wave: First, protect the vulnerable. *CMAJ : Canadian Medical Association journal = journal de l'Association medicale canadienne, 167*(2). pp. 169–169.

White-Newsome, J. L., et al. (2014). Strategies to reduce the harmful effects of extreme heat events: A four-city study. *International Journal of Environmental Research and Public Health, 11*(2), 1960–1988.

Whitehead, D., & Russell, G. (2004). How effective are health education programmes—resistance, reactance, rationality and risk? Recommendations for effective practice. *International Journal of Nursing Studies, 41*(2), 163–172.

Wi, S., et al. (2021). Thermal, hygric, and environmental performance evaluation of thermal insulation materials for their sustainable utilization in buildings. *Environmental Pollution, 272*, 116033.

Wight, D., et al. (2016). Six steps in quality intervention development (6SQuID). *Journal of Epidemiology and Community Health, 70*(5), 520–525.

Wu, X., et al. (2016). Impact of climate change on human infectious diseases: Empirical evidence and human adaptation. *Environment International, 86*, 14–23.

Xiang, J., et al. (2014). Health impacts of workplace heat exposure: An epidemiological review. *Industrial Health, 52*(2), 91–101.

Xu, Z., et al. (2016). Impact of heatwave on mortality under different heatwave definitions: A systematic review and meta-analysis. *Environment International, 89–90*, 193–203.

Yang, J., et al. (2019). Heatwave and mortality in 31 major Chinese cities: Definition, vulnerability and implications. *Science of the Total Environment, 649*, 695–702.

Yu, Z., et al. (2021). Flame-retardant PNIPAAm/sodium alginate/polyvinyl alcohol hydrogels used for fire-fighting application: Preparation and characteristic evaluations. *Carbohydrate Polymers, 255*, 117485.

Zhang, Y., et al. (2017). Risk factors for deaths during the 2009 heat wave in adelaide, Australia: A matched case-control study. *International Journal of Biometeorology, 61*(1), 35–47.

Zhang, Y., Murray, A. T., & Turner, B. L. (2017). Optimizing green space locations to reduce daytime and nighttime urban heat island effects in Phoenix, Arizona. *Landscape and Urban Planning, 165*, 162–171.

Zhao, Y., et al. (2017). Evaluating the physiological and perceptual responses of wearing a newly designed cooling vest for construction workers. *Annals of Work Exposures and Health, 61*(7), 883–901.

Zhao, Q., et al. (2021). Global, regional, and national burden of mortality associated with non-optimal ambient temperatures from 2000 to 2019: A three-stage modelling study. *The Lancet Planetary Health, 5*(7), e415–e425.

Index

Note: 'Page numbers followed by "*f*" indicate figures and "*t*" indicate tables'.

D

E

F

G

H

U

V

W

FSC
www.fsc.org
MIX
Papier | Fördert gute Waldnutzung
FSC® C083411

Kolibri 360 GmbH
Ferdinand-Jühlke-Straße 7
99095 Erfurt, Deutschland
produktsicherheit@kolibri360.de